This book describes advances in the field of superplasticity. This is the ability of certain materials to undergo very large tensile strains, a phenomenon that has increasing commercial applications, but also presents a fascinating scientific challenge in attempts to understand the physical mechanisms that underpin it. Breakthroughs include the development of superplasticity in metallic materials at very high strain rates that are of interest to the automobile industry.

The authors emphasize the materials aspects of superplasticity. They begin with a brief history of the phenomenon. This is followed by a description of the two major types of superplasticity – fine-structure and internal-stress superplasticity – together with a discussion of their operative mechanisms. In addition, microstructural factors controlling the ductility and fracture in superplastic materials are presented. The observations of superplasticity in metals (including alloys of aluminium, magnesium, iron, titanium and nickel), ceramics (including monolithic alloys and composites), intermetallics (including iron, nickel, and titanium base), and laminates are thoroughly described. The technological and commercial applications of superplastic forming and diffusion bonding are presented and examples given.

This book will be of interest to graduate students and researchers in materials science and engineering, especially those working in the aerospace and automobile companies.

T0213389

Cambridge Solid State Science Series

Superplasticity in metals and ceramics

Superplasticity in metals and ceramics

T.G. Nieh
Lawrence Livermore National Laboratory

J. Wadsworth
Lawrence Livermore National Laboratory

O.D. Sherby
Stanford University

CAMBRIDGE
UNIVERSITY PRESS

CAMBRIDGE UNIVERSITY PRESS
Cambridge, New York, Melbourne, Madrid, Cape Town, Singapore, São Paulo

Cambridge University Press
The Edinburgh Building, Cambridge CB2 2RU, UK

Published in the United States of America by Cambridge University Press, New York

www.cambridge.org
Information on this title: www.cambridge.org/9780521561051

First published 1997
This digitally printed first paperback version 2005

A catalogue record for this publication is available from the British Library

Library of Congress Cataloguing in Publication data

Nieh, T. G.
    Superplasticity in metals and ceramics / T.G. Nieh, J. Wadsworth,
O.D. Sherby.
      p.   cm. – (Cambridge solid state science series)
    Includes bibliographical references and index.
    ISBN 0 521 56105 1
    1. Superplasticity.   2. Ceramic materials – Plastic properties.
I. Wadsworth, Jeffrey.   II. Sherby, Oleg D.   III. Title.
IV. Series.
    TA418. 14.N54   1996
    620.1′633–dc20   96-13372 CIP

ISBN-13 978-0-521-56105-1 hardback
ISBN-10 0-521-56105-1 hardback

ISBN-13 978-0-521-02034-3 paperback
ISBN-10 0-521-02034-4 paperback

Contents

Contents

Preface

Superplasticity, the ability of certain materials to undergo very large tensile strains, was first described in 1912. It became the subject of intense research in the early 1960s following a review of Soviet work and the illustration of the potential commercial applications of superplasticity.

There have been enormous advances in the field, of superplasticity since that time. The field has clear commercial applications, but also retains fascinating scientific challenges in understanding the underpinning physical mechanisms. Recent breakthroughs include the development of superplasticity in polycrystalline ceramics, composites and intermetallics, and also the observation of superplasticity in metallic materials at high strain rates. Superplasticity at high strain rates, in particular, is expected to have a significant technological impact on promoting the commercial applications of superplastic materials.

This book emphasizes the materials aspects of superplasticity and thus was written from the materials point of view. A brief history of the development of superplasticity is first introduced. Then, the two major types of superplasticity, i.e. fine-structure and internal-stress superplasticity, and their operative mechanisms are discussed. Other possible superplastic mechanisms, such as Class I solid solutions and superplasticity at dynamic high strain rates are also described. In addition, microstructural factors controlling the ductility and fracture in superplastic materials are presented. The observations of superplasticity in metals (including Al, Mg, Fe, Ti, Ni), ceramics (including monolithics and composites), intermetallics (including Ni-, Ti-, Fe- aluminides), metal-matrix composites (including Al-, Mg- base), and laminates are thoroughly described. Finally, the technological and commercial applications of superplastic forming and diffusion bonding are presented and many examples are given. Over eight hundred literature citations are included in this book.

The book is recommended as a useful reference source for the practicing engineer involved in the design, processing, and manufacture of engineering structural materials. In addition, the book is ideally suited as a text for Superplasticity courses or as supplementary use for Materials Processing, Manufacturing, High Temperature Deformation, and Mechanical Berhavior courses. The materials should be of most interest to Departments of Materials Science and Engineering, Metallurgy, Ceramics, Mechanical Engineering, Aerospace Engineering, Manufacture and Processing Engineering.

T.G. Nieh, J. Wadsworth, O.D. Sherby
Livermore, California

Chapter 1

Introduction

Interest in superplasticity is extremely high. The major areas include superplasticity in metals, ceramics, intermetallics, and composites. Superplasticity at very high strain rates (i.e., approximately $0.1-1\ s^{-1}$) is an area of strong emphasis that is expected to lead to increased applications of superplastic-forming technology.

Historically, there has been no universally accepted definition for superplasticity. After some debate, the following version was proposed and accepted at the 1991 International Conference on Superplasticity in Advanced Materials (ICSAM-91) held in Osaka, Japan [1]:

> Superplasticity is the ability of a polycrystalline material to exhibit, in a generally isotropic manner, very high tensile elongations prior to failure.

It is anticipated that there will continue to be some modifications to this definition, but it should serve as a working definition for a phenomenon that was scientifically reported in 1912 [2] and, indeed, may have a far longer history, as described in the following chapter.

During the course of the ICSAM-91 Conference [1], many different superplastic materials were described. A list of those mentioned is presented in Table 1.1 [3]. It is reasonable to infer from the broad range of superplastic materials listed that there is now a good basic understanding of the requirements for developing superplastic structures. This extensive list also indicates that research on superplastic ceramics and intermetallics has increased dramatically in recent years since the first superplastic ceramic was not observed until 1985 and the first superplastic intermetallic until 1987. The first paper [4] to appear on superplastic ceramics was presented at the 1985 conference in Grenoble, France (though no

Table 1.1   Superplastic materials described during ICSAM–91 [1]

Metallic alloys and composites			Intermetallics	Ceramics and ceramic composites
Al–Ca–Si	IN 905XL	Tool steel	Ni_3Al	YTZP
Al–Ca–Zn	IN 9051, 52	UHC steel	Ni_3Si	$YTZP/Al_2O_3$
Al–Cu	IN 100	Superdux 64	Ti_3Al	Hydroxyapatite $(Ca_{10}(PO_4)_6(OH)_2)$
Al–Cu–Mn	IN 625 LCF	Fe–Cr–Ni	TiAl	Si_3N_4/SiC
Al–Cu–Si	MA6000	SKD11.PM steel	α–2	Al_2O_3
Al–Cu–Zr	MA754	Stainless steels	Super α–2	$3Al_2O_3 \cdot 2SiO_2$
Supral 100	Cu–Al–Ni	T15 PM HSS	$Fe_3(Si,Al)$	α' sialon $(Si_{6-X}Al_XO_YN_{8-X})$
Supral 200	Cu–42Zn	HPb59–1 brass	Nb_3Al	β' sialon $M_{Z/N}Si_{6-X-Z}Al_{X+2}O_XN_{8-X}$
Al–Li	Coronze 328	Pb–62Sn	$Ni_3(Si,Ti)$	Si–Al–M–N–O
Al 8090	Cu–P	Zn–22Al	Ni–9Si	Al_2O_3:Pt (95:5)
Al 2090	Cu–Zn–Ni	Zn–Cu–Ti	Ti–34Al–2Mo	$BaTiO_3$
Weldalite	Nb–Hf–Ti	α/β brass	Ni–Si–Ti(B)	ZnS
Al–Mg–Mn	Ti–Mo–Sn–Zr	SiCp/7475 Al		ZnS/diamond
Al–Mg–Cr	Ti–9V–Mo–Al	α SiC$_w$/2024Al		$PbTiO_3$
Al–Mg–Zr	Ti–6Al–4V	α SiC$_w$/2124Al		Fe_3C/Fe
Al 5083	Ti SP700	α SiC$_w$/6061Al		WC/Co
Al–Zn–Mg	Ti–36Al	SiC$_w$/7075Al		$YBa_2Cu_3O_{7-X}$
Al 7475	Ti–Al–Mo	α Si$_3$N$_{4(w)}$/2124Al		$YBa_2Cu_3O_{7-Y}X$+Ag
Al 7064	Ti IMI843	α Si$_3$N$_{4(w)}$/7064Al		
IN 9021	Mg–Mn–Ce	β Si$_3$N$_{4(w)}$/2024Al		
IN 90211	Mg–Li	β Si$_3$N$_{4(w)}$/6061Al		
IN 905XL	Mg–Al–Zr	SiC$_p$/6061Al		
		SiC$_w$/Zn–22Al		

evidence for large tensile elongations was given in that particular paper); however, an elongation to failure of about 100% in polycrystalline MgO was reported in 1965 by Day and Stokes [5]. The 1991 Osaka conference was the first in the ICSAM series with papers on superplastic intermetallics.

An important conclusion that can be drawn from Table 1.1 is that historically simple concepts regarding key characteristics of superplastic materials are no longer appropriate [3]. For example, it was formerly believed that a superplastic material was a 'metallic two-phase material with a uniform, fine, equiaxed, grain (phase) size.' Clearly, the materials now shown to exhibit superplasticity are much broader than would be covered by this definition, and a more appropriate, although more complex, description is: 'Metallic, ceramic, intermetallic, or composite multiphase materials with uniform or nonuniform, relatively coarse (20 μm) to ultrafine (30 nm) grain sizes that have isotropic or anisotropic grain (phase) shape, size, or orientation.'

Because the development of superplasticity is concerned with achieving high tensile ductility, the maximum elongations that may be attained in different types of materials is of great interest. The maximum elongation achievements to date are summarized in Table 1.2 [3], and this list covers not only the maximum elongations in different classes of materials, but other achievements related to material characteristics, fabrication of parts by superplastic forming, and activities within the scientific community. This list may be viewed as a guide against which to gage future trends in superplasticity research and technology.

References

1. S. Hori, M. Tokizane, and N. Furushiro, *Superplasticity in Advanced Materials*. The Japan Society of Research on Superplasticity, Osaka Japan, 1991.

2. G.D. Bengough, 'A Study of the Properties of Alloys at High Temperatures' *J. Inst. Metals*, 7 (1912), pp. 123–174.

3. T.G. Langdon and J. Wadsworth, 'Summary and Topics of ICSAM-93,' in *International Conference on Superplasticity in Advanced Materials (ICSAM-91)*, pp. 847–852, ed. S. Hori, M. Tokizane, and N. Furushiro, The Japan Society for Research on Superplasticity, Osaka, Japan, 1991.

4. C. Carry and A. Mocellin, ed. *High Ductilities in Fine Grained Ceramics*, pp. 16.1–16.19, ed. C. Carry and A. Mocellin, Centre National de la Recherche Scientifique, Paris, Grenoble, France, 1985.

5. R.B. Day and R.J. Stokes, 'Mechanical Behavior of Polycrystalline Magnesium Oxide at High Temperatures,' *J. Am. Ceram. Soc.*, 49(7) (1966), pp. 345–354.

6. K. Higashi, unpublished research, University of Osaka Prefecture, 1992.

7. G. Gonzales-Doncel, S.D. Karmarkar, A.P. Divecha, and O.D. Sherby, 'Influence of Anisotropic Distribution of Whiskers on the Superplastic Behavior of Aluminum in a Back-Extruded 6061 Al–20% SiC$_w$ Composite,' *Comp. Sci. Technol.*, 35 (1989), pp. 105–120.

8. K. Kajihara, Y. Yoshizawa, and T. Sakuma, 'Superplasticity in SiO$_2$-Containing Tetragonal Zirconia Polycrystal,' *Scr. Metall. Mater.*, 28 (1993), pp. 559–562.

9. T.G. Nieh and J. Wadsworth, 'Superplasticity in Fine-Grained 20%Al$_2$O$_3$/YTZ Composite,' *Acta Metall. Mater.*, 39 (1991), pp. 3037–3045.

10. K. Higashi, T. Okada, T. Muka, and S. Tanimura, 'Postive Exponent Strain-Rate Superplasticity in Mechanically Alloyed Aluminum IN 9021,' *Scr. Metall. Mater.*, **25** (1991), pp. 2503–2506.

11. S.L. Hwang and I.W. Chen, 'Superplastic Forming of SiAlON Ceramics,' *J. Am. Ceram. Soc.*, 77(10) (1994), pp. 2575–2585.

12. T.G. Langdon. 'Superplasticity: a Historical Perspective,' in *International Conference on Superplasticity in Advanced Materials (ICSAM–91)*, pp. 3–12, ed. S. Hori, M. Tokizane, and N. Furushiro, The Japan Society for Research on Superplasticity, Osaka, Japan, 1991.

Table 1.2   Achievements in superplasticity

Property	Level of achievement	Material or contributor
Maximum superplastic elongation in a metal	8000%	Commercial bronze [6]
Maximum superplastic elongation in an intermetallic	>800%	Ti_3Al (super α–2)
Maximum superplastic elongation in a metallic composite	1400%	SiC_w/Al 6061 (thermal cycling) [7]
Maximum superplastic elongation in a ceramic	1038%	YTZP [8]
Maximum superplastic elongation in a ceramic composite	625%	YTZP/Al_2O_3 [9]
Maximum superplastic elongation at a very high strain rate	1250% at 50 s^{-1}	Mechanically alloyed aluminum IN 9021 [10]
Highest strain rate sensitivity	$m=2$	α' sialon, β'' sialon, two-phase sialons [11]
Finest grain size in a consolidated material	30–40 nm	Al_2O_3/SiO_2 (9:1) and ZrO_2 (Y_2O_3-stabilized)
Structural components manufactured per year	265 000	Superform, Ltd.
Longest continuous research career on superplasticity	over 33 years	Professor Presnyakov, Russia [12]
Largest (and only) institute for superplasticity	400–person level	Ufa, Russia

Chapter 2

Key historical contributions

2.1 **Before 1962**

It is often thought that superplasticity is a relatively recent discovery. It is intriguing to speculate, however, that the phenomenon may have had its first application in ancient times. Geckinli [1], for example, has raised the possibility that ancient arsenic bronzes containing up to 10 wt% arsenic, which were used in Turkey in the early Bronze Age, could have been superplastic. This is because the materials are two-phase alloys that may have developed the required stable, fine-grained structure during hand-forging of intricate shapes. Furthermore, the ancient steels of Damascus, in use from 300 B.C. to the late nineteenth century, are similar in composition to modern ultrahigh-carbon steels that have recently been developed, in large part, for their superplastic characteristics [2, 3]. A perspective of the historical context of superplastic studies with respect to the modern time frame is shown in Figure 2.1.

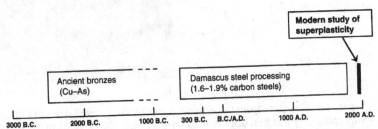

Figure 2.1 Historical perspective of the development of superplasticity from its possible ancient origins to the present time.

A paper published by Bengough in 1912 [4] is believed to contain the first recorded description of superplasticity in a metallic material. Bengough describes how 'a certain special brass ... pulled out to a fine point, just like glass would do, having an enormous elongation.' The quote (which even today is a useful definition containing the essential feature of superplasticity) was made by Bengough in a written discussion following a paper by Rosenhain and Ewen [5] on the *amorphous cement* theory. Examination of Bengough's original work shows that the special brass was an $\alpha+\beta$ brass and exhibited a maximum elongation of 163% at 700 °C; the original data from Bengough's paper are shown in Figure 2.2(a). Considering the crude equipment (Figure 2.2(b)), poor temperature control, large gage length, and relatively high strain rate that was probably used in Bengough's experiment, his results are really rather remarkable. It is interesting to note that materials similar to those studied by Bengough have been processed to be superplastic in recent times. For example, Cu–40% Zn brasses have been developed for superplasticity at temperatures ranging from 600 to 800 °C [6].

Other observations of viscous-like behavior in fine-grained metals can occasionally be found in early literature. Jenkins, for example, achieved elongations of 300 to 400% in Cd–Zn and Pb–Sn eutectics after thermomechanical processing and in 1928 [7] provided the first photographic evidence of superplastic

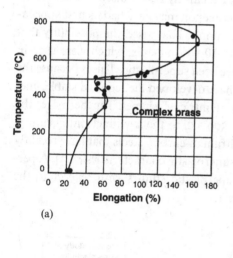

(a)

(b)

Figure 2.2   (a) Data from the original paper by Bengough [4] in 1912 in which superplasticity was first documented scientifically. The elongation to failure of an $\alpha+\beta$ brass is plotted as a function of temperature and a maximum elongation of 163% is noted. (b) The equipment used by Bengough.

behavior in a tested sample as shown in Figure 2.3. Interestingly, Jenkins did not find the observation to be important enough to include in either the synopsis or the conclusions of his paper. In 1934, Pearson [8] dramatically demonstrated, using a Bi–Sn sample that had been deformed to nearly 2000% and then coiled (as shown in Figure 2.4), that unusually large elongations could be achieved in certain fine-structured, two-phase materials. As a result, he is often credited with having first demonstrated superplasticity.

At a later date in East Germany, Sauerwald [9] reported that a number of aluminum- and zinc-based alloys exhibited good tensile ductility, and he patented several compositions for their formability. At the same time, or perhaps slightly earlier, work in the Soviet Union was underway to specifically address the phenomenon, and Bochvar and Sviderskaya [10] coined the term *sverhplastichnost* (ultrahigh plasticity) in their 1945 paper on superplastic alloys. Apparently the term *superplasticity* was used for the first time in the English language in Chemical Abstracts in 1947, but first appears in a 1959 technical paper by Lozinsky and Simeonova [11] on the subject of *Superhigh Plasticity of Commercial Iron under Cyclic Fluctuations of Temperature*. Some of the key observations and discoveries in superplasticity through this time period are summarized in Figure 2.5.

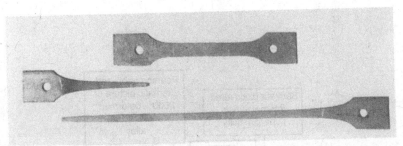

Figure 2.3 The first published photograph of a superplastically deformed sample by Jenkins [7] in 1928. The material is either a cadmium–zinc alloy or a lead–tin alloy (Jenkins does not specify which it is) and has undergone about 300% elongation.

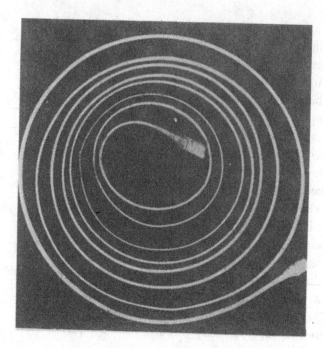

Figure 2.4 Pearson's famous photograph [8] in 1934 of a Bi–Sn alloy that has undergone 1950% elongation.

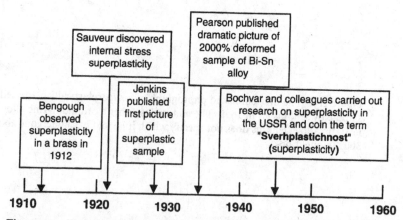

Figure 2.5 Key discoveries in superplasticity in the early and middle part of the twentieth century prior to major development of interest in the West during the 1960s.

2.2 From 1962 to 1982

Although papers on superplasticity occasionally appeared after 1945, the major increase in interest came in 1962 with a review article by Underwood [12] on work in the Soviet Union. A graph (Figure 2.6) was included in this review that illustrated the ductility of Zn–Al alloys after quenching from 375 °C. A maximum ductility of 650% was achieved at 250 °C for a 20% Al and 80% Zn alloy (the monotectoid composition). This remarkable result, achieved on a sample that required only a quenching treatment, attracted the attention of Backofen at the Massachusetts Institute of Technology. His research group [13] studied details of the Zn–Al monotectoid alloy and a Pb–Sn eutectic composition material.

Of great significance is that Backofen and his colleagues showed that the superplastic Zn–Al alloy could be formed into a practical shape by a simple air-pressure operation, as in glassblowing [14]. An example of such an article is shown in Figure 2.7. This practical example, showing the spectacular formability of a superplastic metallic alloy, was probably the single biggest initiator of the rapid growth in the field of superplasticity that took place after the Backofen paper was published in 1964.

By 1968, just four years after the Backofen demonstration of superplastic forming, the first review paper on the subject was published by Chaudhari [15], and in the next year, the first book entiled *Superplasticity of Metals and Alloys* was published by Presnyakov [16]. After that, monographs by Western [2, 6, 17, 18], Soviet [19–21], and Japanese [22] researchers were published, as were numerous review articles [12, 23–25].

2.3 From 1982 to the present

A 1982 international conference entitled *Superplastic Forming of Structural Alloys* [26] showed how great the interest was in the subject of fine-structure superplasticity from both academic and commercial viewpoints. It was also the first conference fully dedicated to the subject of superplasticity. At this conference, the feasibility of the commercial application of superplasticity was reviewed for alloys based on titanium [27], nickel [28], aluminum [29, 30], and iron [31].

Since the 1982 conference, other significant international symposia have been held. Superplastic forming was the topic of a 1984 symposium held in Los Angeles [32]. A conference on superplastic aerospace aluminum alloys was held in Cranfield, United Kingdom, in July 1985 [33]. The second international conference on superplasticity was held in Grenoble, France, in September 1985 [34]. NATO/AGARD selected superplasticity as one of their lecture series in 1987 [35] and again in 1989 [36]; bilateral symposia on superplasticity between

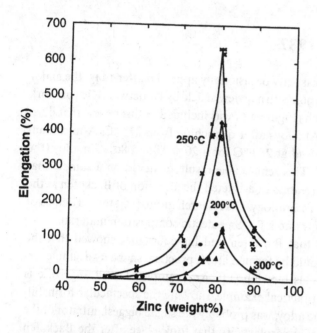

Figure 2.6 The ductility of Zn–Al alloys taken from the 1962 Underwood review article [12].

Figure 2.7 A salt-and-pepper shaker super-plastically formed by a simple gas-pressure operation in the Zn–Al alloy from Backofen *et al.* [13] in 1964.

China and Japan have been held in Beijing in 1985 [37] and in Yokohama in 1986 [38]; in the first Chino–Japanese symposium, 32 separate papers were presented with considerable emphasis placed on superplastic ferrous- and aluminum-based alloys. Both the Chinese and Japanese governments selected superplasticity in 1980 for intense national research and development studies. They envision superplasticity 'as a future technology into the next century.'

The third international conference on superplasticity in advanced materials was held in Blaine, Washington, United States, in August 1988 [39] and attracted a large number of delegates from the Communist bloc countries, including representatives from the Soviet Union. The fourth international conference on superplasticity was held in Osaka, Japan, in June 1991 [40]; a brief summary of this conference, including the materials studied and major technical achievements, has been presented in Chapter 1. Interest in the field continues unabated as seen by the data in Figure 2.8, which shows the number of publications in fine-structure superplasticity from 1960 to 1991. The fifth international conference on superplasticity was held in Moscow, Russia, in May 1994 [41].

Most of the discussion in the preceding sections has dealt with the history of fine-structure superplasticity. The other type of superplasticity is known as internal-stress superplasticity. Internal-stress superplasticity seems to have been first noted by Sauveur in 1924 [42]. He indicated that concurrent phase transformation during thermal cycling lead to exceptional weakness in iron when under a small applied stress. Early researchers who attempted to study this effect were Koref [43] in 1926 and Wasserman [44] in 1937. These investigators coined the term *amorphous plasticity* to explain this type of weakening. Koref studied enhanced plasticity of tungsten coils during recrystallization. Wasserman studied enhanced plasticity during phase transformation (austenite to ferrite) in a nickel steel under stress during low-temperature cooling. Wasserman's work is the predecessor to the modern TRIP (TRansformation Induced Plasticity) steel studies of Zackay et al. [45]. The enhanced plasticity in steels at low temperatures is now more clearly understood to be related to a strain hardening effect than to a strain rate sensitivity effect.

The modern study of phase transformation plasticity can be considered to have started with the work of de Jong and Rathenau [46], who in 1961 established quantitative relations between stress and elongation during thermal cycling plasticity of iron. Early Soviet work in this field was led by Lozinsky and his colleagues [11] who considered that the weakening effect was a result of the 'the transient breaking of atomic bonds.' In the United States, an elongation of 700% was achieved in a 52100 tool steel by Oelschlagel and Weiss [47] under thermal cycling conditions with an imposed stress of 17 MPa.

It should be pointed out that pressure-induced phase changes have also been cited as a source of superplastic flow in geological materials. For example, there is a transformation, because of pressure, in the earth's upper mantle from

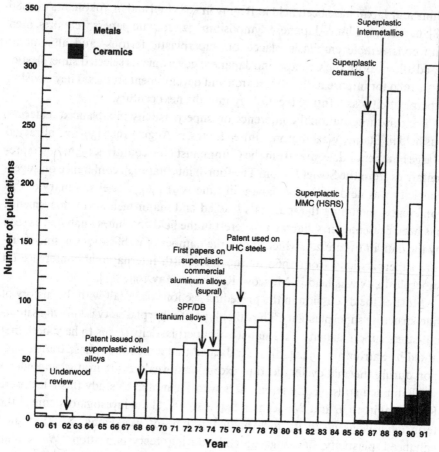

Figure 2.8   The increase in the number of published papers per year since 1960 in the area of superplasticity, as well as some of the highlights since that time.

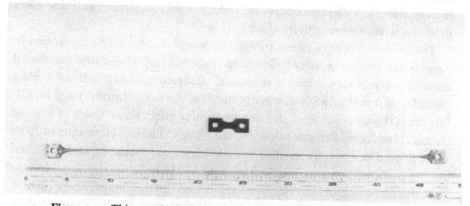

Figure 2.9   This sample shows the extraordinary elongation (8000%) achieved in a Cu-Al alloy by Higashi [56].

orthorhombic olivine to a spinel phase at a depth of about 400 km below the earth's surface [48].

Another common type of internal-stress superplasticity is that developed in polycrystalline materials that exhibit thermal expansion anisotropy. Lobb et al. [49] were among the first to show examples of the high ductility achievable in polycrystalline metals with large anisotropic thermal expansion properties. In their 1972 paper, they showed that a sample of α uranium elongated to about 300% under thermal cycling conditions in contrast to an elongation of about only 50% under optimal isothermal conditions. Their results were recently analyzed using an internal-stress superplasticity model developed by Wu et al. [50]. Internal-stress superplasticity has also been used to show that whisker- and particle-reinforced metal matrix composites can be made to flow superplastically, exhibiting Newtonian-viscous behavior, under thermal cycling conditions [51–53]. A record elongation of 1400% has been found in a SiC whisker-reinforced 6061 aluminum alloy [52].

As discussed in Chapter 1, superplasticity is the ability of a polycrystalline material to exhibit very high tensile elongations prior to failure. In certain fine-grained metal alloy systems, tensile elongations of thousands of percent have been documented. In fact, the subject is of sufficient general scientific interest that a world record has been recognized for the phenomenon in metal alloys [54]. The initial world record of 4850% in a Pb–62 wt% Sn alloy was exceeded in a commercial aluminum bronze (Cu–10 wt% Al-based alloy) by a value reported finally as 8000%, as shown in Figure 2.9 [55, 56].

In the field of ceramics, extensive efforts were made in the West during the 1950s to hot-fabricate ceramics using conventional metallurgical processes such as extrusion, rolling, and forging [57–61]. Interestingly, there was also some evidence indicating that, during a similar time period, a comprehensive effort was underway in the Soviet Union to improve ductility and fabricability in complex ceramics and ceramic composites [62, 63]. The goal was to produce near-net-shape parts to avoid the expensive machining of ceramics. A number of structural oxides, including CaO, MgO, SiO_2, ZrO_2, BeO, ThO_2, and Al_2O_3, were studied [58]. As a result of this work, an improved understanding of ceramic deformation was developed, but certain problems – in particular, the requirement for relatively high forming temperatures – still existed. Day and Stokes [64], for example, have observed a large tensile elongation (100%) in MgO, but the temperature was quite high (about 1800 °C). Also, the temperature required for hot-forging Al_2O_3 was found to be about 1950 °C (Figure 2.10), which is extremely high from a practical standpoint [58]. Subsequently, the concept of thermomechanical processing of ceramics was more-or-less abandoned.

In 1986, a major breakthrough occurred: a 3-mol % yttria-stabilized tetragonal zirconia (YTZP) and its composite (i.e., YTZP-containing alumina, Al_2O_3/YTZP) were shown to be superplastic in tension tests. Wakai et al. [65, 66]

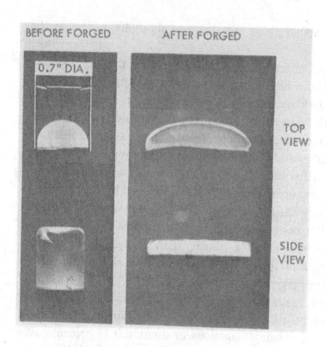

Figure 2.10 Before and after views of a split boule of sapphire forged at 1950 °C. (from Ref. [58]).

showed that the YTZP with a grain size of 0.3 μm, was superplastic at 1450 °C with up to 120% elongation and a strain-rate-sensitivity exponent of 0.5. The elongation of YTZP was further improved to 800% [67] and an elongation of 625% was also recorded in a 20% Al_2O_3/YTZP composite [68].

Since the discovery of superplastic YTZP, superplasticity has been observed in many ceramics, including alumina [69, 70], hydroxyapatite [71], silicon nitride [72], and iron carbide [73]. The superplastic properties of these ceramics apparently do have some characteristics dissimilar to metallic alloys. In contrast to superplastic metals, for example, the tensile ductility of superplastic ceramics does not directly relate to the strain-rate-sensitivity exponent, but rather to the flow stress, as will be discussed in Chapter 10. Also, the grain sizes in all the superplastic ceramics are noted to be less than about 1 μm, which is much smaller than those usually found in superplastic metals (typically, <10 μm). Because of the fineness of the microstructure, grain growth – in particular, dynamic grain growth – during superplastic deformation of ceramics always occurs.

Since its initial report in 1986, the science of ceramic superplasticity has steadily advanced. A knowledge of fundamental issues in ceramic superplasticity has advanced to the stage that the technological application of superplastic deformation, and specifically biaxial gas-pressure deformation, has now been demonstrated [74, 75].

Superplasticity studies in intermetallics began in 1987. Materials such as nickel aluminides [76–81], nickel silicides [82–86], and titanium aluminides [87–93] have been shown to be superplastic. One interesting observation was that the grain size in some superplastic intermetallics, such as Fe_3Al and FeAl [94,

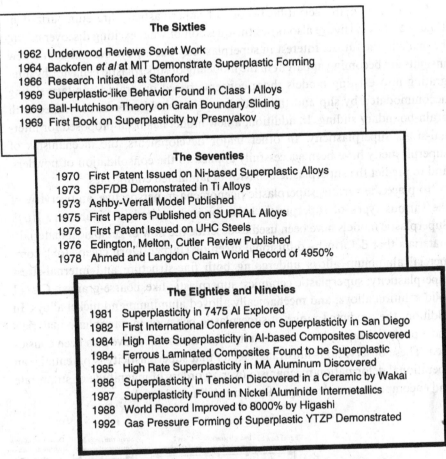

The Sixties

1962 Underwood Reviews Soviet Work
1964 Backofen *et al* at MIT Demonstrate Superplastic Forming
1966 Research Initiated at Stanford
1969 Superplastic-like Behavior Found in Class I Alloys
1969 Ball-Hutchison Theory on Grain Boundary Sliding
1969 First Book on Superplasticity by Presnyakov

The Seventies

1970 First Patent Issued on Ni-based Superplastic Alloys
1973 SPF/DB Demonstrated in Ti Alloys
1973 Ashby-Verrall Model Published
1975 First Papers Published on SUPRAL Alloys
1976 First Patent Issued on UHC Steels
1976 Edington, Melton, Cutler Review Published
1978 Ahmed and Langdon Claim World Record of 4950%

The Eighties and Nineties

1981 Superplasticity in 7475 Al Explored
1982 First International Conference on Superplasticity in San Diego
1984 High Rate Superplasticity in Al-based Composites Discovered
1984 Ferrous Laminated Composites Found to be Superplastic
1985 High Rate Superplasticity in MA Aluminum Discovered
1986 Superplasticity in Tension Discovered in a Ceramic by Wakai
1987 Superplasticity Found in Nickel Aluminide Intermetallics
1988 World Record Improved to 8000% by Higashi
1992 Gas Pressure Forming of Superplastic YTZP Demonstrated

Figure 2.11 Key steps in the history of superplasticity.

95], is much coarser (about 100 μm) than that generally required for both metals and ceramics. Because intermetallics are considered to be the basis for the next-generation, high-temperature, structural materials, the growing interest in this area is certainly warranted.

In the area of composites, superplasticity of a fine-grained, aluminum composite reinforced with silicon carbide whiskers was first discovered in 1984 [96]. Interestingly, it was observed at a very high strain rate (about 10^{-1} s^{-1}). Since this discovery, high-strain-rate superplasticity has been quickly extended to other aluminum-matrix composites containing various forms of ceramic reinforcements and to many of the most advanced aluminum alloys, such as mechanically alloyed materials [97–103] and zirconium-modified conventional alloys [104–106]. Since one of the major drawbacks of conventional superplastic forming is that the forming rate is found at relatively low strain rates (typically about 10^{-4} to 10^{-3} s^{-1}), the development of high-strain-rate superplastic materials is certainly an economically attractive one because of potential energy and cost savings.

Key events in the scientific history of superplasticity are summarized in Figure 2.11. This figure also poses the question of what exciting discoveries will be made in the future. Interest in superplasticity has always been high and new insights are becoming apparent in the modeling of superplastic behavior by integrating into existing models those concepts based on grain-boundary sliding accommodated by slip and the influence of threshold stresses associated with grain-boundary sliding. In addition, a new model has been proposed for internal-stress superplasticity. In other major developments, the mechanisms of superplasticity have been successfully applied to the consolidation of powders, and to predict the superplastic behavior of laminated composites.

To place the various superplastic phenomena in perspective, an overview of the various types of superplastic behavior is presented in Figure 2.12 [107]. Superplastic models have been used to describe the flow of geological materials; materials that deform by diffusional creep or Harper–Dorn creep [108]; commercial aluminum alloys undergoing both fine-structure and internal-stress superplasticity; superplastic ceramics; superplastic-like, coarse-grained, Class I solid solution alloys; and mechanically alloyed aluminum and nickel alloys. In addition, the predicted range of superplastic behavior in nanophase materials and a potential superplastic area at ultrahigh strain rates have also been considered. These observations illustrate the fact that superplasticity is potentially an operational deformation mechanism over 20 orders of magnitude of strain rate and encompasses a wide range of polycrystalline materials.

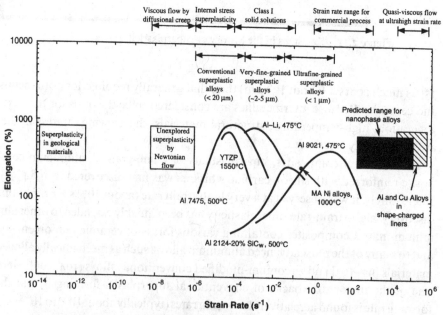

Figure 2.12 Overview of superplasticity in a wide range of materials.

References

1. E. Geckinli, Private communication, Technical University of Istanbul, Turkey, 1987.

2. J. Wadsworth and O.D. Sherby, 'On the Bulat – Damascus Steels Revisited,' *Prog. Mater. Sci.*, **25** (1980), pp. 35–68.

3. D.W. Kum, T. Oyama, O.D. Sherby, O.A. Ruano, and J. Wadsworth in 'Superplastic Ultrahigh Carbon Steels', in *Superplasticity*, pp. 32–42, ed. S. Agrawal, ASM Technical Publications No. 8401–002, ASM, Metals Park, OH, 1984.

4. G.D. Bengough, 'A Study of the Properties of Alloys at High Temperatures,' *J. Inst. Metals*, **7** (1912), pp. 123–174.

5. W. Rosenhain and D. Ewen, 'Intercrystalline Cohesion in Metals,' *J. Inst. Metals*, **8** (1912), pp. 149–185.

6. J.W. Edington, K.N. Melton, and C.P. Cutler, 'Superplasticity,' *Prog. Mater. Sci.*, **21**(2) (1976), pp. 63–170.

7. C.H.M. Jenkins, 'Strength of Cd-Zn and Sn-Pb Alloy Solder,' *J. Inst. Metals*, **40** (1928), pp. 21–32.

8. C.E. Pearson, 'Viscous Properties of Extruded Eutectic Alloys of Pb-Sn and Bi-Sn,' *J. Inst. Metals*, **54** (1934), pp. 111–123.

9. F. Sauerwald, 'Uber den Hochductilen Zustand von Legierungen auf Aluminum-Zink Basis 1,' *Archiv Metallkunde*, **3**(5) (1949), pp. 165–175.

10. A.A. Bochvar and Z.A. Sviderskaya, 'Superplasticity in Zinc-Aluminum Alloys,' *Izv. Akad. Nauk SSSR, Otdel. Tekh. Nauk*, **9** (1945), pp. 821–827.

11. M.G. Lozinsky and I.S. Simeonova, 'Superhigh Plasticity of Commercial Iron under Cyclic Fluctuations of Temperature,' *Acta Metall.*, **7** (1959), pp. 709–715.

12. E.E. Underwood, 'A Review of Superplasticity and Related Phenomenon,' *J. Metals*, **14** (1962), pp. 914–919.

13. W.A. Backofen, I.R. Turner, and D.H. Avery, 'Superplasticity in an Al-Zn Alloy,' *Trans. ASM*, **57** (1964), pp. 980–990.

14. T.H. Thomsen, D.L. Holt, and W.A. Backofen, 'Forming Superplastic Sheet Metal in Bulge Dies,' *Met. Eng. Quart.*, **2** (1970), pp. 1–12.

15. P. Chaudhari, *Superplasticity, Science and Technology*. 42–75, Springer-Verlag, Berlin, Germany, 1968.

16. A.A. Presnyakov, *Sverkhplastichnost' Metallov i Splavov (Superplasticity of Metals and Alloys)*, The British Library, Wetherby, England, 1976.

17. S. Tang, *Mechanics of Superplasticity*, Kriegar Publishing Co., New York, 1979.

18. K.A. Padmanabhan and C.J. Davies, *Superplasticity – Mechanical and Structural Aspects, Environmental Effects, Fundamentals and Applications*, Springer-Verlag, Berlin, Germany, 1980.

19. A.S. Tichonov, *Superplastic Effect in Metals and Alloys*. Izdatelsvo, 'Nauk', Moscow, 1978.

20. O.M. Smirnov, *Working of Metals under Pressure in the Superplastic State*, Machinostroyenia, Moscow, 1979.

21. I.I. Novikov and V.K. Portnoi, *Superplasticity in Alloys with Ultrafine Grains*. Metallurgiya, Moscow, USSR, 1981.

22. O. Izumi, 'Fine Grain Superplasticity,' *J. Jpn Soc. Tech. Plast.*, **16** (1975), pp. 1015–1021.

23. A.K. Mukherjee, 'Deformation Mechanisms in Superplasticity,' *Ann. Rev. Mater. Sci.*, **9** (1979), pp. 191–217.

24. R.H. Johnson, 'Superplasticity,' *Metall. Rev.*, **15** (1970), pp. 115–134.

25. T.G. Langdon, 'The Mechanical Properties of Superplastic Materials,' *Metall. Trans.*, **13A** (1982), pp. 689–701.

26. *Superplastic Forming of Structural Alloys*, edited by N.E. Paton and C.H. Hamilton, The Metallurgical Society of AIME, Warrendale, PA, 1982.

27. C. Hammond, 'Superplasticity in Titanium Base Alloys,' in *Superplastic Forming of Structural Alloys*, pp. 131–146, ed. N.E. Paton and C.H. Hamilton, The Metallurgical Society of AIME, Warrendale, PA, 1982.

28. H.F. Merrick, 'Superplasticity in Nickel-base Alloys,' in *Superplastic Forming of Structural Alloys*, pp. 209–224, ed. N.E. Paton and C.H. Hamilton, The Metallurgical Society of AIME, Warrendale, PA, 1982.

29. D.J. Lloyd and D.M. Moore, 'Aluminum Alloy Design for Superplasticity,' in *Superplastic Forming of Structural Alloys*, pp. 147–172, ed. N.E. Paton and C.H. Hamilton, The Metallurgical Society of AIME, Warrendale, PA, 1982.

30. C.H. Hamilton, C.C. Bampton, and N.E. Paton, 'Superplasticity in High Strength Aluminum Alloys,' in *Superplastic Forming of Structural Alloys*, pp. 173–190, ed. N.E. Paton and C.H. Hamilton, The Metallurgical Society of AIME, Warrendale, PA, 1982.

31. N. Ridley, 'Superplasticity in Iron Base Alloys,' in *Superplastic Forming of Structural Alloys*, pp. 191–224, ed. N.E. Paton and C.H. Hamilton, The Metallurgical Society of AIME, Warrendale, Pennsylvania, 1982.

32. S.P. Agrawal, ed. *Superplastic Forming*, American Society for Metals, Metals Park, OH, 1985.

33. *Superplasticity in Aerospace Aluminum*, edited by R. Pearce and L. Kelly, Ashford Press, Curdridge, England, 1985.

34. *Superplasticity*, edited by B. Baudelet and M. Suery, Centre National de la Recherche Scientifique, Paris, France, 1985.

35. D. Stephen, 'Superplasticity,' in *NATO/AGARD Lecture Series, No. 154*, pp. 7-1-7-35, National Technical Information Services, Springfield VA, 1987.

36. O.D. Sherby and J. Wadsworth, 'Advances and Future Directions in Superplastic Materials,' in *Superplasticity – AGARD Lecture Series No. 168*, pp. 3-1-3-24, NATO/AGARD, 1989.

37. J.-T. Hai and Z.-R. Wang, 'Research and Development of Superplasticity in China,' in *Proceedings of Cino-Japan Joint Symposium on Superplasticity*, pp. 1–7, Chinese Academic Society for Superplasticity, and Japan Society for Research on Superplasticity, Research Group for Forming Process of Metals in the Japan Society for Technology of Plasticity, and Japan Society for Research on Superplasticity, 1985.

38. M. Miyagawa and M. Kobayashi, 'Research and Development of Superplasticity in Japan,' in *Proceedings of Cino-Japan Joint Symposium on Superplasticity*, pp. 8-13, Chinese Academic Society for Superplasticity, and Japan Society for Research on Superplasticity, Research Group for Forming Process of Metals in the Japan Society for Technology of Plasticity, and Japan Society for Research on Superplasticity, 1985.

39. *Superplasticity and Superplastic Forming*, edited by C.H. Hamilton and N.E. Paton, The Minerals, Metals & Materials Society, Warrendale, PA, 1988.

40. *Superplasticity in Advanced Materials*. edited by S. Hori, M. Tokizane, and N. Furushiro, The Japan Society of Research on Superplasticity, Osaka Japan, 1991.

41. *Superplasticity in Advanced Materials – ICSAM-94*, edited by T.G. Langdon, Trans Tech Publications Ltd, Switzerland, Moscow, Russia, 1994.

42. A. Sauveur, 'What is Steel? Another Answer,' *Iron Age*, **113** (1924), pp. 581–583.

43. F. Koref, *Z. Technm. Physik.*, 7 (1926), p. 544.

44. G. Wasserman, 'Untersuchungen an Eisen-Nickel-Legierung uber die Verformbarkeit wahrend der $\gamma-\alpha$ Unwandlung,' *Archiv Fur der Eisenhutt.*, 7 (1937), pp. 321–325.

45. V.F. Zackay, E.R. Parker, D. Fahr, and R. Bush, 'The Enhancement of Ductility in High-Strength Steels,' *Trans. ASM*, **60** (1967), pp. 252–259.

46. M. de Jong and G.W. Rathenau, 'Mechanical Properties of an Iron-Carbon Alloy During Allotropic Transformation in Pure Iron,' *Acta Metall.*, **9** (1961), pp. 714–720.

47. D. Oelschlagel and V. Weiss, 'Superplasticity of Steels During the Ferrite-Austenite Transformation,' *Trans. Am. Soc. Metals*, **59** (1966), pp. 143–154.

48. C.M. Sung and R.G. Burns, 'Kinetics of High Pressure Phase Transformation: Implications to the Evolution of the Olivine-Spinel Transformation in the downgoing Lithosphere and its Consequences on the Dynamics of the Mantle,' *Tectonophy.*, **31** (1976), pp. 1–31.

49. R.C. Lobb, E.C. Sykes, and R.H. Johnson, 'The Superplastic Behavior of Anisotropic Metals Thermally-Cycled under Stress,' *Met. Sci.*, **6** (1971), pp. 33–39.

50. M.Y. Wu, J. Wadsworth, and O.D. Sherby, 'Internal Stress Superplasticity in Anisotropic Polycrystalline Zinc and Uranium,' *Metall. Trans.*, **18A** (1987), pp. 451–462.

51. M.Y. Wu, J. Wadsworth, and O.D. Sherby, 'Superplasticity in a Silicon Carbide Whisker Reinforced Aluminum Alloy,' *Scr. Metall.*, **18** (1984), pp. 773–776.

52. G. Gonzales-Doncel, S.D. Karmarkar, A.P. Divecha, and O.D. Sherby, 'Influence of Anisotropic Distribution of Whiskers on the Superplastic Behavior of Aluminum in a

Back-Extruded 6061 Al–20% SiC$_w$ Composite,' *Comp. Sci. Technol.*, **35** (1989), pp. 105–120.

53. M.Y. Wu, J. Wadsworth, and O.D. Sherby, 'Elimination of the Threshold Stress for Creep by Thermal Cycling in Oxide-Dispersion-Strengthened Materials,' *Scr. Metall.*, **21** (1987), pp. 1159–1164.

54. *The Guiness 1983 Book of World Records*, pp. 144, ed. N. McWhirter and M. Registrada, Bantam Books, Inc., 666 Fifth Avenue, New York, NY, 10103., 1984.

55. Y. Nakatani, T. Ohnishi, and K. Higashi, 'Superplastic Behavior of Commercial Aluminum Bronze,' *Japan Inst. of Metals*, **48** (1984), pp. 113–114.

56. K. Higashi, unpublished research, University of Osaka Prefecture, 1992.

57. R.M. Fulrath, 'A Critical Compilation of Ceramic Forging Methods: III, Hot Forging Processes,' *Am. Ceram. Soc. Bull.*, **43**(12) (1964), pp. 880–885.

58. R.W. Rice, 'Hot-Working of Oxides,' in *High Temperature Oxides, Chapter 3*, pp. 235–280, ed. A.M. Alper, Academic Press, New York, 1970.

59. R.W. Rice, 'Hot Forming of Ceramics,' in *Ultrafine-Grain Ceramics*, pp. 203–250, ed. J.J. Burke, N.L. Reed, and V. Weiss, Syracuse University Press, Syracuse, NY, 1970.

60. P.E.D. Morgan, 'Superplasticity in Ceramics,' in *Ultrafine-Grain Ceramics*, pp. 251–271, ed. J.J. Burke, N.L. Reed, and V. Weiss, Syracuse University Press, New York, 1970.

61. R.C. Bradt, 'Thermal Cycling Deformation in Bi_2O_3–Sm_2O_3 System,' in *Advances in Deformation Processing*, pp. 405–423, ed. J.J. Burke and V. Weiss, Plenum Press, New York, 1978.

62. W.L. Frankhouser, *A Working Paper: Soviet Developments in Regard to Superplasticity in Ceramic Materials.* pp. 1–16, Report SPC 849, System Planning Corporation, 1500 Wilson Boulevard, Arlington, VA, 22209, 1982.

63. 'Conference on High-Melting Compounds, Dedicated to Academician G.V. Samsonov (in Russian),' *Investiya Akademii Nauk SSR, Neorganicheskie Materialy*, **15**(4) (1979), p. 549.

64. R.B. Day and R.J. Stokes, 'Mechanical Behavior of Polycrystalline Magnesium Oxide at High Temperatures,' *J. Am. Ceram. Soc.*, **49**(7) (1966), pp. 345–354.

65. F. Wakai, S. Sakaguchi, and Y. Matsuno, 'Superplasticity of Yttria-Stabilized Tetragonal ZrO_2 Polycrystals,' *Adv. Ceram. Mater.*, **1** (1986), pp. 259–263.

66. F. Wakai, S. Sakaguchi, and H. Kato, 'Compressive Deformation Properties and Microstructures in the Superplastic Y-TZP,' *J. Ceram. Soc. Japan (In Japanese)*, **94** (1986), pp. 72–75.

67. T.G. Nieh, C.M. McNally, and J. Wadsworth, 'Superplastic Behavior of a Yttria-Stabilized Tetragonal Zirconia Polycrystal,' *Scr. Metall.*, **22** (1988), pp. 1297–1300.

68. T.G. Nieh, C.M. McNally, and J. Wadsworth, 'Superplastic Behavior of a 20% Al_2O_3/YTZ Ceramic Composite,' *Scr. Metall.*, **23** (1989), pp. 457–460.

69. P. Gruffel, P. Carry, and A. Mocellin, 'Effects of Testing Conditions on Superplastic Creep of Alumina Doped with Ti and Y,' in *Science of Ceramics, Volume 14*, pp. 587–592, ed. D. Taylor, The Institute of Ceramics, Shelton, Stoke-on-Trent, UK, 1987.

70. L.A. Xue, X. Wu, and I.W. Chen, 'Superplastic Alumina Ceramics with Grain Growth Inhibitors,' *J. Am. Ceram. Soc.*, **74**(4) (1991), pp. 842–845.

71. F. Wakai, Y. Kodama, S. Sakaguchi, and T. Nonami, 'Superplasticity of Hot Isostatically Pressed Hydroxyapatite,' *J. Am. Ceram. Soc*, **73**(2) (1990), pp. 257–260.

72. F. Wakai, Y. Kodama, S. Sakaguchi, N. Murayama, K. Izaki, and K. Niihara, 'A Superplastic Covalent Crystal Composite,' *Nature (London)*, **334**(3) (1990), pp. 421–423.

73. W.J. Kim, J. Wolfenstine, O.A. Ruano, G. Frommeyer, and O.D. Sherby, 'Processing and Superplastic Properties of Fine-Grained Iron Carbide,' *Metall. Trans.*, **23**A (1992), pp. 527–535.

74. J.P. Wittenauer, T.G. Nieh, and J. Wadsworth, 'A First Report on Superplastic Gas-Pressure Forming of Ceramic Sheet,' *Scripta Metall. Maters.*, **26** (1992), pp. 551–556.

75. T.G. Nieh and J. Wadsworth, 'Biaxial Gas-Pressure Forming of a Superplastic Al_2O_3/YTZP,' *J. Mater. Eng. Performance*, **3**(4) (1994), pp. 496–500.

76. V.K. Sikka, C.T. Liu, and E.A. Loria,' Processing and Properties of Powder Metallurgy Ni_3Al-Cr-Zr-B for Use in Oxidizing Environments,' in *Processing of Structural Metals by Rapid Solidification*, pp.

417–427, ed. F.H. Froes and S.J. Savage, Am. Soc. Metals, Metals Park, OH, 1987.

77. R.N. Wright and V.K. Sikka, 'Elevated Temperature Tensile Properties of Powder Metallurgy Ni_3Al Alloyed with Chromium and Zirconium,' *J. Mater. Sci.*, **23** (1988), pp. 4315–4318.

78. M.S. Kim, S. Hanada, S. Wantanabe, and O. Izumi, 'Superplasticity in a Recrystallized Ni_3Al Polycrystal Doped with Boron,' *Mater. Trans. JIM*, **30**(1) (1989), pp. 77–85.

79. A. Choudhury, A.K. Muhkerjee, and V.K. Sikka, 'Superplasticity in an Ni_3Al Base Alloy with 8wt%Cr,' *J. Mater. Sci.*, **25** (1990), pp. 3142–3148.

80. J. Mukhopadhyay, G.C. Kaschner, and A.K. Muhkerjee, 'Superplasticity and Cavitation in Boron Doped Ni_3Al,' in *Superplasticity in Aerospace II*, pp. 33–46, ed. T.R. McNelly and C. Heikkenen, The Minerals, Metals & Materials Society, Warrendale, PA, 1990.

81. J. Mukhopadhyay, G. Kaschner, and A.K. Muhkerjee, 'Superplasticity in Boron Doped Ni_3Al,' *Scr. Metall.*, **24** (1990), pp. 857–862.

82. T.G. Nieh and W.C. Oliver, 'Superplasticity of a Nickel Silicide,' *Scr. Metall.*, **23** (1989), pp. 851–854.

83. S.L. Stoner and A.K. Muhkerjee, 'Superplasticity in Fine Grain Nickel Silicide,' in *International Conference on Superplasticity in Advanced Materials (ICSAM–91)*, pp. 323–328, ed. S. Hori, M. Tokizane, and N. Furushiro, The Japan Society for Research on Superplasticity, Osaka, Japan, 1991.

84. T. Takasugi, S. Rikukawa, and S. Hanada, 'Superplastic Deformation in $Ni_3(Si,Ti)$ Alloys,' *Acta Metall. Mater.*, **40** (1992), pp. 1895–1906.

85. T. Takasugi, S. Rikukawa, and S. Hanada, 'Superplasticity in $L1_2$ Type $Ni_3(Si,Ti)$ Intermetallics,' in *International Conference on Superplasticity in Advanced Materials (ICSAM–91)*, pp. 329–338, ed. S. Hori, M. Tokizane, and N. Furushiro, The Japan Society for Research on Superplasticity, Osaka, Japan, 1991.

86. T.G. Nieh, 'Superplasticity in $L1_2$ Intermetallic Alloys,' in *Superplasticity in Metals, Ceramics, and Intermetallics, MRS Proceeding No.196*, pp. 343–348, ed. M.J. Mayo, J. Wadsworth, and M. Kobayashi, Materials Research Society, Pittsburgh, PA, 1990.

87. S.C. Cheng, J. Wolfenstine, and O.D. Sherby,

'Superplastic Behavior of Two-Phase Titanium Aluminides,' *Metall. Trans.*, **23A** (1992), pp. 1509–1513.

88. R.M. Imayev and V.M. Imayev, 'Mechanical Behavior of TiAl Submicrocrystalline Intermetallic Compound at Elevated Temperatures,' *Scr. Metall. Mater.*, **25** (1991), pp. 2041–2046.

89. R.M. Imayev, V.M. Imayev, and G.A. Salishchev, 'Formation of Submicrocrystalline Structure in TiAl Intermetallic Compound,' *J. Mater. Sci.*, **27** (1992), pp. 4465–4471.

90. R.M. Imayev, O.A. Kaibyshev, and G.A. Salishchev, 'Mechanical Behavior of Fine Grained TiAl Intermetallic Compound – I. Superplasticity,' *Acta Metall. Mater.*, **40**(3) (1992), pp. 581–587.

91. T. Maeda, M. Okada, and Y. Shida, 'Superplasticity in Ti-Rich TiAl,' in *International Conference on Superplasticity in Advanced Materials (ICSAM–91)*, pp. 311–316, ed. S. Hori, M. Tokizane, and N. Furushiro, The Japan Society for Research on Superplasticity, Osaka, Japan, 1991.

92. T. Tsujimoto, K. Hashimoto, and M. Nobuki, 'Alloy Design for Improvement of Ductility and Workability of Alloys Based on Intermetallic Compound TiAl,' *Mater. Trans., JIM*, **33**(11) (1992), pp. 989–1003.

93. W.B. Lee, H.S. Yang, Y.-W. Kim, and A.K. Muhkerjee, 'Superplastic Behavior in a Two-Phase TiAl Alloy,' *Scr. Metall. Mater.*, **29** (1993), pp. 1403–1408.

94. D. Lin, A. Shan, and D. Li, 'Superplasticity in Fe_3Al-Ti Alloy with Large Grains,' *Scr. Metall. Mater.*, **31**(11) (1994), pp. 1455–1460.

95. D. Li, A. Shan, Y. Liu, and D. Lin, 'Study of Superplastic Deformation in an FeAl Based Alloy with Large Grains,' *Scr. Metall. Mater.*, **33** (1995), pp. 681–685.

96. T.G. Nieh, C.A. Henshall, and J. Wadsworth, 'Superplasticity at High Strain Rate in SiC–2124 Al Composite,' *Scr. Metall.*, **18** (1984), pp. 1405–1408.

97. T.G. Nieh, P.S. Gilman, and J. Wadsworth, 'Extended Ductility at High Strain Rates in a Mechanically Alloyed Aluminum Alloy,' *Scripta Metall.*, **19** (1985), pp. 1375–1378.

98. K. Higashi, T. Okada, T. Mukai, S. Tanimura, T.G. Nieh, and J. Wadsworth, 'Superplastic Behavior in a Mechanically-Alloyed Aluminum Composite Reinforced with SiC

Particulates,' *Scr. Metall. Mater.*, **26**(2) (1992), pp. 185–190.

99. K. Higashi, T. Okada, T. Mukai, and S. Tanimura, 'Superplastic Behavior at High Strain Rates of a Mechanically-Alloyed Al-Mg-Li Alloy,' *Scr. Metall. Mater.*, **26** (1992), pp. 761–766.

100. K. Higashi, T. Okada, T. Mukai, S. Tanimura, T.G. Nieh, and J. Wadsworth, 'Superplasticity in Very Fine Grained Al-Based Alloys Produced by Mechanical Alloying,' *Mater. Trans. JIM*, **36**(2) (1995), pp. 317–322.

101. K. Higashi, T.G. Nieh, and J. Wadsworth, 'A Comparative Study of Superplasticity and Cavitation in Mechanically-Alloyed IN9021 and a SiC$_p$/IN9021 Composite,' *Mater. Sci. Eng.*, **188A** (1994), pp. 167–173.

102. Y.W. Kim and L.R. Bidwell, 'Tensile Properties of a Mechanically-Alloyed Al–4.0Mg Powder Alloy,' *Scr. Metall.*, **16** (1982), pp. 799–802.

103. K. Shinagawa, T. Morioka, and K. Osakada, 'Superplasticity of Mechanically-Alloyed Aluminum at High Strain Rates,' in *International Conference on Superplasticity in Advanced Materials (ICSAM–91)*, pp. 581–586, ed. S. Hori, M. Tokizane, and N. Furushiro, The Japan Society for Research on Superplasticity, Osaka, Japan, 1991.

104. T.G. Nieh and J. Wadsworth, 'Effects of Zr on the High Strain Rate Superplasticity of 2124 Al,' *Scr. Metall. Mater.*, **28** (1993), pp. 1119–1124.

105. N. Furushiro, S. Hori, and Y. Miyake, 'High Strain Rate Superplasticity and its Deformation Mechanism in Aluminum Alloys,' in *International Conference on Superplasticity in Advanced Materials (ICSAM–91)*, pp. 557–562, ed. S. Hori, M. Tokizane, and N. Furushiro, The Japan Society for Research on Superplasticity, Osaka, Japan, 1991.

106. N. Furushiro and S. Hori, 'Significance of High Rate Superplasticity in Metallic Materials,' in *Superplasticity in Metals, Ceramics, and Intermetallics, MRS Proceeding No. 196*, pp. 385–390, ed. M.J. Mayo, J. Wadsworth, and M. Kobayashi, Materials Research Society, Pittsburgh, Pennsylvania, 1990.

107. O.D. Sherby and J. Wadsworth, 'Superplasticity-Recent Advances and Future Directions,' *Prog. Mater. Sci.*, **33** (1989), pp. 166–221.

108. J. Harper and J.E. Dorn, 'Viscous Creep of Aluminum near its Melting Temperature,' *Acta Metall.*, **5** (1957), pp. 654–665.

Chapter 3

Types of superplasticity

Phenomenologically, superplasticity is the capability of certain alloys to undergo extensive, neck-free, tensile deformation prior to fracture. Superplastic materials generally exhibit high values of the strain-rate-sensitivity exponent m during tensile deformation, which is characterized by the constitutive equation:

$$\sigma = k\dot{\varepsilon}^m,$$

(3.1)

where σ is the true flow stress, k is a constant, and $\dot{\varepsilon}$ is the true strain rate. Ideal Newtonian viscous behavior is found in materials, such as glass and molasses, for which $m=1$. Most metals and alloys normally exhibit $m<0.2$ whereas superplastic alloys typically have values of $m>0.33$. There are two well-established types of superplastic behavior in polycrystalline solids: fine-structure superplasticity (FSS) and internal-stress superplasticity (ISS).

3.1   Fine-structure superplasticity (FSS)

The first type of superplastic behavior, and the best known and studied, is FSS. The structural prerequisites for developing superplastic materials have been well established for metal-based, fine-grained materials. They are, however, less clearly developed for fine-grained materials such as intermetallics, ceramics, and composites. In the following sections, a number of prerequisites are given for the development of FSS materials.

3.1.1 Fine grain size

In the case of FSS materials, a strain-rate-sensitivity exponent equal to about 0.5 is usually found and the materials are believed to deform principally by a grain-boundary sliding mechanism. Grain-boundary sliding is often experimentally demonstrated by the offset of scratch lines on a sample surface after testing, as illustrated in Figure 3.1. For grain-boundary sliding to dominate deformation, the grain size should be small. Typically, the grain size for metals should be less than 10 μm, and for ceramics less than 1 μm. A finer grain size for ceramics is related to the grain-boundary tensile-fracture stress (Chapter 10). The strain rate for superplasticity increases as grain size decreases when grain-boundary sliding is the rate-controlling process. Normally, the strain rate is inversely proportional to the grain size raised to the second or third power [1], i.e.

$$\dot{\varepsilon} \propto d^{-p} \tag{3.2}$$

where $p=2$ or 3. Thus, grain-size refinement represents a powerful method of increasing the strain rate for superplastic forming of materials. Another impor-

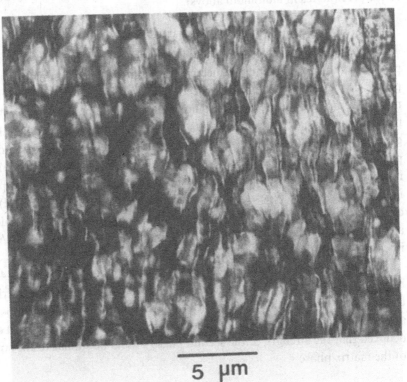

5 μm

Figure 3.1 Offset of surface scratch lines on a superplastically deformed specimen, illustrating the grain boundary sliding process (from Ref. [2]).

tant attribute of achieving a fine grain size is that, for a given rate of deformation, the flow stress decreases as the grain size is decreased. Hence, only low applied forces are required during superplastic forming, thereby reducing costs in energy and, in the case of bulk superplastic forming, die wear.

3.1.2   Second phases

Almost invariably, it is difficult to observe superplasticity in very fine-grained, single-phase materials because grain growth is too rapid at temperatures at which grain-boundary sliding occurs. Therefore, to maintain a fine grain size in the superplastic forming range, the presence of a second phase, or particles at grain boundaries, is required. For this reason, many of the early studies on superplastic materials were based on eutectoid (e.g., $Fe–Fe_3C$), eutectic (e.g., Al–Ca), or monotectoid (e.g., Zn–Al) compositions. All of these classes of materials can be thermomechanically processed to develop fine, equiaxed, two-phase structures. Inhibition of grain growth is usually improved if the quantity of the second phase is increased, provided that the size of the second phase is fine and its distribution is uniform (e.g., by Fe_3C particles in hypereutectoid $Fe–Fe_3C$ alloys or Al_3Zr precipitates in aluminum alloys).

3.1.2.1 Strength of the second phase particle

There is evidence to suggest that the relative strengths of the matrix and second phase constitute important parameters in the control of cavitation during superplastic flow in some systems. Evidence for cavitation at matrix/hard particle or matrix/second phase interfaces is quite extensive. The matrix/hard particle or matrix/second phase interfaces act as stress concentration sites where heterogeneous deformation takes place. For example, the deliberate addition of hard particles of Ag_3Sn to the superplastic Pb–Sn alloy (which does not normally cavitate) was found to cause cavitation nucleated exclusively at the matrix/particle interfaces [3, 4]. This type of behavior has also been observed for Ti(C,N) particles in microduplex stainless steels [5–7], for inclusions in α/β brass [8], at deliberate additions of coarse Fe particles to α/β brass [9], and at the β-phase/matrix interfaces in Cu–Zn–Ni alloys [8, 10].

On the other hand, it should be pointed out that fine-grained Ti–6Al–4V and hypereutectoid $Fe–Fe_3C$ alloys [11] do not exhibit cavitation. This is attributed to the nearly similar strengths of the two phases in the alloys at the temperature at which superplastic flow is found. From these and other observations, it can be deduced that the strength of the second phase should be similar to the strength of the matrix phase.

3.1.2.2 Size, morphology, and distribution of second phase

If the second phase is considerably harder than the matrix phase it should then be distributed uniformly and in fine particle form within the matrix. In the case of fine but hard particles, cavitation during superplastic flow can actually be inhibited by various recovery mechanisms occurring in the vicinity of the particle. For example, Chung and Cahoon [12] have shown how hard but fine silicon particles can minimize cavitation during superplastic flow of a fine-grained Al–Si eutectic alloy. Coarse particles, however, can lead to cavitation. In another material, white cast iron, cavitation occurs at the interface of coarse eutectic carbides (>10 μm) and the matrix during superplastic flow at 700 °C. These large carbides cannot be refined by traditional thermomechanical working; however, it is possible, through rapid solidification technology and powder metallurgy, to obtain fine carbide particles distributed in the iron matrix in white cast iron. It has been shown [13], for example, that a 3% C white cast iron produced by rapid solidification revealed virtually no cavitation after superplastic deformation, even after large tensile elongations in excess of 1000%. It is pointed out that grain-boundary particles may, sometimes, block boundary sliding [14].

Maehara [15] has a unique view on the influence of hard particles on enhancing superplasticity in stainless steels. He investigated a number of stainless steels and concluded that the optimal superplastic condition is achieved when a fine distribution of hard σ phase exists in the austenitic matrix. He obtained over 2000% elongation at 950 °C and at a strain rate of 2×10^{-3} s^{-1} in a stainless steel composed of (by weight) 25% Cr, 7% Ni, 2.8% Mo, 0.85% Mn, 0.5% Si, 0.5% Cu, 0.3% W, and 0.14% N. Maehara considers that optimal superplastic behavior is achieved when recrystallization is occurring in the soft matrix during deformation. He contends that recrystallization will only occur when a sufficiently large amount of hard, small particles exists (about 30% of σ phase). The σ phase particles act as sites for recrystallization because of severe heterogeneous deformation around the hard particles.

Maehara then concluded that soft particles distributed in a hard matrix will not exhibit superplastic behavior. His model has considerable merit. However, it may not be generally applicable to other two-phase systems. For instance, Kim et al. [16] and Kum et al. [17] have shown that iron carbide alloys of eutectic and hypereutectic composition consisting of a continuous phase of relatively hard cementite, with a discontinuous soft phase of iron, are superplastic from 700 to 1025 °C.

3.1.3 Nature of the grain-boundary structure

3.1.3.1 Grain-boundary orientation

The grain boundaries between adjacent matrix grains should be high energy (i.e., high angle, or disordered) because grain-boundary sliding is generally the predominant mode of deformation during superplastic flow. Low-energy (i.e., low-angle) boundaries, such as those often obtained during warm-working, do not readily slide under the appropriate shearing stresses; this is illustrated in the sliding of aluminum [18], copper [19], and zinc [20] bicrystals. The results from the sliding of aluminum bicrystals are given in Figure 3.2. Notably, structures containing low-angle grain boundaries in a eutectoid composition steel are not superplastic, but can be made superplastic by converting the low-angle boundaries to high-angle boundaries by appropriate thermal or thermomechanical treatment. Similar findings have been reported in other polycrystalline solids, for example, a tool steel [21] and an Al–Li-based alloy [22].

In a study of the sliding of Cu–Fe–Co bicrystals, Monzen et al. [14] found that the higher the disorder in the atomic arrangement of a boundary, the less the resistance of the boundary to slide. Monzen et al. [23] further demonstrated that the presence of grain-boundary precipitates can greatly reduce the sliding mobility of the boundary. Other boundary irregularities, such as grain-boundary ledges, may also affect the sliding properties of the boundary [24].

3.1.3.2 Homophase and heterophase

In addition to structure, chemical composition of a boundary can also influence its sliding mobility. A heterophase boundary (a boundary between two grains with dissimilar chemical compositions), for example, is noted to slide more readily than a homophase boundary (a boundary between two grains with similar chemical compositions) [25, 26]. Specifically, the sliding velocity of an austenite–ferrite (α/γ) boundary has been estimated to be about 200 times faster

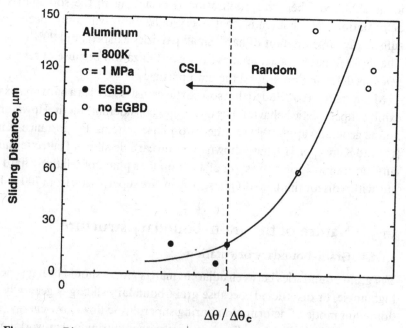

Figure 3.2 Bicrystal sliding displacement as a function of boundary misorientation angle in aluminum at 800 K and $\sigma = 1$ MPa (data from Ref. [18]).

than that of a ferrite–ferrite (α/α) boundary [25]. In addition, the sliding velocity (or rate) of a heterogeneous interface was found to be:

$$v = k\tau^2, \tag{3.3}$$

where v is the boundary sliding velocity, k is a material constant, and τ is the
applied shear stress. This is shown in Figure 3.3. Interestingly, Equation (3.3) has
a similar mathematical form (Equation 3.1) to those normally observed in superplastic polycrystalline materials. It suggests that strain accommodation during
superplastic deformation of some two-phase alloys (e.g., α/β brass [26] and α/γ
stainless steel [25]) may play an unimportant role in determining sliding kinetics.

3.1.4 Texture and shape of grains

The shape of grains should be equiaxed so that the grain boundary can experience a shear stress allowing grain-boundary sliding to occur. Polycrystalline
materials with elongated cylindrical grains, even though fine-grained in a transverse direction, would be expected to exhibit limited grain-boundary sliding
when tested longitudinally. Strain accommodation at triple points is relatively
difficult in materials with textured grains. Testing in a transverse direction could
also lead to extensive creep cavitation along grain boundaries and, therefore,
result in nonsuperplastic behavior.

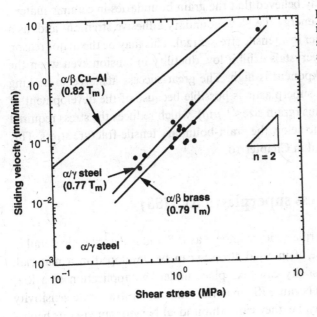

Figure 3.3 Sliding veloc-
ity of α/β brass and α/γ
stainless steel bicrystals as
a function of shear stress
(data from Refs. [25, 26]).

3.1.5   Mobility of grain boundaries

Grain boundaries in superplastic alloys should be mobile. During grain-boundary sliding, stress concentrations develop at triple points, as well as at other obstructions along the grain boundary. The ability of grain boundaries to migrate during grain-boundary sliding permits reduction of these stress concentrations. In this manner, grain-boundary sliding can continue as the major deformation process. The fact that grains remain equiaxed after extensive superplastic deformation is indirect evidence that grain-boundary migration is occurring. One possible explanation for the limited ductility observed in most fine-grained ceramic polycrystals at high temperatures (even despite the high value of strain-rate-sensitivity exponent) is that the grain boundaries are not very mobile. In this case, the lack of boundary mobility can lead to high stress concentrations at triple points during grain-boundary sliding, resulting in crack nucleation and early failure. Also, the presence of grain-boundary impurities in metals, and an excessive amount of glassy phases in ceramics is expected to promote the fracture process.

3.1.6   Grain boundaries and their resistance to tensile separation

Grain boundaries in the matrix phase should not be prone to ready tensile separation. It is generally believed that the grain boundaries in ceramic materials have a high surface energy (i.e., low boundary adhesive strength) and, as a result, will separate under low tensile stresses [27]. This may be the major reason why many ceramic polycrystals exhibit low ductility in tension even when the strain-rate-sensitivity exponent is high. The great success achieved in making polycrystalline zirconia superplastic is possible because of the development of extraordinarily fine grains (grain size<1 μm), which reduces the stress required for plastic flow to a value below the grain-boundary tensile-fracture stress. This will be discussed in detail in Chapter 10.

3.2   Internal-stress superplasticity (ISS)

The second type of superplasticity is known as ISS and is described in detail in Chapter 11. In this case, if internal stresses can be developed in a material, considerable tensile plasticity can take place under the application of a low, externally applied, stress because ISS materials can have a strain-rate-sensitivity exponent as high as unity, i.e. they can exhibit ideal Newtonian viscous behavior. Such superplastic materials deform by a slip-controlled process, and a fine grain size is not a necessary condition. In most cases, internal stresses are generated by (1) thermal cycling or pressure cycling of polymorphic materials through a phase change, (2) thermal cycling of pure metals or single phase alloys that have anisotropic thermal expansivity coefficients, and (3) thermal cycling of composite materials in which the constituents have different thermal expansivity coefficients. Interestingly, pressure-induced phase changes have been cited as a source of superplastic flow in geological materials [28]. It is also possible to achieve Newtonian viscous flow under isothermal and constant pressure conditions from the presence of internal stresses.

The concepts and principles described in FSS and ISS have been technologically applied to enhanced powder consolidation through superplastic flow and the development of superplasticity in laminated composites containing at least one superplastic component.

3.3   High-strain-rate superplasticity (HSRS)

High-strain-rate superplasticity (HSRS) is, in fact, an extension of FSS. It is treated as a separate topic because of the great interest in this subject within the superplasticity community. This area is important because conventional superplasticity is only found at relatively low strain rates, typically about 10^{-4} to 10^{-3} s^{-1}. In contrast, HSRS occurs at much faster rates of about 10^{-1} to 10^{1} s^{-1} which are similar to the rates for conventional forging. These rates will result in a great reduction of forming time; HSRS is expected to have a significant technological impact on the commercial applications of superplastic materials. Grain size is obviously an important microstructural parameter contributing to HSRS. However, Nieh and Wadsworth [29, 30] have pointed out that a fine grain size is a necessary but insufficient condition for the observation of HSRS. The observed HSRS phenomenon is apparently related to the presence of liquid phases at grain boundaries. We will address HSRS in detail in Chapter 9.

3.4   Other mechanisms

There are other observations of large tensile elongations in metals that do not fit into the above classifications. These are described in Chapter 12 and include (1) observations of superplastic-like behavior (up to several hundred percent) in coarse-grained, Class I solid solution alloys, (2) the possibility of achieving high elongations in relatively coarse-grained materials exhibiting values of $m=1$ at low strain rates through Coble creep [31] (grain-boundary diffusion controlled), Nabarro–Herring creep [32] (lattice diffusion controlled), and Harper–Dorn creep [33] (slip controlled), and (3) observations of large plastic strains in Cu and Al under the extremely high strain rates such as those found in anti-armor shape-charge liners. These mechanisms are all included in the overview figure shown in Figure 2.12.

References

1. O.D. Sherby and J. Wadsworth, 'Development and Characterization of Fine Grain Superplastic Material,' in *Deformation, Processing and Structure*, pp. 355–389, ed. G. Krauss, ASM, Metal Park, Ohio, 1984.

2. T.R. Bieler, T.G. Nieh, J. Wadsworth, and A.K. Muhkerjee, 'Superplastic-Like Behavior at High Strain Rates in a Mechanically Alloyed Aluminum,' *Scr. Metall.*, **22** (1988), pp. 81–86.

3. C.W. Humphries and N. Ridley, 'Effect of Relatively Hard Particles on Cavitation of Microduplex Pb-Sn Eutectic During Superplastic Flow,' *J. Mater. Sci.*, **12** (1977), pp. 851–855.

4. D.W. Livesey and N. Ridley, 'Cavitation During Superplastic Flow of Ternary Alloys Based on Microduplex Pb-Sn Eutectic,' *J. Mater. Sci.*, **13** (1978), pp. 825–832.

5. C.I. Smith, B. Norgate, and N. Ridley, 'Superplastic Deformation and Cavitation in a Microduplex Stainless Steel,' *Metal Sci.*, **10** (1976), pp. 182–188.

6. C.W. Humphries and N. Ridley, 'Cavitation in Alloy Steels During Superplastic Deformation,' *J. Mater. Sci.*, **9** (1974), pp. 1429–1435.

7. C.W. Humphries and N. Ridley, 'Cavitation During the Superplastic Deformation of an α/β Brass,' *J. Mater. Sci.*, **13** (1978), pp. 2477–2482.

8. N. Ridley, C.W. Humphries, and D.W. Livesey, 'Cavitation in Copper-Base Alloys and a Steel during Superplastic Flow,' in *Strength of Metals and Alloys, ICSMA 4*, pp. 433–437, ed. E.N.S.M.I.M., Nancy, France, 1976.

9. S. Sagat and D.M.R. Taplin, 'Fracture of a Superplastic Ternary Brass,' *Acta Metall.*, **24** (1976), pp. 307–315.

10. D.W. Livesey and N. Ridley, 'Superplatsic Deformation, Cavitation, and Fracture of Microduplex Cu-Ni-Zn Alloys,' *Metall Trans.*, **9A** (1978), pp. 519–526.

11. B. Walser and O.D. Sherby, 'Mechanical Behavior of Superplastic Ultrahigh Carbon Steels at Elevated Temperature,' *Metall. Trans.*, **10A** (1979), pp. 1461–1472.

12. J.W. Chung and J.R. Cahoon, 'Superplasticity in Aluminum-Silicon Eutectic,' *Metal Sci.*, **13** (1979), pp. 635–640.

13. O.A. Ruano, L.E. Eiselstein, and O.D. Sherby, 'Superplasticity in Rapidly Solidified White Cast Irons,' *Metall. Trans.*, **13A** (1982), pp. 1785–1792.

14. R. Monzen, M. Futakuchi, K. Kitagawa, and T. Mori, 'Measurement of Grain Boundary Sliding of [011] Twist Boundaries in Copper by Electron Microscopy,' *Acta Metall. Mater.*, **41**(6) (1993), pp. 1643–1646.

15. Y. Maehara, 'Superplastic Deformation Mechanism of δ/γ Duplex Stainless Steels,' *Trans. ISIJ*, **27** (1987), pp. 705–712.

16. W.J. Kim, G. Frommeyer, O.A. Ruano, J.B. Wolfenstine, and O.D. Sherby, 'Superplastic Behavior of Iron Carbide,' *Scr. Metall.*, **23** (1989), pp. 1515–1520.

17. D.W. Kum, G. Frommeyer, N.J. Grant, and O.D. Sherby, 'Microstructures and Superplastic Behavior of Eutectic Fe-C and Ni-Cr White Cast Irons Produced by Rapid Solidification,' *Metall. Trans.*, **18A** (1987), pp. 1703–1711.

18. H. Kokawa, T. Watanabe, and S. Karashima, 'Sliding Behavior and Dislocation Structure in Aluminum Grain Boundary,' *Phil. Mag. A*, **44**(6) (1981), pp. 1239–1254.

19. M. Kato and T. Mori, 'Internal Friction of Copper Bicrystals with [001] Twist Boundaries,' *Phil. Mag. A*, **68**(5) (1993), pp. 939–949.

20. T. Watanabe, M. Yamada, S. Shima, and S. Karashima, 'Misorientation Dependence of Grain Boundary Sliding in <1010> Tilt Zinc Bicrystals,' *Phil. Mag. A*, **40**(5) (1979), pp. 667–683.

21. J. Wadsworth, J.H. Lin, and O.D. Sherby, 'Superplasticity in a Tool Steel,' *Metals Technol.*, **8** (1981), pp. 190–193.

22. J. Wadsworth and A.R. Pelton, 'Superplastic Behavior of a Powder-Source Aluminum-Lithium Based Alloy,' *Scr. Metall.*, **18** (1984), pp. 387–392.

23. R. Monzen, K. Kitagawa, and T. Mori, 'Boundary Sliding and Elastic Distortion in Cu Bicrystals with Boundary Precipitates,' *Acta Metall.*, **37** (1989), pp. 1619–1625.

24. P.J. Samsel and C.S. Nichols, 'A Molecular Dynamics Investigation of a Stepped Metal-Ceramic Interface,' *Scr. Metall. Mater.*, **28** (1993), pp. 943–947.

25. S. Hashimoto, F. Moriwaki, T. Mimaki, and
S. Miura, 'Sliding Along the Interphase
Boundary in Austenitic/Ferritic Duplex
Stainless Steel Bicrystals,' in *International
Conference on Superplasticity in Advanced
Materials (ICSAM–91)*, pp. 23–32, ed. S.
Hori, M. Tokizane, and N. Furushiro, The
Japan Society for Research on Superplasticity,
Osaka, Japan, 1991.

26. A. Eberhardt and B. Baudelet, 'Interphase
Boundary Sliding at High Temperature in
Two-Phase (α/β)-Brass Bicrystals,' *Phil. Mag.*,
41(6) (1980), pp. 843–867.

27. J.J. Gilman, 'Mechanical Behavior of
Crystalline Solids,' *National Bureau of
Standards Monograph*, **59** (1963), pp. 79–104.

28. C.M. Sung and R.G. Burns, 'Kinetics of High
Pressure Phase Transformation: Implications
to the Evolution of the Olivine-Spinel
Transformation in the downgoing Lithosphere
and its Consequences on the Dynamics of the
Mantle,' *Tectonophy.*, **31** (1976), pp. 1–31.

29. T.G. Nieh and J. Wadsworth, 'High Strain
Rate Superplasticity in Aluminum Matrix
Composites,' *Mater. Sci. Eng.*, **A147** (1991),
pp. 229–237.

30. T.G. Nieh, J. Wadsworth, and T. Imai, 'A
Rheological View of High-Strain-Rate
Superplasticity in Metallic Alloys and
Composites,' *Scr. Metall. Mater.*, **26**(5) (1992),
pp. 703–708.

31. R.L. Coble, 'A Model for Boundary Diffusion
Controlled Creep in Polycrystalline Materials,'
J. Appl. Phys., **34** (1964), pp. 1679–1682.

32. F.R.N. Nabarro, 'Steady State Diffusional
Creep,' *Phil. Mag. A*, **16** (1967), pp. 231–237.

33. J. Harper and J.E. Dorn, 'Viscous Creep of
Aluminum near its Melting Temperature,'
Acta Metall., **5** (1957), pp. 654–665.

Chapter 4

Mechanisms of high-temperature deformation and phenomenological relations for fine-structure superplasticity

4.1 Creep mechanisms

Creep is a plastic deformation process that occurs in solids at high temperatures, typically, above approximately 0.5 of the homologous temperature, i.e. T/T_m, where T_m is the absolute melting point of the solid [1]. During creep, a solid deforms permanently, under external forces, initially with negligible formation of cracks or voids. This capacity for plastic flow is associated with three discrete mechanisms that can occur at the atomic level. These mechanisms are (a) slip by dislocation movement, (b) sliding of adjacent grains along grain boundaries, and (c) directional diffusional flow. To a first approximation, the three mechanisms can be generally considered to occur independently of one another. Although, in some cases, one mechanism may be necessary to permit accommodation of another, e.g. diffusional flow or slip may be an accommodation mechanism for grain boundary sliding. For the case of large plastic strains, these mechanisms are all thermally activated and are controlled by the diffusion of atoms. They are, therefore, both temperature- and time-dependent.

Creep is commonly characterized by a strain–time curve, i.e. a creep curve. The creep rate is measured directly from the slope of such a creep curve. There are usually two basic types of creep curves: a metal type and an alloy type. As shown in Figure 4.1, for the metal type, the curve normally starts with a primary regime during which the creep rate decreases with time; this region is usually followed by a steady-state regime during which the creep rate is essentially constant; eventually, cavitation and necking begin to develop in the specimen which results in an acceleration of creep rate and leads to a tertiary region and the final failure.

For the alloy type (Figure 4.2), on the other hand, the primary stage (or primary creep) is absent; there exist only the steady-state stage (or secondary creep) and the final stage (or tertiary creep). It is generally considered that the existence of a steady state is a result of a balance between work hardening and recovery softening. From a microstructural point of view, such a region represents a 'constant' microstructure, and as such is considered very useful for model development and evaluation. Studies in the past were mainly focused on this steady-state behavior and led to the development of constitutive equations to describe each of the above mechanisms of plastic flow [2].

Thus, the flow stress, σ (i.e. the stress to cause plastic flow), can be described as a function of strain rate, $\dot{\varepsilon}$, and absolute temperature, T, in the following mathematical form [3]:

$$\dot{\varepsilon} = A' \frac{DGb}{kT} \left(\frac{\sigma}{G}\right)^n \qquad (4.1)$$

where D is the diffusion coefficient; k is Boltzmann's constant; n is the stress exponent (and is equal to $1/m$, where m is the strain-rate-sensitivity exponent); G is the dynamic, unrelaxed, shear modulus; b is Burgers vector, and A is some function of microstructure (principally reflecting the influences of grain size, subgrain size, and dislocation density). To incorporate the grain size dependence, Equation (4.1) can be rewritten as:

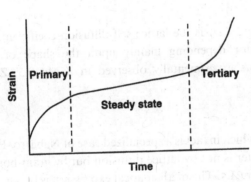

Figure 4.1 A metal-type creep curve showing that the creep rate first decreases in the primary creep regime, is followed by a constant value in the steady-state regime, and accelerates in the tertiary creep regime which leads to the final failure.

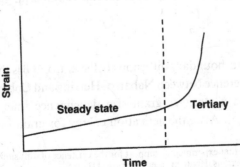

Figure 4.2 An alloy-type creep curve showing the absence of primary creep.

$$\dot{\varepsilon} = A' \frac{DGb}{kT} \left(\frac{b}{d}\right)^{p} \left(\frac{\sigma}{G}\right)^{n}$$ (4.2)

where d is the grain size, and p and A are constants. Equation (4.2) is often referred to as the Dorn equation, or the power-law equation. Each creep mechanism can be described by Equation (4.2) and has specific values of n, p, and D by which the mechanism is defined uniquely. In the following, we summarize briefly creep mechanisms and their specific mathematical forms.

4.1.1 Diffusional creep (n=1)

When high-temperature deformation is a result of the transport of matter by diffusion, rather than dislocation motion, it is named diffusional creep. It occurs typically in fine-grained materials at a very high temperatures ($\sim 0.9\, T_m$) where atom diffusion is rapid. Diffusion is induced by a nonhydrostatic stress and atoms flow from compression sites to tension sites. Diffusional creep can be further divided into Nabarro–Herring creep and Coble creep.

4.1.1.1 Nabarro–Herring creep

In this case, the transport of matter is by diffusion through the grain lattice, as shown in Figure 4.3. The mathematical expression is [4, 5]:

$$\dot{\varepsilon} = \alpha \frac{D_L \sigma \Omega}{d^2 kT}$$ (4.3)

where Ω is the atomic volume, D_L is the lattice self-diffusion coefficient, and α (~ 10) is a numerical factor depending mainly upon the shape of grains. Nabarro–Herring creep was experimentally observed in a Mg–0.5%Zr alloy (Figure 4.4)[*] [7, 12].

4.1.1.2 Coble creep

In the case of Coble creep, which in fact is a specialized case of Nabarro–Herring creep, the transport of matter is not by lattice diffusion but by grain-boundary diffusion, as shown in Figure 4.5. The mathematical expression is [4, 5]:

$$\dot{\varepsilon} = \frac{141 D_{gb} \delta \sigma \Omega}{d^3 kT}$$ (4.4)

where D_{gb} represents the grain-boundary diffusion coefficient, and δ is the grain-boundary thickness. The difference between Nabarro–Herring and Coble creep is that Coble creep ($p=3$) has a stronger grain size dependence than that of Nabarro–Herring creep ($p=2$). Also, the activation energy for grain-boundary

---

[*] It is worth pointing out that experimental support for the existence of diffusional creep has been the source of a serious debate in recent years [6–11].

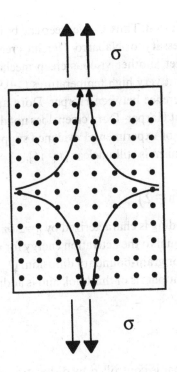

Figure 4.3 Schematic illustration of Nabarro–Herring creep. Atomic flow through the grain lattice leads to plastic deformation.

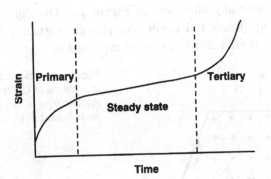

Figure 4.4 Zones denuded of precipitates (magnesium hydride) in polycrystalline magnesium after high-temperature tensile deformation; these denuded zones are believed to be a result of the consequence of plastic deformation by diffusional flow, as illustrated by the schematic diagram in Figure 4.3. (Photograph from Ref. [7].)

diffusion is smaller than that for lattice diffusion. Thus, Coble creep can be important at lower temperatures than those necessary for Nabarro–Herring creep.

It is worth mentioning that there is yet another viscous creep mechanism, Harper–Dorn creep, which exhibits $n=1$ at very high temperatures ($>0.95\ T_m$) and very low stresses [13]. The main difference between Harper–Dorn and the other two diffusional creep processes is that Harper–Dorn creep does not depend on grain size. There exists several pieces of experimental evidence supporting Harper–Dorn creep, but the exact mechanism is still unclear [14–16].

4.1.2   Grain-boundary sliding (n=2)

Plastic deformation by grain-boundary sliding is characterized by $n=2$ ($m=0.5$) and an activation energy which is either equal to the activation energy for lattice diffusion, Q_L, or to the activation energy for grain-boundary diffusion, Q_{gb}. This particular deformation mode is the main emphasis of this book and is addressed in detail in the next section.

4.1.3   Dislocation creep

Dislocation creep refers to deformation that is controlled by dislocation slip in the grain lattice, as schematically illustrated in Figure 4.6. The slip process involves both glide on slip planes and climb over physical obstacles. It is a sequential process and the overall creep rate can be expressed as:

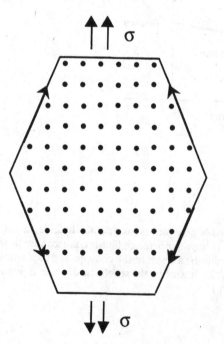

Figure 4.5   Schematic illustration of Coble creep. Atomic flow occurs along grain boundaries.

$$\frac{1}{\dot{\varepsilon}_t} = \frac{1}{\dot{\varepsilon}_g} = \frac{1}{\dot{\varepsilon}_c} \tag{4.5}$$

where $\dot{\varepsilon}_t$ is the overall creep rate, and $\dot{\varepsilon}_g$ and $\dot{\varepsilon}_c$ represent the dislocation glide and climb rates, respectively. It is evident in Equation (4.5) that the slower process controls the overall creep rate.

4.1.3.1 Glide-controlled creep (n=3)

In certain alloys, creep is controlled by the glide step in the glide/climb mechanism mainly because solute atoms impede dislocation motion [1, 17]. The stress exponent in these cases is about 3, i.e.

$$\dot{\varepsilon} = KD_s\sigma^3 \tag{4.6}$$

where D_s is the diffusion coefficient of the solute atom in the alloy. Materials in which deformation is controlled by the glide of dislocations can have relatively large elongations (up to 400%). This group of materials, usually called Class I solid solution alloys, will be discussed in detail in Chapter 12.

4.1.3.2 Climb-controlled creep (n=4–5)

In the majority of alloys and pure metals, dislocation climb is more difficult than glide; creep deformation is controlled by the climb of dislocations over physical obstacles in the glide/climb sequential mechanism. In this case, the creep rate can be described by [18]:

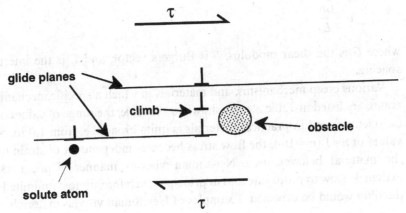

Figure 4.6 Schematic illustration of dislocation creep. Dislocation slip involves both glide on slip planes and climb over physical obstacles.

$$\dot{\varepsilon}=KD_L\sigma^n \tag{4.7}$$

where D_L is the lattice diffusion coefficient, and $n=4$–5. As pointed out by Weertman [18], by making various assumptions on the stress concentration factors in the pile-ups of dislocations against obstacles, it is possible to obtain stress exponent values as high as 6. The creep rates of dislocation slip creep are particularly noted to be independent of grain size.

4.1.4 Dispersion-strengthened alloys ($n>8$)

Many alloys and composites are strengthened by the dispersion of fine, hard particles (e.g., oxides, carbides, nitrides) [19, 20]. In these dispersion-strengthened materials, the creep stress exponent is extremely high ($n\sim25$) and therefore strain-rate-sensitivity values ($m=1/n$) are very low ($m\sim0.04$). The high stress exponents are apparently a result of the existence of a threshold stress [21]. In this case, the stress in Equation (4.2) is replaced by an effective stress

$$\sigma_{eff}=\sigma-\sigma_{th} \tag{4.8}$$

where σ_{eff} and σ_{th} are the effective and threshold stresses, respectively. Although the origin of the threshold stress is sometimes unclear and controversial [22], it is often attributed to Orowan bowing, as shown in Figure 4.7. The bowing stress is expressed as [23]:

$$\sigma\approx\frac{Gb}{L} \tag{4.9}$$

where G is the shear modulus, b is Burgers vector, and L is the interparticle spacing.

Various creep mechanisms, and materials in which a specific mechanism operates, are listed in Table 4.1. As shown in the table, the range of values of stress exponents found in practical materials is quite broad, i.e. from 1.0 to >10. At values of $n=1$ ($m=1/n$), the flow stress becomes independent of strain rate and the material behaves in a Newtonian viscous manner. Thus, necking is extremely slow to propagate and in principle, therefore, almost unlimited tensile ductility would be expected. Examples of Newtonian viscous materials include

Figure 4.7 Dislocation bowing around physical obstacles.

hot glass, tar, and well-masticated chewing gum. Classical polycrystalline, fine-grained metallic superplastic materials typically have values of strain rate sensitivity that are about 0.5. Extremely large values of elongation-to-failure (up to 8000%) have been found in such materials, although typical values are about 400–1000%.

In some metallic alloys that are controlled by the glide step in the glide/climb mechanism of high temperature deformation, the stress exponent is approximately 3, and for materials in this group, relatively large elongations (up to 400%) can be found that are intermediate between superplastic ($n \leq 2$) and non-superplastic ($n \geq 5$) materials. In contrast, most materials that deform by a climb-controlled mechanism exhibit a value of stress exponent of ≥ 5 and total elongations are usually not in excess of 50–80%. In dispersion-strengthened materials, the creep stress exponent is extremely high ($n \geq 10$) and therefore strain-rate-sensitivity values ($m = 1/n$) are very low ($m \sim 0.1$). In such cases, corresponding tensile ductilities are also extremely low ($<5\%$).

Table 4.1 Stress exponent values found in different classes of polycrystalline solids

Stress exponent, n $(1/m)$	Materials
1	— Fine grained ceramics
	— Internal stress superplastic materials (α-U, Zn, Al-based and Zn-based composites, Fe, Ti)
	— Fine grained superplastic materials in which grain boundary sliding is accommodated by glide in a Class I solid solution
	— Materials undergoing viscous creep (Nabarro–Herring, Coble, Harper–Dorn)
2	— Classical fine grained superplastic materials including metallic alloys, intermetallics and ceramics
3	— Class I solid solutions
	— Some composites at high strain rates
	— Some coarse-grained intermetallic alloys
4–5	— Class II solid solutions
	— Many pure metal
≥ 10	— Alloys containing dispersions of hard phases

4.2   Grain-boundary sliding with various accommodation processes

Although the structural prerequisites for fine-structure superplasticity (FSS) appear to be well understood, a detailed understanding of the exact mechanism of superplastic flow in FSS materials has not been thoroughly developed. Three principal modes of deformation have been considered in explaining the creep behavior of FSS materials: (1) diffusional flow (Coble), (2) grain-boundary sliding (GBS) accommodated by diffusional flow (Ashby–Verrall), and (3) GBS accommodated by dislocation slip. The predicted relations between strain rate and stress (by the three different modes of deformation mentioned above) are compared with experimental data in Figure 4.8. The data are for a Zn–Al alloy of monotectoid composition and are taken from 11 separate investigations [24–34]. Creep tests were carried out at intermediate temperatures and at fine grain sizes (d=0.5 to 3.5 μm). Under such conditions the superplastic creep rate is presumably controlled by grain-boundary diffusion. As can be seen in Figure 4.8, the slopes of the creep curves represented by diffusional creep (Coble creep

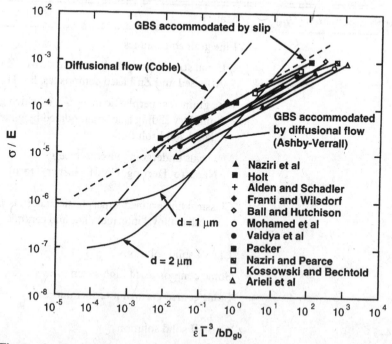

Figure 4.8 Comparison of experimental data and predictions by the three different modes of deformation: diffusional flow (Coble), GBS accommodated by diffusional flow (Ashby–Verrall), and GBS accommodated by slip. (See [38] for references.)

[35, 36]) and by GBS accommodated by diffusional flow (Ashby–Verrall flow [37]) do not agree well with the experimental data. However, GBS accommodated by slip, as represented by the following equation, appears to fit the data rather well:

$$\dot{\varepsilon} = 2 \times 10^9 \left(\frac{b}{d}\right)^2 \frac{D_L}{b^2} \left(\frac{\sigma}{E}\right)^2, \qquad (4.10)$$

where $\dot{\varepsilon}$ is the strain rate, b is Burgers vector, d is the grain size, D_L is the lattice diffusivity, σ is the stress, and E is the dynamic, unrelaxed Young's modulus.

These correlations apparently favor the concept of GBS occurring in a mantle-like region within the region adjacent to the grain boundaries (Figure 4.9), similar to the *core and mantle theory* proposed by Gifkins [39]. Thus, one can consider plastic flow as arising from two independent processes in fine-grained superplastic materials. In one process, GBS accommodated by slip occurs in the mantle region, and in the other, slip occurs within the core of each grain. When the former process dominates deformation, superplasticity can occur, and when the latter dominates deformation, normal ductility is expected. For example, the results obtained by Wyon and coworkers [40–42] provide excel-

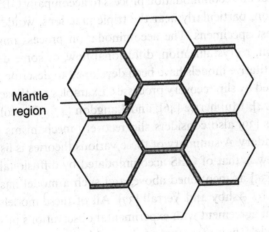

Figure 4.9 A mantle-like region within the region adjacent to the grain boundaries in superplastic materials.

Mantle region

lent support for Gifkins's core and mantle model. These researchers showed how a fine-grained particulate composite of aluminum can be made superplastic at low temperatures by creating weak mantle regions at grain boundaries by the diffusion of gallium along grain boundaries.

Another example is provided by a Zn–Al eutectic alloy, processed to a fine grain size, with a strong crystallographic hexagonal close-packed texture [43]. During superplastic deformation at strain rates for which $m=0.5$, specimens of this alloy experienced a change in cross section from circular to elliptical. Such a directionality in plastic flow indicates that slip occurs near the grain boundary along specific crystallographic planes, contributing to directionality in GBS. The study of an Al–9Zn–1Mg alloy [44] also led to the conclusion that slip processes occur during GBS. Still another example [45] is provided by a sample of fine-grained cadmium deformed at a temperature (150 °C) and strain-rate range (10^{-3} to 10^{-4} s^{-1}) where $m=0.5$. In this case, it was shown that the strength of the fine-grained cadmium polycrystal was dictated by the crystallographic texture existing in the sample. Samples in which grains were oriented for easy basal slip required a lower stress for GBS than did samples in which grains were oriented so that basal slip was difficult.

The most commonly considered mechanisms for superplastic flow involve GBS, and it is necessary for an accommodation process to accompany GBS; otherwise, extensive cavitation, particularly near the triple junctions, would cause premature fracture of test specimens. The accommodation process might be grain-boundary migration, recrystallization, diffusional flow, or some dislocation slip processes. Quantitative models have been developed to describe superplastic flow accommodated by slip recovery processes. Examples are those given by Ball and Hutchison [24], Mukherjee [46], and Langdon [47]. The core and mantle theory of Gifkins [39] also considers slip recovery mechanisms in the vicinity of the grain boundary. A summary of these various theories is listed in Table 4.2. A different view is that of GBS accommodated by diffusional flow, such as by Coble creep [35] as mentioned above, and such a model has been quantitatively developed by Ashby and Verrall [37]. All of these models have certain features that are in agreement with experimental observations in superplastic materials. The models, however, have not been able to predict quantitatively the creep rates actually observed in fine-grained superplastic materials; the predicted rates are usually of two to three orders of magnitude slower than the experiments. In addition, none of the theories is able to predict in one relation the correct stress, temperature, and grain size dependencies.

Generally, the high-temperature deformation (or creep) behavior of fine-grained materials can be schematically represented by the curve shown in Figure 4.10 [55, 56]. The trends shown in Figure 4.10 are based on two competing independent processes during deformation of fine-grained superplastic materials. These processes are (1) GBS with a threshold stress and (2) dislocation-controlled

Table 4.2 Summary of proposed models of GBS

Reference	Equation	Year	Remarks
Slip accommodation (rate controlling)			
Ball–Hutchison [24]	$\dot{\varepsilon}=K_3\left(\dfrac{b}{d}\right)^2 D_{gb}\left(\dfrac{\sigma}{E}\right)^2$	1969	Sliding of group of grains
Langdon [47]	$\dot{\varepsilon}=K_5\left(\dfrac{b}{d}\right) D_L\left(\dfrac{\sigma}{E}\right)^2$	1970	Movement of dislocations adjacent to grain boundary
Mukherjee [46]	$\dot{\varepsilon}=K_3\left(\dfrac{b}{d}\right)^2 D_{gb}\left(\dfrac{\sigma}{E}\right)^2$	1971	Grains slide individually
Hayden *et al.* [48]	$\dot{\varepsilon}=K_7\left(\dfrac{b}{d}\right)^3 D_p\left(\dfrac{\sigma}{E}\right)^2$	1972	$T<T_c$, GBS is rate-controlled by slip
Gifkins [39]	$\dot{\varepsilon}=K_4\left(\dfrac{b}{d}\right)^2 D_{gb}\left(\dfrac{\sigma}{E}\right)^2$	1976	Pile-up at triple points (core–mantle)
Gittus [49]	$\dot{\varepsilon}=K_6\left(\dfrac{b}{d}\right)^2 D_{IPB}\left(\dfrac{\sigma-\sigma_o}{E}\right)^2$	1977	Pile-up at interphase boundary
	$\dot{\varepsilon}=K_7\left(\dfrac{b}{d}\right)^3 D_L\left(\dfrac{\sigma}{E}\right)^2$		$T>T_c$, creep in the grains
Arieli–Mukherjee [50]	$\dot{\varepsilon}=K_8\left(\dfrac{b}{d}\right)^2 D_{gb}\left(\dfrac{\sigma}{E}\right)^2$	1980	Climb of individual dislocations near grain boundary
Sherby–Wadsworth [51]	$\dot{\varepsilon}=6\times10^8\left(\dfrac{b}{d}\right)^3 \dfrac{D_{gb}}{b^2}\left(\dfrac{\sigma}{E}\right)^2$	1984	Phenomenological $T=0.4–0.6T_m$
	$\dot{\varepsilon}=2\times10^9\left(\dfrac{b}{d}\right)^2 \dfrac{D_L}{b^2}\left(\dfrac{\sigma}{E}\right)^2$		Phenomenological $T>0.6T_m$
Kaibyshev *et al.* [52]	$\dot{\varepsilon}=\dfrac{A}{kT}\left(\dfrac{\sigma-\sigma_o}{G}\right)^2\left(\dfrac{b}{d}\right)^2$	1985	Hardening and recovery of dislocations at GBS
Fukuyo *et al.* [53]	$\dot{\varepsilon}=K_{10}\left(\dfrac{b}{d}\right)^2\left(\dfrac{D_{chem}}{b^2}\right)\left(\dfrac{\sigma}{E}\right)^2$	1990	GBS accommodated by dislocation climb

Table 4.2 (*continued*)

Reference	Equation	Year	Remarks
Diffusional accommodation (rate controlling)			
Ashby–Verrall [37]	$\dot{\varepsilon} = K_1 \left(\dfrac{b}{d}\right)^2 D_{eff}\left(\dfrac{\sigma - \sigma_o}{E}\right)$	1973	$D_{eff} = D_L 9\left[1 + \left(\dfrac{3.3w}{d}\right)\left(\dfrac{D_{gb}}{D_L}\right)\right]$
Diffusional accommodation (not rate controlling)			
Padmanabhan [54]	$\dot{\varepsilon} = K_9 \left(\dfrac{b}{d}\right)^2 D\left(\dfrac{\sigma}{E}\right)^2$	1980	D may differ from D_L and D_{gb}

w = Grain-boundary width.

T_c = Critical temperature.

K_1–K_{10} = Materials constants.

σ_o = Threshold stress.

D_{gb}, D_p, and D_{IPB} stand for diffusivities along grain boundary, dislocation pipe, interfacial phase boundary, respectively; D_L and D_{chem} are lattice and chemical diffusivities, respectively.

slip. The dashed line in Figure 4.10 depicts the creep rate of a fine-grained material when GBS with a threshold stress is the deformation process. The solid line in Figure 4.10 represents the creep rate of the material when dislocation creep in the matrix is the rate-controlling process. Because these two processes are considered to be independent, the fastest one is the rate-controlling one. The predicted behavior is given by the thick curve in Figure 4.10. Four regions are depicted: Regions 0, I, II, and III. Regions 0 and III represent deformation controlled by slip, Region I is where a threshold stress is observed and low ductility can be expected, and Region II is where superplastic flow is expected. The data in Figure 4.10 indicate that the threshold stress is a function of temperature, decreasing in magnitude as temperature increases. The actual meaning of threshold stress and its value as a function of temperature, as well as of grain size, is as yet unclear.

Another obstacle to understanding mechanisms of superplastic flow is the existence of Region I creep (described in Figure 4.10). Whereas many researchers associate this low-stress region with low-strain-rate sensitivity, others have observed high-strain-rate sensitivity, even in the same alloy system, inciting fierce debates on this subject [57, 58]. Region I creep has been explained and discussed in various ways: as a deformation process controlled by slip [59], as a process associated with a threshold stress for plastic flow, and as a GBS process with a reduced creep rate from grain growth [60].

Another view, developed by Mayo and Nix [61, 62], uses a modification of

Gifkins's core and mantle model to explain Regions I, II and III in the following way. The core is assumed to have a strain-rate-sensitivity exponent value typical of that observed in coarse-grained materials ($m=0.2$). The mantle region is considered to be weaker than the core region, but otherwise has a similar strain-rate-sensitivity exponent. Mayo and Nix assume that the mantle region, which has a finite width, increases in volume as stress decreases, and it is this change with stress that leads to the transition Region II. Such a model leads to values of m higher than 0.2, although a specific value of m is not predicted for Region II. Experimental work by Mayo and Nix on the surface observations of grains in the superplastic region (involving Pb–Sn and Zn–Al alloys) suggests that the mantle region makes up a large fraction of the total material.

Typical strain-rate-sensitivity exponents observed in FSS materials are at values clustered around 0.5. This is especially true at intermediate temperatures and at fine grain sizes where the activation energy for superplastic flow is equal to that for grain-boundary diffusion [51]. At high temperatures, however, where the activation energy for superplastic flow is equal to that for lattice diffusion, m is equal either to 0.5 or to values greater than 0.5 [51]. Fukuyo *et al.* [53] have developed a model to explain the different results obtained at high temperatures. The model is similar to the Ball–Hutchison, Mukherjee, and Langdon concepts based on a GBS process accommodated by slip (Figure 4.11). As can be seen, the slip accommodation process involves the sequential steps of glide and climb. When

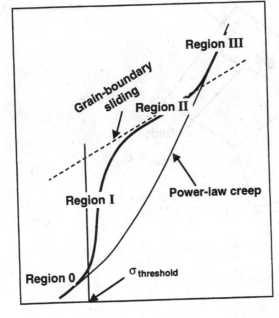

Figure 4.10 High-temperature deformation (or creep) behavior of fine-grained materials.

Log (σ/E)

climb is the rate-controlling step, $m=0.5$ because of the pile-up stress at the head of the climbing dislocation (this, in fact, is the prediction of the Ball–Hutchison,. Mukherjee, and Langdon relations). When glide is the rate-controlling step, however, m is equal to unity because there is no pile-up stress. Because glide and climb processes are sequential, the slower of the two processes is the rate-controlling one. This model predicts that fine-grained, Class I solid-solution alloys can exhibit a high value of m equal to unity because in these alloys the glide step (solute-drag-controlled dislocation creep) is often the slowest process [37, 63]. However, fine-grained, Class II solid-solution alloys in which dislocation climb is the rate-controlling step should exhibit m values no greater than 0.5.

The predictions of Fukuyo *et al.* have been confirmed for a number of fine-grained, solid-solution alloy systems studied at high temperatures, as can be seen in Figures 4.12 and 4.13. In these figures, m is plotted as a function of the strain rate for fine-grained, Class I solid-solution alloys (Figure 4.12) and fine-grained, Class II solid-solution alloys (Figure 4.13). In Region I, the strain rate is normalized to the strain rate at which the strain rate sensitivity $m=0.3$. In this manner, different materials can be assessed from a common base. The Class I solid-solution alloys shown in Figure 4.12 are Fe–10Al–1.25C [64], Ti–6Al–4V [65, 66], Mg–33Al [67], Al–33Cu [68] and Al–25Cu–11Mg [69]. At low strain rates, m is about 0.5, then increases with the increasing strain rate to values as high as 0.8.

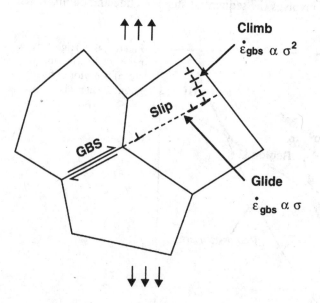

Climb

$\dot{\varepsilon}_{gbs} \propto \sigma^2$

Slip

GBS

Glide

$\dot{\varepsilon}_{gbs} \propto \sigma$

Figure 4.11 A model based on a GBS process accommodated by slip.

This is the trend predicted from the Fukuyo *et al.* model wherein the GBS process is accommodated by climb at low strain rates and accommodated by glide at high strain rates. These two regions are identified as part of Region II of Figure 4.10. At yet higher strain rates, slip deformation becomes the dominant deformation mode and m decreases to about 0.1 to 0.2.

The superplastic materials shown in Figure 4.13 for fine-grained, Class II solid-solution alloys exhibit trends that are quite different from those observed for the fine-grained, Class I solid-solution alloys. The four alloys listed are Ni–39Cr–10Fe–7.5Ti–1Al [70], Cu–39.4Zn [71], Fe–26Cr–6.5Ni–0.05Al [48], and Ag–28Cu [72]. For these fine-grained, solid-solution materials, m is about 0.5 at low strain rates and decreases as strain rates increase. This is the type of creep behavior expected in fine-grained, Class II solid-solution alloys because the accommodation process, in this case, is GBS controlled by dislocation climb where $m = 0.5$.

For superplastic ceramics, the apparent strain-rate-sensitivity values are often observed to be over 0.33 [73–75]. These values were noted to be determined under conditions of concurrent grain growth, commonly taking place during superplastic deformation of ceramics [76–79]. After compensating for the microstructural instability, Nieh and Wadsworth [80] argued that the 'true' m value for superplastic yttria-stabilized tetragonal zirconia (YTZP) is, in fact, greater than

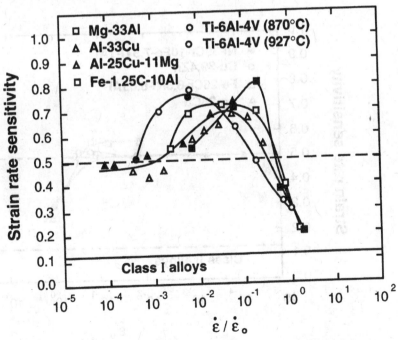

Figure 4.12 Predictions of Fukuyo's model [64] with experimental data from a number of fine-grained, Class I solid-solution alloy systems studied at high temperatures.

0.5. This result, combined with microstructural observations, has led to the general recognition that GBS is also the dominant deformation mechanism in fine-grained, superplastic ceramics. Furthermore, because the deformation temperatures for superplastic ceramics are often relatively low for slip to occur within grains, the accommodation of GBS is unlikely to be caused by slip. Similar to the case of metals, superplastic flow in ceramics can generally be expressed as the following [73]:

$$\dot{\varepsilon} = A d^{-p} \sigma^n \exp\left(-\frac{Q}{RT}\right), \tag{4.11}$$

where $\dot{\varepsilon}$ is the strain rate, A, p, and n are constants, d is the grain size, σ is the flow stress, Q is the activation energy for flow, R is the gas constant, and T is the absolute temperature. The values for p, n, and Q vary according to the microstructure, specific flow/diffusion law, and sometimes the impurity content in a material. It is also noted that, in fine-grained ceramics, dynamic grain growth often takes place during superplastic deformation. Thus, Equation (4.11) is kinematical in nature.

One of the major microstructural features of ceramics, in contrast to metals, is that ceramics usually contain glassy phases at their grain boundaries [81]. Therefore, in addition to the possible diffusional accommodations listed in Table

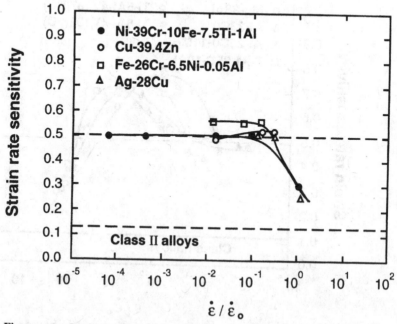

Figure 4.13 Predictions of Fukuyo's model [64] with experimental data from a number of fine-grained, Class II solid-solution alloys systems studied at high temperatures.

4.2, one must consider an accommodation process that is controlled by interface reaction creep. In this case, and when interface reaction and diffusion are sequential processes, creep rate can be expressed by the following equation:

$$\frac{1}{\dot{\varepsilon}} = \frac{1}{\dot{\varepsilon}_d} = \frac{1}{\dot{\varepsilon}_i} \qquad (4.12)$$

where $\dot{\varepsilon}_d$ and $\dot{\varepsilon}_i$ represent the creep rates controlled by diffusion and interface reactions, respectively. The slower process of these two should be the rate-controlling step. In other words, the rate-controlling process for the accommodation of GBS in superplastic ceramics can be either diffusional creep or interface reaction creep [82]. For superplastic YTZP, Wakai and colleagues [83, 84] and Lakki et al. [85] have suggested that the interface reaction is the dominant mechanism. This is not applicable to the observations in some superplastic ceramics that do not contain grain-boundary glassy phases.

In the case of intermetallics, superplasticity was observed in two-phase or quasi-single-phase materials. In addition to the common values of $m=0.5$, a value of $m=0.33$ was often recorded [86–88]. Compared to superplastic ceramics, grain sizes appear to be much coarser in some superplastic intermetallics; thus, the microstructure is stable. In the case of those exhibiting values of $m=0.33$, therefore, viscous dislocation glide has been suggested as the rate-controlling process during superplastic deformation [89]. The drag force was suggested to be generated by the restoration of the lattice order after dislocations glide. If this were the case, the drag force would be expected to be related to the order–disorder energy; however, no data comparison has yet been made.

4.3 Optimizing the rate of superplastic flow in FSS materials

For a given superplastic material, and at a given temperature, there exists a maximum strain rate where superplastic flow by GBS is no longer the dominant process; another mode of deformation becomes important, namely dislocation slip creep. The maximum strain rate at which GBS remains rate-controlling in a conventional superplastic material is typically on the order of 10^{-4} s^{-1}, a rate considerably lower than those used in most commercial forming operations (e.g., 10^{-1} to 1 s^{-1}). From a technological viewpoint, it would be desirable to increase the maximum strain rate for superplastic flow. The approach for attainment of this goal is, in principle, straightforward. One must select structural variables that will enhance GBS but make slip creep more difficult (Figure 4.14).

In Figure 4.14, the logarithm of the stress is plotted as a function of the logarithm of the strain rate. The two separate processes contributing to GBS and slip creep are represented as straight lines. The point of intersection (marked $\dot{\varepsilon}_{SPmax}$) represents the maximum strain rate for superplastic flow for a given set of microstructural conditions. As noted on the figure, the principal equations for GBS (governed by grain-boundary diffusion) and for slip can be written as Equations (4.13) and (4.14), respectively [24, 51]:

$$\dot{\varepsilon}_{gbs} = A_{gbs} \left(\frac{b}{d}\right)^3 \left(\frac{D_{gb}}{b^2}\right) \left(\frac{\sigma}{E}\right)^2 \qquad (4.13)$$

$$\dot{\varepsilon}_{slip} = A_s \left(\frac{\lambda}{b}\right)^3 \left(\frac{D_L}{b^2}\right) \left(\frac{\sigma}{E}\right)^8 \qquad (4.14)$$

where $\dot{\varepsilon}_{gbs}$ and $\dot{\varepsilon}_{slip}$ are the strain rates for GBS and slip, respectively; A_{gbs} and A_s are constants; b is the Burgers vector; d is the mean linear intercept grain size; λ is the minimum barrier spacing governing slip creep (typically the inter-particle spacing or the grain size); D_{gb} is the grain-boundary diffusivity; D_L is the lattice diffusivity; σ is the stress; and E is the dynamic, unrelaxed Young's modulus.

Equation (4.13) is the same as Equation (4.10) except that A_{gbs} is substituted in Equation (4.13) as a more general constant for the average value of 2×10^8 used to describe superplastic fine-grained materials. The point of intersection on Figure 4.14 can be shown by combining Equations (4.13) and (4.14) and rearranging them to be

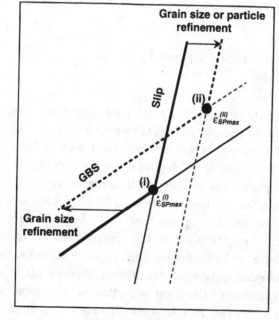

Figure 4.14 Two competing mechanisms of GBS and slip. If the structural state is changed so that GBS is enhanced and slip creep is made more difficult, as shown by the dashed lines, the maximum strain rate for superplasticity flow is increased.

$$\left.\frac{\sigma}{E}\right|_{\dot{\varepsilon}_{SP\,max}} = \left(\frac{A_{gbs}D_{gb}b^6}{A_s D_L d^3 \lambda^3}\right)^{1/6} \tag{4.15}$$

$$\dot{\varepsilon}_{SPmax} = \left[\frac{(A_{gbs}D_{gb})^{4/3}}{(A_s D_L)^{1/3}}\right]\frac{b^3}{d^4 \lambda} \tag{4.16}$$

The generation of a new set of microstructural conditions to make GBS more facile and to inhibit the slip-creep process can increase the maximum strain rate for superplastic flow. Such a change leads to an increase in the maximum rate for superplastic flow from $\dot{\varepsilon}_{SPmax}$(i) to $\dot{\varepsilon}_{SPmax}$(ii), as shown in Figure 4.14 in which the dashed lines represent the new microstructural conditions. The most straightforward structural feature that can be modified to achieve such an enhancement in superplasticity is to decrease the grain size. When the grain size is reduced, the superplastic flow rate (by GBS) is increased and the normal flow rate (by slip creep) is reduced (the Hall–Petch type relation, i.e. $\sigma = \sigma_0 + k\,d^{-1/2}$, where σ and σ_0 are the flow and lattice friction stresses, respectively, d is the grain size, and k is a material constant). In another example, a bimodal distribution of second-phase particles may lead to an ideal superplastic structure. The ultrafine particles can pin dislocation networks (subgrains), yielding a low value of λ, and fine particles can pin the fine grain size, leading to a low value of d.

A series of predictions is shown in Figure 4.15 from Equations (4.13) and (4.14) for the case of a typical high-stacking fault-energy material. In Figure 4.15(a), the logarithm of the maximum strain rate for superplastic flow is shown as a function of the logarithm of grain size for a fixed value of λ (i.e., 1 μm) with homologous temperature indicated as a variable. The figure demonstrates the significant influence of grain size on $\dot{\varepsilon}_{SPmax}$ at all temperatures. By refining the grain size from 100 μm to a submicron value, the strain rate for superplastic flow can be increased from glacial flow rates to values typical of that for high rate forming operations such as extrusion. The influence of temperature is also significant; an increase from 0.55 to 0.72 T_m can improve $\dot{\varepsilon}_{SPmax}$ by an order of magnitude at a fixed grain size. It is important to recognize that the microstructure must remain stable over this increased temperature range for these predictions to be realized. The predictions shown in Figure 4.15 are based on $A_{gbs} = 10^8$ and $A_s = 10^9$. The lattice diffusion coefficient was selected for a typical body-centered cubic metal ($D_L = 10^{-4} \exp[-17T_m/T]$ m²/s and the grain-boundary diffusion coefficient was calculated from $D_{gb} = 10^{-4} \exp[-11T_m/T]$ m²/s).

High-strain-rate superplastic forming may not always be practical. This is because the force required during superplastic forming must be low. For example, in blow forming procedures, there is a limitation to the gas pressure that can be utilized to form a superplastic part. In bulk forming procedures, there may be a limit to the pressure used in filling a die related to strength limitations

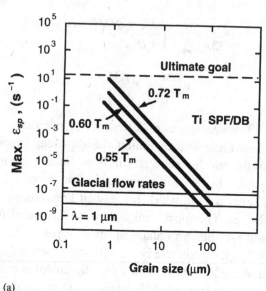

Figure 4.15 Deformation maps for a typical high-stacking fault-energy material showing the influence of temperature (a), grain size (b), and interparticle spacing λ (c) on the maximum strain rate for superplasticity $\dot{\varepsilon}_{SPmax}$.

(a)

of the die. Therefore, a compromise often must to be reached between a high forming rate and a low applied force.

In Figure 4.15(b) and (c), the influence of the ease (or difficulty) of slip creep on the strain rate for superplastic flow is examined for the specific homologous temperatures of 0.60 and 0.72, respectively. The logarithm of $\dot{\varepsilon}_{SPmax}$ is plotted as a function of d, the microstructural feature governing slip creep for a variety of grain sizes. As may be seen, the predominant influence on $\dot{\varepsilon}_{SPmax}$ is that of grain size; however, the role of λ, the microstructural feature that governs slip creep, is also significant. For example, at a grain size of 2 μm at 800 °C, refinement of λ from 1 μm to 0.01 μm is predicted to result in an improvement in $\dot{\varepsilon}_{SPmax}$ by two orders of magnitude from 10^{-3} to 10^{-1} s^{-1}.

A number of recent developments have centered on aluminum alloys with the objective of increasing the strain rate range for superplastic flow by refining grain size. An illustration of the beneficial effect of fine grain size in aluminum alloys is shown in Figure 4.16 [90]. In this figure, an increase in strain rate or a decrease in stress with decreasing grain size is clearly illustrated. It is also noted that, despite the different strain rates resulting from different grain sizes, the mechanisms governing superplastic deformation appear to be similar for all these alloys, namely, they exhibit a similar stress exponent value. The high-strain-rate behavior of superplastic materials is discussed in detail in Chapter 9.

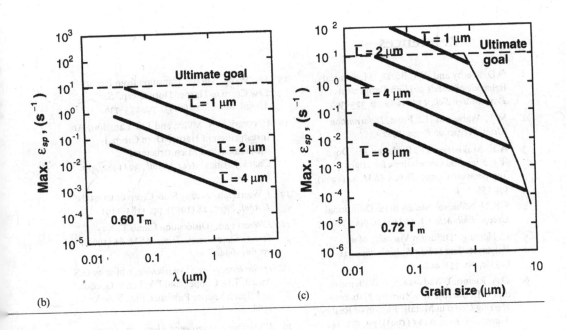

(b)

(c)

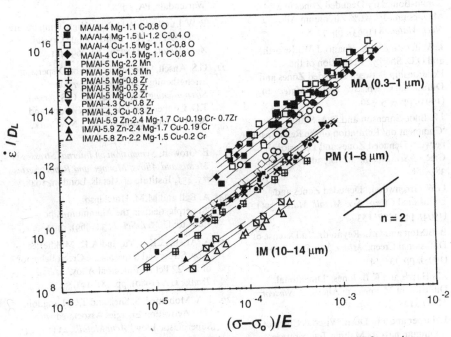

Figure 4.16 Logarithm of $\dot{\varepsilon}_{SPmax}/D_L$ vs logarithm of $(\sigma - \sigma_o)/E$ for various ingot-metallurgy (IM), powder-metallurgy (PM), and mechanically alloyed (MA) aluminum alloys showing grain size effects on superplastic strain rate.

References

1. O.D. Sherby and P.M. Burke, 'Mechanical Behavior of Solids at High Temperatures,' *Prog. Mater. Sci.*, **13** (1967), pp. 325–390.

2. M.F. Ashby and H.J. Frost, *Deformation Maps*. Pergamon Press, 1982.

3. A.K. Mukherjee, J.E. Bird, and J.E. Dorn, 'Experimental Correlations for High-Temperature Creep,' *Trans. ASM*, **62** (1969), pp. 155–179.

4. F.R.N. Nabarro, 'Steady State Diffusional Creep,' *Phil. Mag. A*, **16** (1967), pp. 231–237.

5. C. Herring, 'Diffusion Viscosity of a Polycrystalline Solid,' *J. Appl. Phys.*, **21** (1951), pp. 437–445.

6. O.A. Ruano, J. Wadsworth, J. Wolfenstine, and O.D. Sherby, 'Evidence for Nabarro-Herring Creep in Metals: Fiction or Reality?,' *Mater. Sci. Eng.*, **A165** (1993), pp. 133–141.

7. R.L. Squires, R.T. Weiner, and M. Phillips, 'Grain-Boundary Denuded Zones in a Magnesium–1/2 wt.% Zirconium Alloy,' *J. Nucl. Mater.*, **8** (1963), pp. 77–80.

8. J. Wolfenstine, O.A. Ruano, J. Wadsworth, and O.D. Sherby, 'Refutation of the Relationship Between Denuded Zones and Diffusional Creep,' *Scr. Metall. Mater.*, **29** (1993), pp. 515–520.

9. J.B. Bilde-Sorenson and D.A. Smith, 'Comment on 'Refutation of the Relationship between Denuded Zones and Diffusional Creep',' *Scr. Metall. Mater.*, **30** (1994), pp. 383–386.

10. G.W. Greenwood, 'Denuded Zones and Diffusional Creep,' *Scr. Metall. Mater.*, **30**(12) (1994), pp. 1527–1530.

11. B. Burton and G.L. Reynolds, 'In Defense of Diffusional Creep,' *Mater. Sci. Eng.*, **A191** (1995), pp. 135–141.

12. J.E. Harris and R.B. Jones, 'Directional Diffusion in Magnesium Alloys,' *J. Nuclear Mater.*, **10** (1963), pp. 360–362.

13. J. Harper and J.E. Dorn, 'Viscous Creep of Aluminum near its Melting Temperature,' *Acta Metall.*, **5** (1957), pp. 654–665.

14. J.N. Wang and T.G. Nieh, 'Effects of the Peierls Stress on the Transition from Power Law Creep to Harper-Dorn Creep,' *Acta Metall. Mater.*, **43** (1995), pp. 1415–1419.

15. J.N. Wang, 'On the Transition from Power Law Creep to Harper-Dorn Creep,' *Scr. Metall. Mater.*, **29** (1993), pp. 733–736.

16. P. Yavari, D.A. Miller, and T.G. Langdon, 'An Investigation of Harper-Dorn Creep, I-Mechanical and Microstructural Characteristics,' *Acta Metall.*, **30** (1982), pp. 871–879.

17. J. Weertman, 'Steady State Creep of Crystals,' *J. Appl. Phys.*, **28** (1957), pp. 1185–1191.

18. J. Weertman, 'Dislocation Climb Theory of Steady-State Creep,' *Trans. ASM*, **61** (1968), pp. 681–694.

19. *Oxide Dispersion Strengthening*, edited by G.S. Ansell, T.D. Cooper, and F.V. Lenel, Gordon and Beach Science Publisher, Inc., New York, 1968.

20. *Dispersion Strengthened Aluminum Alloys*, edited by Y.-W. Kim and W.M. Griffith, TMS, Warrendale, PA, 1988.

21. R.W. Lund and W.D. Nix, 'High Temperature Creep of Ni–20Cr–2ThO$_2$ Single Crystal,' *Acta Metall.*, **24** (1976), pp. 469–481.

22. G.S. Ansell, 'The Mechanism of Dispersion Strengthening,' in *Oxide Dispersion Strengthening*, pp. 61–141, ed. G.S. Ansell, T.D. Cooper, and F.V. Lenel, Gordon and Beach Science Publisher, Inc., New York, 1968.

23. E. Orowan, *Symposium on Internal Stresses in Metals and Alloys, Monogr. and Report Series* #5. 451, Institute of Metals, London, 1948.

24. A. Ball and M.M. Hutchinson, 'Superplasticity in the Aluminum-Zinc Eutectoid,' *Met. Sci. J.*, **3** (1969), pp. 1–6.

25. A. Arieli, A.K.S. Yu, and A.K. Mukherjee, 'Low Stress and Superplastic Creep Behavior of Zn–22 Pct Al Eutectoid Alloy,' *Metall. Trans.*, **11A** (1980), pp. 181–191.

26. F.A. Mohamed, S. Shei, and T.G. Langdon, 'The Activation Energies Associated with Superplastic Flow,' *Acta Metall.*, **23** (1975), pp. 1443–1450.

27. H. Naziri, R. Pearce, M. Brown, and K.F. Hale, 'Microstructural-Mechanism Relationship in the Zinc/Aluminum Eutectoid Superplastic Alloy,' *Acta Metall.*, **23** (1975), pp. 489–496.

28. D.L. Holt, 'The Relationship between Superplasticity and Grain Boundary Shear in the Aluminum-Zinc Eutectoid Alloy,' *Trans. TMS-AIME*, **242** (1968), pp. 25–31.

29. T.H. Alden and H.W. Schadler, 'The Influence of Structure on the Flow Stress-Strain Rate Behavior of Zn-Al Alloys,' *TMS-AIME*, **242** (1968), pp. 825–832.

30. M.L. Vaidya, K.L. Murty, and J.E. Dorn, 'High-Temperature Deformation Mechanisms in Superplastic Zn–22Al Eutectoid,' *Acta Metall.*, **21** (1973), pp. 1615–1623.

31. R.J. Prematta, P.S. Venkatesan, and A. Pense, 'Superplasticity in Tin-Zinc Eutectoid System,' *Metall. Trans.*, **7A** (1976), pp. 1235–1236.

32. C.M. Packer, Ph.D. Thesis, Stanford University, (1967).

33. H. Naziri and R. Pearce, 'Behavior of Superplastic Zn-Al Eutectoid,' *J. Inst. Metals*, **101** (1973), pp. 197–202.

34. R. Kossowsky and J.H. Bechtold, 'Structural Changes during Superplastic Deformation in Al–78 Wt Pct Zn Alloy,' *Trans. TMS-AIME*, **242** (1968), pp. 716–719.

35. R.L. Coble, 'A Model for Boundary Diffusion Controlled Creep in Polycrystalline Materials,' *J. Appl. Phys.*, **34** (1964), pp. 1679–1682.

36. J.R. Spingarn, D.M. Barnett, and W.D. Nix, 'Theoretical Description of Climb Controlled Steady State Creep at High and Intermediate Temperatures,' *Acta Metall.*, **27** (1979), pp. 1549–1562.

37. M.F. Ashby and R.A. Verrall, 'Diffusion-Accommodated Flow and Superplasticity,' *Acta Metall.*, **21** (1973), pp. 149–163.

38. O.D. Sherby and J. Wadsworth, 'Superplasticity-Recent Advances and Future Directions,' *Prog. Mater. Sci.*, **33** (1989), pp. 166–221.

39. R.C. Gifkins, 'Grain Boundary Sliding and its Accommodation during Creep and Superplasticity,' *Metall. Trans.*, **7A** (1976), pp. 1225–1232.

40. S.K. Marya and G. Wyon, 'Superplasticite a l'ambiante de l'aluminium a grain fin, en liaison avec l'existence d'un film inter-granulaire de solution solide riche en gallium,' *J. Physique*, **36**(C.4) (1975), pp. 309–313.

41. S.K. Marya and G. Wyon, 'Proprietes Superplasticite d'alliages d'Al et de Composites Al-Al$_2$O$_3$ par Formation in situ d'un film intergranulaire de Solution Solide Al-Ga,' in *Strength of Metals and Alloys, ICSMA 4*, pp. 438–442, ed. E.N.S.M.I.M., Nancy, France, 1976.

42. F. Weill and G. Wyon, 'Superplastic Behavior of Fine Grained Aluminum Alloys whose Grain Boundaries have been Enriched by Gallium,' in *Strength of Metals and Alloys, ICSMA 5*, pp. 387–392, ed. P. Haasen, V. Gerold, and G. Kostorz, Pergamon, Oxford, 1979.

43. R.H. Johnson, C.M. Packer, A. L. Anderson, and O.D. Sherby, 'Microstructure of Superplastic Alloys,' *Phil. Mag.*, **18**(156) (1968), pp. 1309–1314.

44. K. Matsuki, H. Morita, M. Yamada, and Y. Murakami, 'Relative Motion of Grains during Superplastic Flow of an Al–9Zn–1wt%Mg Alloy,' *Metal Sci.*, **11** (1977), pp. 156–163.

45. S.-E. Hsu, G.R. Edwards, and O.D. Sherby, 'Influence of Texture on Dislocation Creep and Grain Boundary Sliding in Fine-Grained Cadmium,' *Acta Metall.*, **31** (1983), pp. 763–772.

46. A.K. Mukherjee, 'The Rate Controlling Mechanism in Superplasticity,' *Mater. Sci. Eng.*, **8** (1971), pp. 83–89.

47. T.G. Langdon, 'Grain Boundary Sliding as a Deformation Mechanism During Creep,' *Phil. Mag.*, **22A**(178) (1970), pp. 689–700.

48. H.W. Hayden, S. Floreen, and P.D. Goodall, 'The Deformation Mechanisms of Superplasticity,' *Metall. Trans.*, **3A** (1972), pp. 833–842.

49. J.H. Gittus, 'Theory of Superplastic Flow in Two-Phase Materials: Roles of Interphase-Boundary Dislocations, Ledges, and Diffusion,' *Trans. ASME – J. Eng. Mater. Tech*, **99** (1977), pp. 244–251.

50. A. Arieli and A.K. Mukherjee, 'A Model for the Rate-Controlling Mechanism in Superplasticity,' *Mater. Sci. Eng.*, **45** (1980), pp. 61–70.

51. O.D. Sherby and J. Wadsworth, 'Development and Characterization of Fine Grain Superplastic Material,' in *Deformation, Processing and Structure*, pp. 355–389, ed. G. Krauss, ASM, Metal Park, Ohio, 1984.

52. O.K. Kaibyshev, R.Z. Valiev, and A.K. Emaletdinov, 'Deformation Mechanisms and the Theory of Structural Superplasticity of Metals,' *Phys. Stat. Sol. (a)*, **90** (1985), pp. 197–206.

53. H. Fukuyo, H.C. Tsai, T. Oyama, and O.D. Sherby, 'Superplasticity and Newtonian-Viscous Flow in Fine-Grained Class I Solid Solution Alloys,' *ISIJ International*, **31**(1) (1991), pp. 76–85.

54. K.A. Padmanabhan, 'A Reply to 'Comments on Theories of Structural Superplasticity',' *Mater. Sci. Eng.*, **40** (1979), pp. 285–292.

55. O.D. Sherby, 'Advances in Superplasticity and in Superplastic Materials,' *ISIJ International*, **29** (1989), pp. 698–716.

56. A. Salama, Ph.D. Thesis, U.S. Naval Postgraduate School, Monterey, CA, (1987).

57. D. Grivas, J.W. Morris, Jr, and T.G. Langdon, 'Observations on the Differences Reported in Region I for the Superplastic Zn–22%Al Eutectoid,' *Scr. Metall.*, **15** (1981), pp. 229–236.

58. A. Arieli and A.K. Mukherjee, 'Reply to Observations on the Differences Reported in Region I for Superplastic Zn–20%Al Eutectoid,' *Scr. Metall.*, **15** (1981), pp. 237–244.

59. O.A. Kaibyshev, 'Contribution of Intragranular Slip to the Total Superplastic Strain in AMg6 Alloy,' *Czech. J. Phys.*, **31B** (1981), pp. 223–227.

60. G. Rai and N.J. Grant, 'On Measurements of Superplasticity in an Al-Cu Alloy,' *Metall. Trans.*, **6A** (1975), pp. 385–390.

61. M.J. Mayo and W.D. Nix, 'A Microindentation Study of Superplasticity in Pb, Sn, and Sn–28%Pb,' *Acta Metall.*, **36** (1988), pp. 2183–2192.

62. M.J. Mayo, Ph.D. Thesis, Stanford University, (1988).

63. W.R. Cannon and O.D. Sherby, 'Creep Behavior and Grain Boundary Sliding in Polycrystalline Al_2O_3,' *J. Am. Ceram. Soc.*, **60** (1970), pp. 44–47.

64. H. Fukuyo, Engineering Dissertation, Dept. of Materials Science and Engineering, Stanford University, (1987).

65. C.H. Hamilton and A.K. Ghosh, 'Strain Rate Sensitivity Index in Superplastic Deformation of Ti–6Al–4V,' *Metall. Trans.*, **11A** (1980), pp. 1494–1496.

66. C.H. Hamilton, A.K. Ghosh, and M.M. Mahoney, 'Superplastic Deformation in Ti Alloys,' in *Advanced Processing Methods for Titanium*, pp. 129–138, ed. D.F. Hasson and C.H. Hamilton, TMS-AIME, Warrendale, PA, 1982.

67. D. Lee, 'The Nature of Superplastic Deformation in the Mg-Al Eutectic,' *Acta Metall.*, **17** (1969), pp. 1057–1069.

68. D.L. Holt and W.A. Backofen, 'Superplasticity in the Al-Cu Eutectic Alloy,' *Trans. ASM*, **59** (1966), pp. 755–765.

69. R. Horiuchi, A.B. El-Sebai, and M. Otsuka, 'Superplasticity in the Ternary Eutectic Alloys Al–33wt%Cu–7wt%Mg and Al–25wt%Cu–11wt%Mg,' *Scr. Metall.*, **7** (1973), pp. 1101–1104.

70. H.W. Hayden, R.C. Gibson, H.P. Merrick, and J.H. Brophy, 'Superplasticity in the Ni-Fe-Cr System,' *Trans. ASM*, **60**(1) (1967), pp. 3–13.

71. M. Suery and B. Baudelet, 'Hydrodynamical Behavior of a Two-Phase Superplastic Alloy: α/β Brass,' *Phil. Mag. A*, **41** (1980), pp. 41–64.

72. H.E. Cline and D. Lee, 'Strengthening of Lamellar vs Equiaxed Ag-Cu Eutectic,' *Acta Metall.*, **18** (1970), pp. 315–323.

73. T.G. Nieh, J. Wadsworth, and F. Wakai, 'Recent Advances in Superplastic Ceramics and Ceramic Composites,' *Inter. Mater. Rev.*, **36**(4) (1991), pp. 146–161.

74. T.G. Langdon, 'Superplastic Ceramics – an Overview,' in *Superplasticity in Aerospace II*, pp. 3–18, ed. T.R. McNelly and C. Heikkenen, The Minerals, Metals & Materials Society, Warrendale, PA, 1990.

75. T. Hermansson, H. Swan, and G. Dunlop, 'The Role of the Intergranular Glassy Phase in the Superplastic Deformation of Y-TZP Zirconia,' in *Euro-Ceramics*, pp. 3.329–3.333, ed. G. de With, R.A. Terstra, and R. Metselaar, Elsevier Applied Science, London, 1989.

76. C. Carry and A. Mocellin, 'Superplastic Forming of Alumina,' *Proc. Brit. Ceram. Soc*, **33** (1983), pp. 101–115.

77. T.G. Nieh and J. Wadsworth, 'Dynamic Grain Growth in Yttria-Stabilized Tetragonal Zirconia during Superplastic Deformation,' *J. Am. Ceram. Soc.*, **72**(8) (1989), pp. 1469–1472.

78. D.J. Schissler, A.H. Chokshi, T.G. Nieh, and J. Wadsworth, 'Microstructural Aspects of Superplastic Tensile Deformation and Cavitation Failure in a Fine-Grained, Yttria Stabilized, Tetragonal Zirconia,' *Acta Metall. Mater.*, **39**(12) (1991), pp. 3227–3236.

79. L.A. Xue and R. Raj, 'Superplastic Deformation of Zinc Sulfide near Transformation Temperature (1020 °C),' *J. Am. Ceram. Soc*, **72** (1989), pp. 1792–1796.

80. T.G. Nieh and J. Wadsworth, 'Superplastic Behavior of a Fine-Grained, Yttria-Stabilized, Tetragonal Zirconia Polycrystal (Y-TZP),' *Acta Metall. Mater.*, **38** (1990), pp. 1121–1133.

81. D.R. Clarke and G. Thomas, 'Grain Boundary Phases in a Hot-Pressed MgO Fluxed Silicon Nitride,' *J. Am. Ceram. Soc.*, **60**(11–12) (1977), pp. 491–495.

82. R. Raj, 'Creep in Polycrystalline Aggregates by Matter Transport Through a Liquid Phase,' *J. Geophys. Res.*, **87**(B6) (1982), pp. 4731–4739.

83. F. Wakai, 'Non-Newtonian Flow and Micrograin Superplasticity in Ceramics,' in *MRS Intl. Meeting on Advanced Materials Vol 7 (IMAM-7, Superplasticity)*, pp. 225–232, ed. M. Kabayashi and F. Wakai, Materials Research Society, Pittsburgh, PA, 1989.

84. F. Wakai and T. Nagano, 'The Role of Interface-Controlled Diffusion Creep on Superplasticity of Yttria-Stabilized Tetragonal ZrO_2 Polycrystals,' *J. Mater. Sci. Lett.*, **7** (1988), pp. 607–609.

85. A. Lakki, R. Schaller, M. Nauer, and C. Carry, 'High Temperature Superplastic Creep and Internal Friction of Yttria Doped Zirconia Polycrystals,' *Acta Metall. Mater.*, **41** (1993), pp. 2845–2853.

86. J. Mukhopadhyay, G.C. Kaschner, and A.K. Muhkerjee, 'Superplasticity and Cavitation in Boron Doped Ni_3Al,' in *Superplasticity in Aerospace II*, pp. 33–46, ed. T.R. McNelly and C. Heikkenen, The Minerals, Metals & Materials Society, Warrendale, PA, 1990.

87. T.G. Nieh, 'Superplasticity in $L1_2$ Intermetallic Alloys,' in *Superplasticity in Metals, Ceramics, and Intermetallics, MRS Proceeding No.196*, pp. 343–348, ed. M.J. Mayo, J. Wadsworth, and M. Kobayashi, Materials Research Society, Pittsburgh, PA, 1990.

88. H.S. Yang, P. Jin, E. Dalder, and A.K. Muhkerjee, 'Superplasticity in a Ti_3Al-base Alloy Stabilized by Nb, V, and Mo,' *Scr. Metall. Mater.*, **25** (1991), pp. 1223–1228.

89. W.B. Lee, H.S. Yang, Y.-W. Kim, and A.K. Muhkerjee, 'Superplastic Behavior in a Two-Phase TiAl Alloy,' *Scr. Metall. Mater.*, **29** (1993), pp. 1403–1408.

90. O.D. Sherby, unpublished work, Stanford University, 1993.

Chapter 5

Fine-structure superplastic metals

Most superplastic metal alloys exhibit large tensile elongations of about 500% to over 1000%. For the most advanced structures, however, the forming strains are typically less than 200 to 300%. Thus, these elongation values are sufficient to make extremely complex shapes using superplastic forming technology. In so doing, large cost and weight savings (through redesign) have provided the driving force for the change from conventional forming to superplastic forming technology. The principal, fine-structured alloy systems that have been commercially exploited for superplastic forming are those based on aluminum, magnesium, iron, titanium, and nickel alloys [1–3]. Other alloy systems, e.g., Zn–Al, Cu–Al, and Pb–Sn, have also been widely explored. The study of these alloys, however, is usually for achieving basic understanding rather than for structural applications. Many reviews already exist to cover these alloys, so in the following sections, we will only discuss those that are important for structural applications.

5.1 Aluminum-based alloys

It is instructive to review the evolution of superplastic aluminum alloys to gain a basic understanding of how a structural alloy group is developed. For this purpose, an overview of the development of superplastic aluminum alloys from 1966 to 1984 is presented in Figure 5.1, where each box represents an individual publication. The description within each of the boxes refers to the nominal alloy composition (in wt%) or to the commercial designation, if appropriate. There is also information in each box regarding whether or not the probability exists that

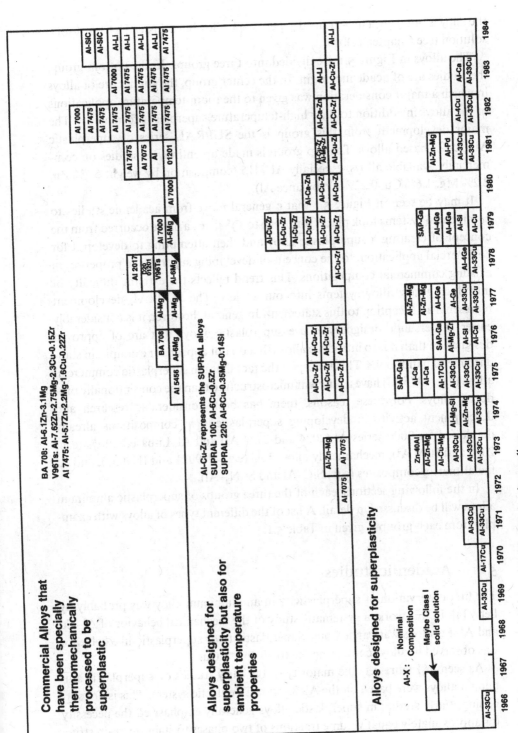

Figure 5.1 The evolution of superplastic aluminum alloys.

the alloy is not truly a fine-grained superplastic alloy but is instead a Class I solid solution (see Chapter 12).

The alloys in Figure 5.1 are divided into three groups. In the bottom group, the studies are of academic origin. In the center group, the studies are of alloys in which a major consideration was given to the room-temperature applications of the alloys in addition to their high-temperature superplastic formability. The major development from this group is the SUPRAL series of superplastic Al–Cu–Zr based alloys. The top group is made up entirely of studies on commercially available alloys, especially Al 7475 (composition by weight 5.7% Zn, 2.2% Mg, 1.6% Cu, 0.22% Cr, balance Al).

It may be seen in Figure 5.1 that a general move from academic studies to commercial systems took place from 1966 to 1984; i.e., a move occurred from the concept of creating a superplastic alloy and then attempting to develop it for commercial application, to the concept of developing superplastic properties in existing commercial compositions. This trend reflects the extreme difficulty of introducing new alloy systems into our society. The SUPRAL development represents an exception to this statement. In general, however, it is considerably easier to persuade designers to use superplastic alloys that are of approved designation than it is to introduce alloys that do not appear, for example, in standard metal handbooks. This is despite the fact that the superplastic commercial alloys (e.g., 7475 Al) have a different microstructure than the conventionally produced alloys. For these reasons, there has been considerable research and development activity in developing superplasticity in compositions already developed for 2000 series Al (2024 and 2124 Al) and Al–Li-based alloys (e.g., 2090 and 8090 Al), mechanically alloyed Al (e.g., IN9021 and IN9052), and Al metal-matrix composites (e.g., SiC–Al and Si_3N_4–Al).

In the following sections, each of the three groups of superplastic aluminum alloys will be discussed in detail. A list of the different types of alloys with examples from each group is given in Table 5.1.

5.1.1   Academic studies

The first observation of superplasticity in an aluminum alloy was probably by Petty [4]. In this work, a systematic study of the mechanical behavior of Al–Cu and Al–Fe alloys was carried out. Superplasticity or superplastic-like behavior was observed in alloys containing 7.9 to 40.4 wt% Cu.

As seen in Figure 5.1, the majority of the early studies on superplastic aluminum alloys were based on the Al–33 wt% Cu eutectic system. This is not surprising. Early studies in superplastic alloys generally emphasized the necessity for approximately equal volume fractions of two phases to stabilize grain structure and prevent rapid grain growth. The relative ease of thermomechanically processing lamellar eutectic structures into fine-grained, spheroidized, two-phase mixtures led to their popularity as model superplastic systems (e.g., Pb–Sn, Bi–Sn, Al–Ca, Al–Si, and Zn–Al). A typical microstructure of the processed Al–33 wt% Cu eutectic is given in Figure 5.2. As may be seen in Figure 5.1, the popularity of the Al–33Cu eutectic alloy as a model system has not diminished over the years.

Although academic studies on aluminum alloys were largely centered on eutectic structures (Al–Cu, Al–Ca, Al–Si, and Al–Pd) other significant academic studies on noneutectic structures were also carried out. These alloys included Al–Cu–Mg, Al–Zn–Mg, Al–Mg–Si, and Al–Mg–Zr. Some of these studies, in particular the work on Al–Ca and Al–Mg–Zr, were directly responsible for the

Table 5.1 Different types of aluminum alloys with specific examples of compositions that have been processed to possess superplastic or superplastic-like behavior

Alloy type	Example (wt%)
Binary eutectics	Al–33Cu
	Al–7.6Ca
	Al–12Si
	Al–23Pd
Binary hypoeutectic	Al–17Cu
Ternary eutectic	Al–Ca–Zn
	Al–Cu–Mg
Grain-size-stabilized, nominally single-phase solid solutions	Al–Mg–(Zr)
	Al–Zn–Mg–(Zr)
Two-phase or multiphase alloys stabilized by a specific element addition	Al–Cu–(Zr)
	7000 series Al (Cr or Zr)
	Al–Li–(Zr)
	Al–Cu–Li–Mg–(Zr)
Fine-grained, two-phase alloys (metal-matrix composites)	Al–Al_2O_3–(Ga)
	Al–SiC
	Al–Si_3N_4
Mechanically alloyed materials	Al–Cu–Mg–O–C
	Al–Mg–O–C
Class I solid-solution alloys	Al–5Mg
	Al–5456
	Al–4Ge
	Al–Mg–Sc

development work in alloys (e.g., Al 5083) combining superplasticity with acceptable room-temperature properties.

In the above examples, the microstructure developed for superplasticity varies significantly. In the eutectic type of compositions, the microstructure is a two-phase one in which a significant volume fraction of coarse (~ 15 μm), second-phase particles is present, which prevents the growth of the matrix grains. In the other type of alloy (e.g., Al–Mg–Zr), a relatively small volume fraction (less than 1 or 2%) of very fine Al_3Zr precipitates pins matrix grains and prevent growth.

An unusual study was carried out on Al–Al_2O_3 particulate alloy (also known as SAP, sintered aluminum powder) by Marya and Wyon [6]. In this work, the aluminum particulate composite was made superplastic near room temperature by diffusion of gallium into the grain-boundary regions. Aluminum containing 10% alumina particles was heavily worked and recrystallized to develop a fine grain size (~ 10 μm). This material exhibited about 8% tensile elongation at 50 °C. As expected at this temperature, the superplastic properties are poor because slip processes dominate deformation. If the Al–Al_2O_3 specimen was immersed in liquid gallium at 50 °C, the gallium diffused along the grain boundaries, forming a very thin film at such regions. The specimen was then soaked for 50 h at 50 °C, allowing gallium to diffuse into a narrow region adjoining the grain boundary, forming an Al–Ga solid-solution mantle.

As a result of the strong temperature dependence of the Al–Ga solidus at a

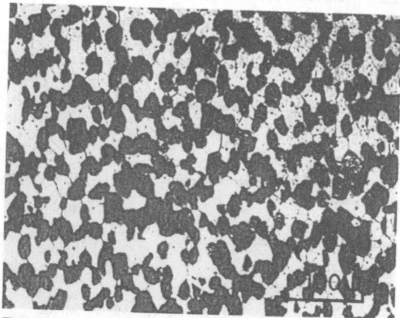

Figure 5.2   A typical microstructure of processed Al–33 wt% Cu eutectic (from Ref. [5]).

temperature of 50 °C, the mantle region is at nonetheless a high homologous temperature. In this condition the material behaves superplastically, exhibiting 300% elongation at 50 °C. In this structural state, grain-boundary sliding (GBS) is possible, because the grain boundary is embedded in the low-melting-point Al–Ga solid-solution region. Thus, even though the temperature is only 50 °C (T/T_m is 0.35 for pure aluminum), the homologous temperature at the grain-boundary region is probably above 0.5 T_m. After superplastic forming, the material can be strengthened by a high-temperature heat treatment that disperses the gallium throughout the aluminum matrix. The above results were further illustrated in several Al–Ti alloys containing 10 wt% Ga [7].

5.1.2 Alloys designed for room-temperature properties as well as superplasticity

The center section of Figure 5.1 consists of alloys that were designed to be superplastic but equal weight was given to the consideration of final use, i.e., room-temperature properties. The most important alloy in this section is the commercial SUPRAL Al–Cu–Zr alloy [8–14]; however, work was also carried out on the Al–Ca–Zn-based alloys [11, 15, 16] and Al–Li-based alloys.

The breakthrough for superplastic Al–Cu alloys was made by Stowell, Watts and Grimes in 1969 when the first of several dilute aluminum alloys (Al–6% Cu–0.5% Zr) was rendered superplastic with the introduction of relatively high levels of zirconium in solution via specialized casting techniques and subsequent thermal treatment to create extremely fine ZrAl$_3$ precipitates. This work was not reported until 1976 [9]. These dilute alloys containing zirconium, later to be known by the trade name SUPRAL, were heavily cold worked to sheet and dynamically recrystallized to a fine stable grain size, typically 4–5 μm, during the initial stages of hot deformation. The dispersion of ZrAl$_3$ particles effectively inhibited grain growth by exerting a drag on the grain boundaries – the Zener effect. Superplastic tensile elongations to failure approaching 2000% have been measured in SUPRAL alloys.

As previously described, the introduction of new alloys into structural applications is sufficiently difficult that work on aluminum alloys has been centered on developing superplasticity in existing commercial alloys (7000 series) rather than adopting new alloys such as the SUPRAL and FORMALL alloys. Work on 7000 series alloys and SUPRAL alloys are described in several reviews [1, 2, 14, 17–19] and for this reason will not be described in detail here; however, some of the superplastic properties of SUPRAL and FORMALL alloys are listed in Table 5.2.

It is interesting to note the studies in 1972 and 1973 on the commercial 7000 series type of compositions manufactured by a noncommercial, rapid-solidification processing (RSP) approach. Although limited superplastic properties were observed in these studies, the work foreshadowed the tremendous interest in developing superplasticity in commercial 7000 series alloys by thermomechanical processing of ingot alloys. Also, since 1985, the work on Al–Li-based alloys has largely centered on potentially commercial compositions, such as 2090 and 8090 Al.

5.1.3 Commercial alloys

The top section of Figure 5.1 consists of commercial alloys that have been thermomechanically processed to develop superplasticity. The main effort has been on the Al 7000 series alloys, Al–Li alloys, Al-based metal-matrix composites, and mechanically alloyed materials [1–3, 18]. A list of the room-temperature mechanical properties of some commercial superplastic aluminum alloys is presented in Table 5.3. The work on aluminum metal-matrix composites and mechanically alloyed materials will be discussed further in Chapter 9, as they exhibit superplasticity at unusually high strain rates. In this section, superplastic monolithic aluminum alloys are explored. In particular, Al–Li-based alloys are used as an example to illustrate the general principles for producing fine-structure superplasticity in aluminum alloys.

Aluminum–lithium alloys are primarily of great interest for the aircraft and aerospace industries because Li is one of only three elements that significantly increases the elastic modulus and simultaneously decreases the density of aluminum; the other elements are beryllium and boron. For aerospace and aircraft applications, improvements in specific modulus and specific strength can lead directly to weight savings [20–22].

To produce a fine-grained structure for superplastic alloys, thermomechanical deformation processes (also known as particle stimulated nucleation, PSN [23])

Table 5.2 Superplastic properties of SUPRAL and FORMALL alloys

Alloy	Temperature (°C)	Strain rate (s^{-1})	Elongation (%)
SUPRAL 100*	470	2.0×10^{-3}	1165
SUPRAL 220	500	4.5×10^{-3}	1060
SUPRAL 150	470	4.5×10^{-3}	890
SUPRAL 5000	500	4.5×10^{-3}	230
FORMALL 700	510	1.0×10^{-3}	240
FORMALL 545	530	1.0×10^{-3}	290
FORMALL 548	530	1.0×10^{-3}	430

* International alloy designation 2004 Al.

are usually used, followed by a recrystallization process that is either static (discontinuous) or dynamic (continuous). In static recrystallization, a heavily deformed microstructure containing a uniform dispersion of relatively coarse (~ 1 μm) and very fine precipitates (e.g., Al_3Zr) is heated to the superplastic forming temperature. Recrystallization is initiated at heavy deformation zones, e.g., at regions adjacent to the coarse (~ 1 μm) second-phase precipitates that are often created by a deliberate over-age practice; see Figure 5.3. Fine grain sizes in aluminum alloys developed by the above over-age practice are quite stable even at temperatures as high as about 500 °C; grain growth is prevented by the dispersion of fine particles, i.e., Al_3Zr. The fine-grained, recrystallized microstructure is very stable in the absence of stress but will usually coarsen under stress during superplastic forming. Examples of commercial type alloys that undergo static recrystallization are Al–Cu–Mg–Li–Zr alloys [24] and the Al 7475 alloy [23].

In the case of dynamic recrystallization, a heavily deformed microstructure undergoes gradual recrystallization under the presence of an applied stress, often during superplastic forming at relatively high strain rates (as opposed to static recrystallization taking place prior to superplastic forming). A fine, steady-state grain size is usually obtained during superplastic deformation [25]. Examples of dynamically recrystallized alloys include the commercial SUPRAL alloys [8] as well as some of the Al–Li-based alloys [26]. Other superplastic aluminum alloys that undergo dynamic or static recrystallization are described elsewhere [27].) The precise roles of the second-phase size and distribution, as well as the thermomechanical processing history, in the subsequent recrystallization behavior of an alloy vary, depending upon the composition of the alloy. For example, data for Al–Cu alloys indicate that the tensile elongation of the alloys is closely related to their Cu content; this is shown in Figure 5.4 [8, 28].

Table 5.3   Room-temperature properties of some commercial superplastic aluminum

Alloys	Modulus (GPa)	Yield strength (MPa)	Tensile strength (MPa)	Elongation (%)	Density (g/cm³)
SUPRAL 100 (2004 Al)	73.8	300	420	8	2.84
SUPRAL 220	73.8	450	510	6	2.84
FORMALL 570	—	417	482	7.9	2.69
5083 Al	72	150	300	15	2.67
7475 Al	70.0	460	525	12	2.80
2090 Al–Li	80.0	455	490	6	2.61
8090 Al–Li	80.0	350	470	6	2.58

It is interesting to note that as long ago as 1970, an Al–17Cu alloy was produced by Beghi *et al.* [29] by RSP in the belief that the Wne grain size thus produced would render the alloy superplastic; however, this may not always be true. For example, an alloy of composition of Al–3Cu–2Li–1Mg–0.2Zr (similar to Al 2090 alloy) has been produced by rapidly solidified, powder metallurgy (PM) processing [26]. Despite the fact that the PM alloy appeared to be fine-grained, it was not superplastic in the as-extruded condition, because in this condition, the fine microstructure consists of medium-angle (3–10°) subgrains and not true (high-angle) grain boundaries. It is difficult for these medium-angle boundaries to undergo GBS. After appropriate processing, involving static recrystallization, a fine-grained structure (less than 2–4 μm) consisting of high-angle grain boundaries (>10° misorientation) is developed and superplastic behavior is observed (Figure 5.5).

A direct comparison of the properties between the as-extruded and thermomechanically processed Al–3Cu–2Li–1Mg–0.2Zr alloy is given in Figure 5.6. The strain-rate-sensitivity value increases from 0.3 for the as-extruded material to 0.4 for the thermomechanically processed material. Also, the rate-controlling process is noted to change from a lattice diffusion to a more rapid grain-boundary diffusion process. Therefore, the exact nature of grain boundaries in an alloy is also important in determining whether an alloy is superplastic. For commercial alloys, proper subsequent thermomechanical processing is often necessary to produce high-angle grain boundaries.

In aluminum alloys, it has been generally recognized that the fine grain sizes necessary for superplasticity can be produced and maintained by small but deliberate additions of elements such as Zr. For SUPRAL alloys, an amount of

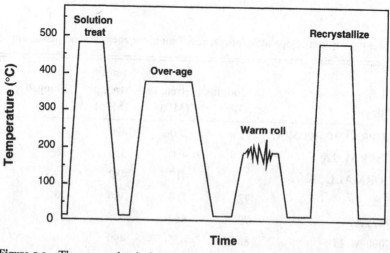

Figure 5.3   Thermomechanical treatment to control grain size in heat-treatable Al alloys.

Zr from about 0.4 to 0.5 wt% was considered to be necessary. This is in excess of that usually found in large-scale ingot alloys (less than 0.2 wt%), therefore, these alloys need to be chill-cast in production. It is now clear that amounts smaller than 0.4 wt% Zr are adequate, at least in some aluminum alloys. The processing steps required, however, have to be modified and the precise physical mechanism leading to fine grain sizes may be different, i.e., in some cases static recrystallization is observed, in others dynamic recrystallization occurs [30]. For example, Al–Cu–Li–Mg–Zr alloys manufactured by ingot metallurgy were restricted to levels of Zr from 0.12 to 0.18% [31]. The alloys were processed by an over-age practice and were also shown to be superplastic. In these cases, the alloys exhibit static recrystallization. In contrast, another Al–Li–Mg–Cu alloy containing only 0.18% Zr, which was manufactured by a PM method, has been demonstrated to be superplastic via dynamic recrystallization [32]. It would appear, therefore, that the thermomechanical processing is as important as the Zr content in an alloy.

Some examples of the superplastic aluminum alloys undergoing either static or dynamic recrystallization, are listed in Table 5.4. The most recent effort in the development of superplastic aluminum alloys is to increase the strain rate for superplasticity to overcome the relatively slow strain rates in conventional superplastic materials and to improve the economic feasibility of superplastic forming. The principal approach is to add an appropriate amount of Zr to Al to reduce its grain size. Alloys such as 7475 [33] and 2124 Al [34] have been studied by using this approach. This subject will be addressed in Chapter 9.

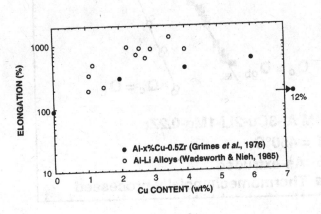

Figure 5.4 Elongation of aluminum alloys as a function of Cu content.

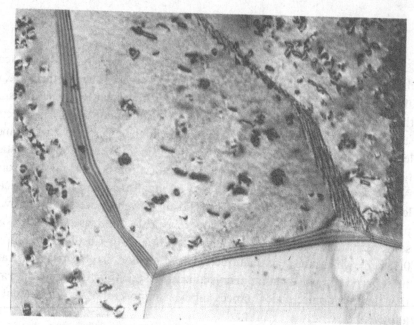

Figure 5.5 A single, high-angle grain from thermomechanically processed, powder metallurgy Al–3Cu–2Li–1Mg–0.2Zr alloy that contains a low-angle boundary.

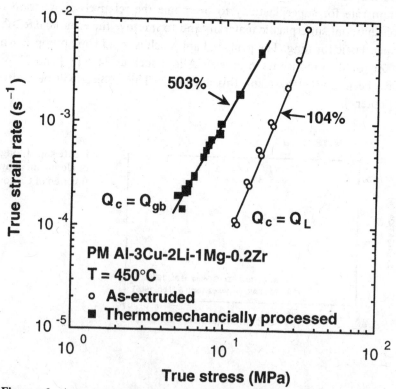

Figure 5.6 A summary of the difference in properties between as-extruded and thermomechanically processed, powder metallurgy Al–3Cu–2Li–1Mg–0.2Zr alloy.

5.2 Magnesium-based alloys

Magnesium alloys have the inherent advantage of being light. As shown in Figure 5.7, they are less dense than fiberglass and just over half as dense as aluminum; magnesium is also an abundant element. Thus, there are many potential opportunities for the use of magnesium alloys in automotive drive train components. This is a consequence not only of magnesium's relatively low density, which can directly and substantially reduce vehicle weight, but also is a result of its good damping characteristics, dimensional stability, machinability, and low casting costs. These desirable attributes enable magnesium to economically replace many zinc and aluminum die castings, as well as cast iron and steel components and assemblies in automotive drive trains [35, 36]. Magnesium metal-matrix composites have yet greater potential to be used in high-performance aerospace and automobile applications because of their superior dimensional stability, strength, and wear resistance compared to competitive materials. The major limitation in using magnesium for structural applications is not its strength *per se*, but its poor corrosion resistance [37].

The study of superplasticity in magnesium alloys was recorded as early as the 1960s [38–42]. These studies were primarily on Mg–6Zn–0.5Zr (ZK60) alloy. One of the important findings was that the grain size of the alloy must be finer than 10 μm to observe superplasticity. The strain-rate-sensitivity exponent of the alloy as a function of strain rate for different grain sizes at the optimum temperature of 270 °C is shown in Figure 5.8. This figure indicates that not only are the

Table 5.4 Superplastic properties of some aluminum alloys

Alloys	Temperature (°C)	Strain rate (s^{-1})	m	Elongation (%)
Statically recrystallized				
Al–33Cu	400–500	8×10^{-4}	0.8	400–1000
Al–4.5Zn–4.5Ca	550	8×10^{-3}	0.5	600
Al–(6–10)Zn–1.5Mg–0.2Zr	550	1×10^{-3}	0.9	1500
Dynamically recrystallized				
Al–6Cu–0.5Zr (SUPRAL 100)	470	2×10^{-3}	0.3	1165
Al–6Cu–0.35Mg–0.14Si (SUPRAL 220)	500	4.5×10^{-3}	0.3	1060
Al–4Cu–3Li–0.5Zr	450	5×10^{-3}	0.5	900
Al–3Cu–2Li–1Mg–0.2Zr	500	1.3×10^{-3}	0.4	878

0.8- and 4-μm materials superplastic, but also the finer the grain size, the faster the superplastic strain rate. This view was further demonstrated by Mabuchi *et al.* [43]. Shown in Figure 5.9 is a log–log plot of the superplastic strain rate of serveral Mg alloys as a function of the inverse of grain size. It is readily seen that the strain rate is a power-law function of the grain size and, specifically,

$$\dot{\varepsilon} \propto d^{-3} \tag{5.1}$$

The third power dependence is often attributed to a grain-boundary sliding mechansim controlled by grain-boundary diffusion [44].

In the case of extrusion, elongation values of over 600% are obtainable for AZ61A of 6 to 8 μm grain size when deformed at 290 °C with an initial strain rate of $3.3 \times 10^{-4}\,\mathrm{s}^{-1}$. More than a 700% elongation was recorded for ZK60A of 4 to 5 μm grain size when deformed at 310 °C and at a higher initial strain rate of $3.3 \times 10^{-3}\,\mathrm{s}^{-1}$. Concurrent grain growth during superplastic deformation occurred in AZ61A (grain size increases from 5 to 17 μm), but occurred only slightly in ZK60A (grain size increases from 4 to 5 μm), suggesting that the latter alloy was thermally more stable than the former. The mechanical properties of these alloys after superplastic deformation were also evaluated; room-temperature tensile properties are maintained near values for extruded stock after superplastic deformation of 100% for AZ61A and 200% for ZK60A. ZK60A appears to be a more practical alloy for superplastic forming of parts than AZ61A because (1) maximum superplastic deformation is possible at higher strain rates

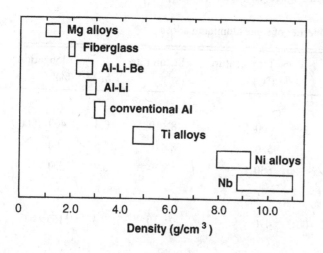

Figure 5.7 Relative densities of common alloys and materials.

for ZK60A and (2) ZK60A retains more strength than AZ61A when both are superplastically deformed to the same extent. In fact, closed-die forming of a ZK60A part was successfully demonstrated in 1 min at 300 °C and 42 MPa. Die-fill capability, as revealed by the reproduction of fine details on the die surface, appears to be good.

For sheet forms, in addition to ZK60A and AZ61A, studies have been carried out on two British alloys ZW1 (Mg–1Zn–0.5Zr) and ZW3 (Mg–3Zn–0.5Zr), AZ31, and Mg–12Al [45]. These alloys can be thermomechanically processed into fine-grained structures, and they are all superplastic under the appropriate testing conditions. A summary of the superplastic characteristics of these alloys is listed in Table 5.5. Tensile properties of these alloys are reduced slightly following superplastic elongation of 100%, except for the experimental composition Mg–12Al, which is brittle after 100% elongation.

Work in the Former Soviet Union on superplastic Mg was extensive (see, for example, Ref. [51]). Alloys studied include MA8 (Mg–1.8Mn–0.25Ce), MA14 (ZK60), MA15 (Mg–1.6Cd–3.1Zn–0.65Zr–0.95La), and MA21 (Mg–5.2Al–4.7Cd–8.1Li–0.21Mn–1.38Zn). Most of these alloys are noted to be single-phase or pseudo-single-phase solid solutions, except MA21, which forms two phases (Mg-rich, hexagonal close-packed, α-solid solution and Li-rich, body-centered cubic, β-solid solution). Grains in the single-phase materials are probably stabilized by various fine precipitates.

In the 1990s, reasearch efforts have been devoted to developing low-density, superplastic Mg–(6.5–9)Li alloys, in particular, by Sherby and colleagues [48,

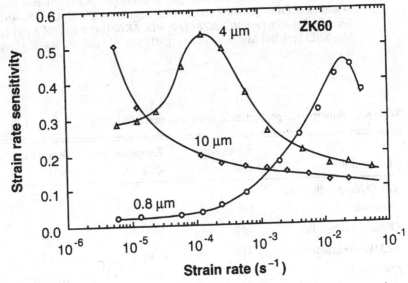

Figure 5.8 Dependence of strain-rate-sensitivity exponent on strain rate for ZK60 magnesium alloy. (Data from Ref. [41].)

52–54], although a limited study was also carried out on commercial AZ91 [46]. For the Mg–Li system, a fine-grained (~1.5 μm) Mg–9Li binary alloy was fabricated. Interestingly, the material exhibits superplasticity (450% elongation) at 100 °C, which is only 0.43 T_m, where T_m is the melting point of the alloy. Even at room temperature a high elongation of 60% was achieved. These superior

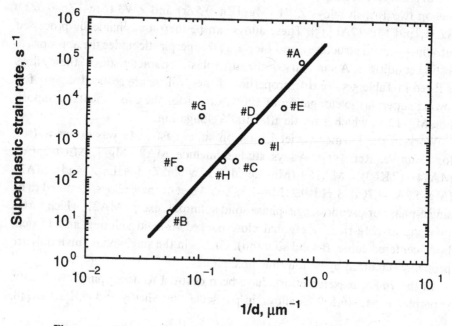

Figure 5.9 The variation of superplastic strain rate as a function of the inverse of grain size for magnesium alloys. (Data sources: #A: RS Mg–11Si–4Al [43], #B: Mg–4Si–4Zn [43], #C: AZ61 [45], #D: ZK60 [45], #E: P/M AZ91 [46], #F Mg–8.5Li [47], #G Mg–8.5Li–1Y [47], #H Mg–9Li [48], #I Mg–9Li [49].)

Table 5.5 Superplastic properties of some commercial Mg alloys

Alloy/condition	Elongation (%)	Temperature (°C)	Strain rate (s⁻¹)	m
AZ61A/cold-rolled and annealed	213	—	3.3×10^{-4}	0.69
ZK60A/warm-rolled	289	290	1.7×10^{-4}	0.45
AZ31B/annealed	319	350	1.7×10^{-4}	0.53
ZW3/aged	195	300	1.7×10^{-4}	—
Mg–12Al/hot-rolled	162	360	8.5×10^{-4}	0.70

mechanical properties were apparently attributed to the alloy's fine grain size. Superplastic flow in this Mg–9Li was postulated to be GBS controlled by grain-boundary diffusion and accommodated by slip. The elongation data of this Mg–9Li are consistent with those measured from the Russian MA21 alloy containing 8.1 wt% Li (see Figure 5.10). In addition to Mg–Li alloys, some studies have been conducted with Mg-based composites [49, 55, 56]; this will be addressed in Chapter 8.

5.3   Iron-based alloys

Superplasticity in iron-based materials has been reviewed by Ridley [57], Walser and Ritter [58], Sherby and Wadsworth [59], and Maehara and Langdon [60]. It is interesting to note that, although observed in a wide range of alloy systems, the number of iron-based alloys that exhibit superplasticity is surprisingly small. These groups are (1) based on the Fe–C system – hypoeutectoid and eutectoid steels, hypereutectoid steels, and white cast irons; (2) low- and medium-alloy-content steels; and (3) microduplex stainless steels.

5.3.1   Plain carbon steels

5.3.1.1   Hypoeutectoid and eutectoid steels

As described in the review by Ridley [57], early studies on plain carbon steels were not particularly successful. After extensive heat treatments to produce microstructures of fine ferrite grains pinned by spheroidized cementite particles, compositions containing between 0.2 and 1.0 wt% C were shown to exhibit elongations typically of only about 130% [61]. A problem with hypoeutectoid and eutectoid steels is that low-angle dislocation boundaries are usually gener-

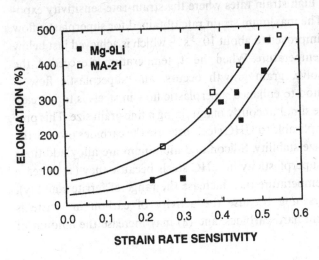

Figure 5.10 Elongation to failure vs strain-rate-sensitivity exponent for the Mg–9Li laminates and for MA–21 (from Ref. [48]).

ated during warm-working procedures. Low-angle boundaries do not permit GBS, which results in materials with low strain-rate-sensitivity exponents. Thermal cycling of these steels after mechanical processing, however, changed fine-grained, low-angle boundaries to fine-grained, high-angle boundaries. The m value increased to 0.5, and tensile ductility of the order of 1000% were thereby achieved [44].

5.3.1.2 Hypereutectoid steels

Fine ferrite/spheroidized cementite microstructures in steels containing a large amount of carbon (i.e., 1 to 2.1 wt% C) can be readily prepared by hot- and warm-working procedures. These hypereutectoid steels, termed ultrahigh-carbon (UHC) steels, had traditionally been ignored because of the generally held belief that they would be brittle at room temperature.

The original work on UHC steels demonstrated that superplasticity could be developed over the range of composition from about 0.8 to 2.1 wt% C and temperatures ranging from 650 to 800 °C. This temperature range is from just below the A_1 temperature (727 °C) to just above it, as shown in Figure 5.11 and, therefore, includes ferrite–cementite as well as austenite–cementite structures. Furthermore, superplasticity was observed at intermediate strain rates (10^{-5} to 10^{-3} s^{-1}). Two major developments have expanded the usefulness of superplasticity in UHC steels [62]. The first is a novel alloying concept that extends the temperature and strain-rate range for superplastic flow of UHC steels. The second is novel processing methods that have increased the range of carbon composition over which superplasticity can be observed. An overview of the expansion in composition and temperature ranges now available for superplasticity in UHC steels and cast irons is also shown in the iron–cementite phase diagram in Figure 5.11.

Generally, the maximum rate of superplastic flow is limited by the introduction of slip processes at high strain rates where the strain-rate-sensitivity exponent is small ($m \approx 0.1$). The maximum strain rate obtained for superplastic flow in UHC steels has been improved to about 10^{-3} s^{-1}, which is achieved just below the A_1 transformation temperature. When the A_1 temperature is exceeded, the eutectoid carbides dissolve, grain growth occurs, and superplastic flow is reduced. Thus, one method to enhance superplastic flow in steels is to increase the A_1 temperature, while simultaneously maintaining a fine grain size. This processing route is ideally applicable to UHC steels because the carbides provide the pinning sites for grain-size stability. Silicon and aluminum are alloy additions that favorably enhance superplasticity in UHC steels because they (1) increase the A_1 transformation temperature (i.e., increase the range of ferrite stability), (2) inhibit carbide coarsening because the activity of carbon in ferrite is increased, (3) do not form hard carbides, and (4) may increase the amount of proeutectoid carbides.

Figure 5.12 illustrates the influence of aluminum and silicon on the maximum strain rate for superplastic flow of several UHC steels with a grain size of about 2 μm, as a function of temperature. It is seen that a UHC steel containing 3 wt% Si or 1.6 wt% Al exhibits superplastic flow at a maximum strain rate of about 10^{-2} s^{-1} at temperatures ranging from about 770 to 800 °C. Figure 5.12 also illustrates the predicted superplastic behavior of a UHC steel–12Al alloy of 2 μm grain size, which should exhibit superplastic flow at a strain rate of 0.3 s^{-1} at 950 °C. This is an impressive strain rate, and the predicted results would indicate that commercial superplastic forging of UHC steel–Al alloys is economical, feasible, and worthy of extensive development studies. An additional benefit of the developed UHC steels containing aluminum or silicon is that they have a low resistance to plastic flow in the superplastic range at low strain rates (e.g., 14 MPa at a strain rate of 10^{-4} s^{-1} at 800 °C). This means that a UHC steel, prepared as a fine-grained sheet, can be gas-pressure-formed superplastically with existing blow-forming equipment.

At temperatures above 800 °C, the UHC–Al–Si steels undergo transformation to austenite(γ) as the ferrite(α) and cementite phases gradually disappear. Because austenite is stronger than ferrite, a remarkably constant flow stress is achieved in the three-phase region consisting of $\alpha + \gamma + Fe_3C$. For a UHC–3Si steel, the flow stress remains at about 14 MPa ($\pm$1.5 MPa) in a wide temperature

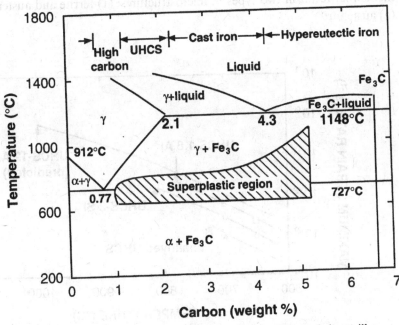

Figure 5.11 Iron–carbon (cementite) phase diagram. The shaded area illustrates the temperature and composition range over which UHC steels and white cast irons have been made superplastic.

range of 750 to 920 °C. These results suggest that superplastic forming of such a UHC steel in this temperature range does not require accurate temperature control.

5.3.1.3 White cast irons

Superplasticity has been developed in compositions containing even more carbon than UHC steels, namely white cast irons [63, 64]. These white cast irons have been made fine-grained by several PM techniques, including liquid atomization or RSP. Such techniques lead to powders that, on annealing at a low temperature (600–700 °C), have the desired fine microstructure for superplasticity. The powders are readily compacted into fully dense compacts by warm-pressing at temperatures below A_1 or by multiple phase transformations, under pressure, through thermal cycling. Because the compaction temperature is low, ultrafine structures (1–2 μm ferrite grain size) are found in densified compacts. These white cast irons are superplastic at intermediate temperatures. A maximum tensile elongation of 1410% has been found for a 3.0 wt% C–1.5 wt% Cr white cast iron [65].

5.3.2   Low- and medium-alloy-content steels

Superplastic studies have been carried out on alloy steels containing alloying elements that result in two types of microstructures: (1) ferrite and austenite and (2) austenite.

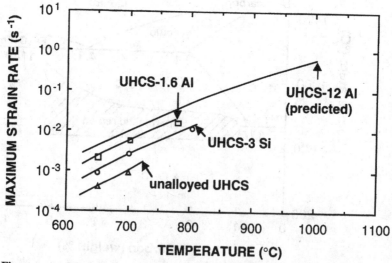

Figure 5.12   Maximum strain rate for superplastic flow of UHC steels is enhanced by increasing the range of ferrite stability through silicon and aluminum additions; arrows indicate maximum temperatures for ferrite stability.

5.3.2.1 Ferrite and austenite

In steels $(\alpha+\gamma)$ containing about 1 to 2 wt% Mn and 0.1 to 0.4 wt% C, superplastic elongations of about 300 to 500% were initially recorded in microstructures of grain size 1 to 2 μm [66]. These observations were further expanded by Smith *et al.* [67] to develop improved superplastic behavior (up to 600%) in Fe–4Ni–3Mo–(1–2)Ti alloys. The reason for retaining a fine grain size in these alloys is to achieve a temperature condition at which about 50% ferrite–50% austenite coexist, each phase inhibiting the other from growing. Difficulties with these superplastic steels include the narrow temperature range (800 °C$\pm$50 °C) in which superplastic flow occurs. Within this temperature range the rapid growth of the two phases can also occur, especially of the more mobile ferrite grains.

5.3.2.2 Austenite

Superplastic behavior in austenitic steels has been initially investigated in low C–Mn steels containing minor Nb and Al additions [68]. Moderate elongations of 200% were observed. Additional work by Lee *et al.* [69] has centered on the development of fine grains in modified Hadfield manganese steels (13 wt% Mn and 1.2–1.7 wt% C). The method of preparation was to mechanically work the steels in the austenite–cementite region to precipitate the carbides at austenite grain boundaries, followed by a cold-working and recrystallization treatment. Such processing led to Hadfield steels with grain sizes of the order of 3 to 4 μm. Superplasticity tests at 800 °C revealed $m=0.5$ with elongations on the order of 500%. Extensive cavitation occurred during superplastic flow, presumably because of the strength differential between austenite and the carbides (cementite); the strength of austenite is higher than that of cementite.

5.3.3 Microduplex stainless steels

Microduplex stainless steels are of great interest to the chemical industry because of their high strength and resistance to both stress corrosion and general corrosion. These steels are commercially available under a range of different names, the most prominent of which are alloy IN 744 and SUPERDUX64. These complex steels have compositions of Fe–(18–26) wt% Cr–5.8 wt% Ni and can also contain Mo, Ti, Cu, Si, Mn, C, and N. They are called 'microduplex' steels because the basis of their high plasticity is the presence of two phases (ferrite and austenite) in fine-grained form, with grain size of about 2 to 3 μm.

Extensive studies have been devoted to the characterization and understanding of this group of materials [67, 70–81]. Hayden et al. [82], for example, in studying the superplastic properties of a Fe–39 wt% Cr–10 wt% Ni–1.75 wt% Ti–1 wt% Al alloy, revealed that a high m value of 0.5 was observed over four orders of magnitude of strain rate (6×10^{-6} to 4×10^{-2} s^{-1}) at 980 °C, giving strong credence to the presence of a single discrete mechanism for superplastic flow. This mechanism was believed to be GBS accommodated by slip in the vicin-ity of the grain boundary [83]. Cavitation during extensive superplastic flow was noted in these microduplex steels and is probably related to the higher strength of the austenite component compared to that of the ferrite component at high temperatures [67]; this is similar to behaviour in austenitic steels.

Maehara [84] added 0.14 wt% solute nitrogen as an austenite stabilizer to a 25 wt% Cr–7 wt% Ni–3 wt% Mo duplex (δ-ferrite/austenite) stainless steel and studied the superplastic properties of the alloy. A maximum elongation of greater than 2500% was recorded in a solution-treated and then cold-worked material at a strain rate of $4\times10^{-3}\,\mathrm{s}^{-1}$ at 950 °C. Two ductility maxima were also observed at the optimum strain rate (Figure 5.13), which were attributed to the microstructural dynamics [72]. Specifically, at temperatures above about 1000 °C, γ grains broke into spherical particles during initial deformation, resulting in a homogeneous dispersion of γ particles within a δ-ferrite matrix. Further deformation caused the preferential recrystallization of the soft δ-ferrite matrix and eventually led to the formation of the steady-state, equiaxed δ/γ duplex microstructure. At low temperatures (<1000 °C), however, a metastable σ phase precipitation occurred during initial deformation as a result of the eutec-toid decomposition of δ ferrite into γ and σ phases. Upon further deformation, plastic strain induced the recrystallization of the soft γ grains adjacent to σ par-ticles and resulted in the final γ/σ equiaxed duplex structure. One of the nice fea-tures of superplastic microduplex stainless steels is the fact that superplastic forming can be performed in a relatively wide range of temperature (850–1050 °C); this is clearly illustrated in Figure 5.13.

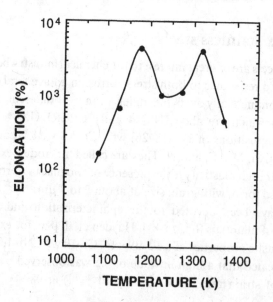

Figure 5.13 Elongation as a function of temperature for a 25 wt% Cr–7 wt% Ni–3 wt% Mo–0.14 wt% N duplex stainless steel deformed at $4\times10^{-3}\,\mathrm{s}^{-1}$.

In an attempt to further improve the superplastic strain rate of microduplex stainless steels, Maehara [74] developed thermomechanical processes that essentially involved solution treatment at temperatures in the δ single-phase region with subsequent heavy cold-rolling. The processed alloy at 1000 °C exhibited elongation values of greater than 300% at an initial strain rate of 1 s^{-1}. Again, dynamic recrystallization of δ grains adjacent to hard γ particles was proposed to be the dominant deformation mechanism.

5.3.4 Nonsuperplastic steels made superplastic by lamination

A number of studies have been conducted on developing superplastic properties in nonsuperplastic steels by lamination with superplastic steels [85–88]. This special group of materials will be discussed in Chapter 8.

5.3 Titanium-based alloys

Superplasticity in some of the titanium alloys has been known for over 20 years [89]. Particular interest has developed for titanium alloys because (1) they can be quite superplastic as a natural consequence of conventional processing procedures, and (2) they are expensive and difficult to fabricate into structural shapes by the more traditional forming and machining methods. Superplastic forming has been shown to offer the potential for significant cost savings for the construction of structural titanium alloy parts, particularly when the design is developed for the capabilities and limitations of the superplastic process. Many commercial applications of the superplastic forming process for titanium (usually the Ti–6Al–4V alloy) have been successfully demonstrated. Further improvements in the process and a reduction of costs have been envisioned if improved superplastic titanium alloys could be developed [90]. This general interest in superplastic titanium alloys has stimulated the study of fundamental mechanisms and microstructural features of superplasticity. Several reviews on superplasticity in titanium-based materials are available [91–93].

Most of the experimental studies as well as commercial applications have focused on the Ti–6Al–4V alloy. This alloy in particular has found wide-spread use because of its desirable structural properties. These are exclusive of its superplasticity, but the observation that it is superplastic in conventional wrought forms has provided the incentive to concentrate on this alloy. Subsequent studies, however, have shown that a range of titanium-based alloys can also be superplastic. These alloys include Ti–6Al–4V [94, 95], Ti–6Al–5V [93], Ti–6Al–2Sn–4Zr–2Mo [96], Ti–6Al–4V–2Ni [97], Ti–6Al–4V–2Co [97], Ti–6Al–4V–2Fe [97], Ti–5.8Al–4.0Sn–3.5Zr–0.7Nb–0.5Mo–0.3Si–0.05C [98],

Ti–5Al–2.5Sn [89], Ti–4.5Al–5Mo–1.5Cr [99], Ti–4Al–4Mo–2Sn–0.5Si [100], Ti–3Al–15V–3Sn–3Cr [93], Ti–3Al–11V–13Cr [101], Ti–3Al–10V–2Fe [99], Ti–8Mn [101, 102], and Ti–15Mo [101]. All these highly superplastic titanium alloys are primarily two-phase alloys, containing approximately an equal volume fraction of α and β phases. Superplastic properties of some of these α/β titanium alloys are summarized in Table 5.6.

For superplastic α/β alloys, the average grain size is about 5 to 10 μm and the flow stress is of the order of 1 to 10 MPa [94]. As with all superplastic alloys, the microstructure, and particularly the grain size and the phase concentration in titanium alloys, is crucial to the development of superplasticity. The effect of grain size on the high-temperature properties of superplastic titanium [94, 95] (see Figure 5.14) is qualitatively similar to that of other superplastic alloys. As the grain size increases, the flow stress in the superplastic region increases, and the maximum m value decreases and shifts to a lower strain rate. The grain-size exponent p for Ti–6Al–4V in the equation $\dot{\varepsilon} \propto d^{-p}$, where $\dot{\varepsilon}$ is the strain rate and d is the grain size, has been reported by several investigators [89, 94] and found to have a substantial variation. Such variability makes it difficult to establish a constitutive relation based on fundamental microstructural parameters, and sometimes causes confusion in deducing the deformation mechanism(s) involved. The exponent p is expected to be 2 if the grain-boundary accommodation mechanism is dislocation climb processes and 3 if the mechanism is grain-boundary diffusional creep [44].

Regarding phase concentration in superplastic α/β titanium alloys, the two

Table 5.6   Superplastic properties of several α/β titanium alloys

Alloys	Temperature (°C)	Strain rate (s^{-1})	m	Elongation (%)
Ti–6Al–4V	840–870	1.3×10^{-4} to 1.3×10^{-3}	0.75	750–1170
Ti–6Al–5V	850	8×10^{-4}	0.70	700–1100
Ti–6Al–2Sn–4Zr–2Mo	900	2×10^{-4}	0.67	538
Ti–4.5Al–5Mo–1.5Cr	871	2×10^{-4}	0.63–0.81	>510
Ti–6Al–4V–2Ni	815	2×10^{-4}	0.85	720
Ti–6Al–4V–2Co	815	2×10^{-4}	0.53	670
Ti–6Al–4V–2Fe	815	2×10^{-4}	0.54	650
Ti–5Al–2.5Sn	1000	2×10^{-4}	0.49	420
Ti–5.8Al–4Sn–3.5Zr–0.7Nb–0.5Mo–0.3Si–0.05C	990	1×10^{-4}	0.82	>300

phases are notably different in their high-temperature deformation-related characteristics. The α phase has fewer slip systems than the β phase, and the self-diffusivity is about two orders of magnitude slower. Both of these features would suggest that the α is the *harder* phase and β the *softer* at superplastic temperatures and strain rates. Work by Leader *et al.* [97] has shown evidence for plastic deformation in the β phase with little or no deformation in α for the Ti–6Al–4V alloy at 900 °C. These data suggest that the superplasticity in titanium alloys may involve at least some plastic flow concentrated in the β phase. This is similar to that observed during the superplastic deformation of microduplex δ/γ stainless steel [74]. In the latter case, dynamic recrystallization of soft δ (ferrite) grains adjacent to hard γ particles was suggested to be the dominant factor contributing to deformation.

The maximum strain-rate-sensitivity and total elongation values for a number of titanium alloys for which the phase content was known are shown as functions of volume % β phase in Figures 5.15 and 5.16, respectively [93]. As can be seen in these figures, a maximum is indicated for both m and elongation at inter-

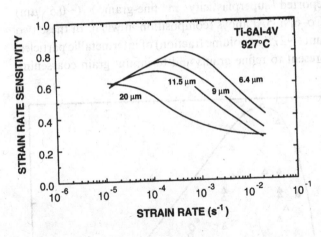

Figure 5.14 Strain-rate-sensitivity value as a function of strain rate for Ti–6Al–4V with various grain sizes.

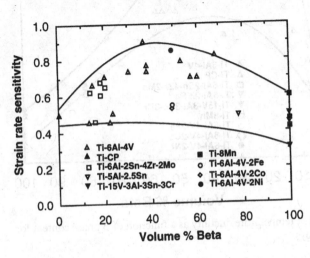

Figure 5.15 Strain-rate sensitivity as a function of β phase content for several titanium alloys.

mediate β contents, indicating that a microstructure with a presence of both α and β in significant concentrations (i.e., greater than about 10%) is important in achieving superplasticity in titanium alloys.

This characteristic superplastic behavior of titanium alloys is, in fact, a result of a balance between grain-size control and the ease of deformation. While the β phase has high diffusivity, it also undergoes extremely rapid grain growth at superplastic temperatures, making it unsuitable for superplastic flow. A finite concentration of the α phase significantly restricts the grain growth because of the long-range diffusion necessary to transfer highly partitioned alloy elements (e.g., vanadium in β phase). Without the presence of a second phase, the β phase will rapidly grow at temperatures above about 760 °C. In addition to microstructural stabilization, the presence of a second phase can affect boundary sliding processes. For example, it is generally recognized that a homogeneous boundary slides with much more difficulty than does a heterogeneous boundary [103, 104].

Liu *et al.* [105] reported superplasticity in fine-grained (~0.5 μm) Ti–12Co–5Al and Ti–6Co–6Ni–5Al alloys (composition in wt%). In these two alloys, a significant amount (~27% in volume fraction) of intermetallic particles, Ti$_2$Co and Ti$_2$Ni, was present to refine grains and to inhibit grain coarsening

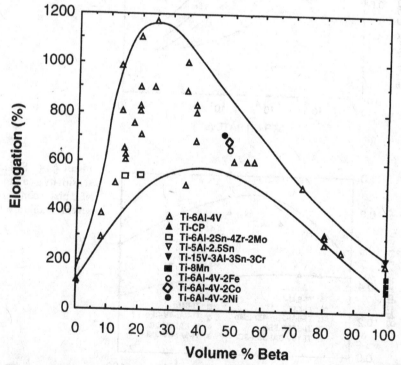

Figure 5.16   Elevated temperature ductility as a function of β phase content for several titanium alloys.

during superplastic deformation. A maximum superplastic elongation of over 2000% was observed at a strain rate as high as 3×10^{-2} s^{-1} and at a temperature as low as 700 °C. Evidently, the relatively high strain rate and low temperature were attributed to finer grain sizes in these two alloys as compared to those in the other superplastic titanium alloys, which are typically about 5 to 10 μm [94]. The temperature at which superplasticity is found in titanium alloys is usually between about 0.4 T_m (about 800 °C) [99] to 0.61 T_m (about 1030 °C). At temperatures higher than this range, the grain growth kinetics degrade the microstructural stability necessary to retain a fine grain size, and thus, superplasticity is no longer observed [94]. Also, the microstructure of titanium alloys undergoes dynamic changes, i.e., rapid grain growth, during superplastic deformation.

5.5   Nickel-based alloys

The study of superplastic, nickel-based alloys began as early as the 1960s [82]. Up to the present time, superplasticity has been observed in a variety of nickel-based alloys, including cast and wrought alloys (e.g., Inconel 718 [106–108], a Ni–30Cr–5Al alloy [109]), PM IN–100 [110–113]), and oxide-dispersion-strengthened materials (MA 754 and MA 6000 [114, 115]). The well-known, commercial gatorizing process was, in fact, developed mainly for the nickel-based superalloys [116]. In a similar situation to aluminum alloys, there is relatively little interest in developing new superplastic nickel alloys; work on superplastic nickel was primarily emphasized on existing alloys. For a comprehensive review on superplastic nickel-base alloys the reader is referred to Ref. [115].

The microstructural requisites, e.g., fine grain size (< 10 μm) and multi-phase structure, for superplastic nickel-based alloys are similar to those for the other superplastic metallic materials. The second phase is mainly the ordered face-centered cubic γ'. The γ' phase is considered to have a similar strength to the γ matrix at the superplastic deformation temperature (typically, 950 °C). The desired microstructures are usually achieved via thermomechanical processing, such as microduplex processing [117] and MINIGRAIN* processing [118].

The essential features of microduplex processing are described by Figure 5.17 for operations that involve hot-working alone (hot-work cycle) and those that involve cold-working (cold-work cycle). In the hot-work cycle, after initial ingot breakdown through rolling or forging in the single-phase region (Figure 5.17(a)), hot-working of the billet is continued so that the finish temperature is in the two-phase region. If precipitation of the second phase is sluggish, it may be advantageous to reheat the billet in the two phase region and then continue

* MINIGRAIN is a trademark of United Technologies Corp.

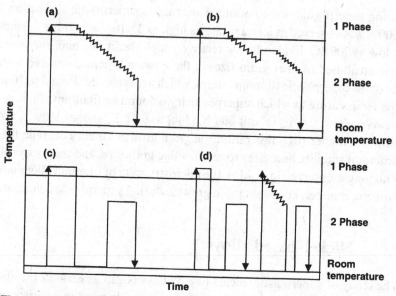

Figure 5.17 Processing cycles for producing a microduplex structure in two-phase Ni–Cr–Fe alloys (from Ref. [115]).

hot-working (Figure 5.17(b)). In the MINIGRAIN process, heat treatment in the two phase region is applied following the initial hot working. Final working is conducted at or below the recrystallization temperature. These controlled thermomechanical processes to derive fine-grain structures have been applied to a variety of nickel-based alloys.

The most significant progress in the area of Ni-base alloys is probably the development of superplastic Inconel 718 alloy (composition by weight%: 50–55 Ni, 17–21 Cr, 4.8–5.25 Nb, 2.8–3.3 Mo, 0.651.15 Ti, 0.2–0.8 Al, 0.35 Mn, 0.03 C, 0.015 S, bal Fe) [108, 119]. Inconel 718 was developed by Inco Alloys International Inc. in the early 1960s. It is widely used in jet engine parts such as compressor and turbine disks and rings, turbine shafts, exhaust sections, hot air ducting, and fasteners. The alloy is made by vacuum induction melting and electroslag remelting. It can be conventionally hot-worked, then cold processed to make a sheet product with a grain size of 5 μm. The grain size is sufficiently stable at temperatures about 980 °C or less to permit superplastic forming. The superplastic behavior of fine-grained Inconel 718 was found to be strongly dependent upon the strain rates, as shown in Figure 5.18 [108]. An elongation value of over 700% can be obtained at a strain rate of 1.33×10^{-5} s^{-1}. In another study [119], a heavily rolled IN–718 exhibited a maximum elongation to failure of over 1000% at 920 °C at a strain rate of about 10^{-3} s^{-1}. In this case, the microstructure dynamically recrystallized into a fine structure during the early stage of deformation and led to superplastic behavior.

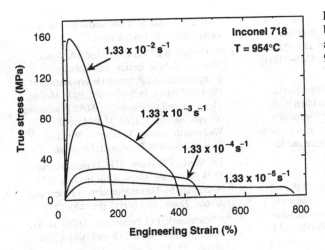

Figure 5.18 Stress–strain behavior of Inconel 718 as a function of strain rate at 954 °C (from Ref. [108]).

Superplastic deformation mechanisms in nickel alloys involve also GBS accommodated by dislocation slip [83, 112] in a manner that is similar to other metals. It is worth noting that the results obtained from oxide-dispersion-strengthened materials MA 754 and MA 6000 [114] were unusual. Specifically, while conventional superplasticity occurs at strain rates of about 10^{-4} to $10^{-3}\,s^{-1}$, superplasticity in MA 754 and MA 6000 occurs at relatively high strain rates ($\sim 10^{-1}\,s^{-1}$). This result will be discussed in detail in Chapter 9.

References

1. *Superplastic Forming of Structural Alloys*, edited by N.E. Paton and C.H. Hamilton, The Metallurgical Society of AIME, Warrendale, PA, 1982.

2. *Superplasticity and Superplastic Forming*, edited by N.E. Paton and C.H. Hamilton, The Metallurgical Society of AIME, Warrendale, PA, 1988.

3. S. Hori, M. Tokizane, and N. Furushiro, *Superplasticity in Advanced Materials*. The Japan Society of Research on Superplasticity, Osaka, Japan, 1991.

4. E.R. Petty, 'The Deformation Behavior of Some Aluminum Alloys Containing Intermetallic Compounds,' *J. Inst. Metals*, **91** (1962–63), pp. 274–279.

5. M.J. Stowell, J.L. Robertson, and B.M. Watts, 'Structural Changes during Superplastic Deformation of the Al-Cu Eutectic Alloy,' *Metal Sci.*, **3** (1969), pp. 41–45.

6. S.K. Marya and G. Wyon, 'Superplasticite a l'ambiante de l'aluminium a grain fin, en liaison avec l'existence d'un film intergranulaire de solution solide riche en gallium,' *J. Physique*, **36**(C.4) (1975), pp. 309–313.

7. F. Weill and G. Wyon 'Superplastic Behavior of Fine Grained Aluminum Alloys whose Grain Boundaries have been Enriched by Gallium,' in *Strength of Metals and Alloys, ICSMA 5*, pp. 387–392, ed. P. Haasen, V. Gerold, and G. Kostorz, Pergamon, Oxford, 1979.

8. R. Grimes, M.J. Stowell, and B.M. Watts, 'Superplastic Aluminum-based Alloys,' *Metals Technol.*, **3** (1976), pp. 154–160.

9. B.M. Watts, M.J. Stowell, B.L. Baikie, and D.G.E. Owen, 'Superplasticity in Al-Cu-Zr Alloys Part I: Material Preparation and Properties,' *Metal Sci.*, **10** (1976), pp. 189–197.

10. B.M. Watts, M.J. Stowell, B.L. Baikie, and D.G.E. Owen, 'Superplasticity in Al-Cu-Zr Alloys Part II: Microstructural Study,' *Metal Sci.*, **10** (1976), pp. 198–206.

11. D.J. Lloyd and D.M. Moore, 'Aluminum Alloy Design for Superplasticity,' in *Superplastic Forming of Structural Alloys*, pp. 147–172, ed. N.E. Paton and C.H. Hamilton, TMS-AIME, Warrendale, PA, 1982.

12. A.J. Shakesheff, 'The Effect of Superplastic Deformation on the Post-Formed Mechanical Properties of the Commercially Produced Supral Alloys,' in *Superplasticity in Aerospace-Aluminum*, pp. 36–54, ed. R. Pearce and L. Kelly, Ashford Press, Curdridge, Southampton, Hampshire, 1985.

13. B. Geary, J. Pilling, and N. Ridley, 'Effect of Strain Rate and Grain Size on Cavitation in Supral 220,' in *Superplasticity in Aerospace-Aluminum*, pp. 127–135, ed. R. Pearce and L. Kelly, Ashford Press, Curdridge, Southampton, Hampshire, 1985.

14. R. Grimes, M.J. Stowell, and B.M. Watts, 'The Forming Behavior of Commercially Available Superplastic Aluminum Alloys,' in *Superplasticity in Aerospace*, pp. 97–113, ed. C. Heikkenen and T.R. McNelly, TMS-AIME, Warrendale, PA, 1988.

15. D.M. Moore and L.R. Morris, 'A New Superplastic Aluminum Sheet Alloy,' *Mater. Sci. Eng.*, **43** (1980), pp. 85–92.

16. Q. Wu, H. Ma, and L. Ma, 'Research on Superplasticity of Al-Ca-Zn Alloys,' in *Superplasticity in Aerospace-Aluminum*, pp. 168–179, ed. R. Pearce and L. Kelly, Ashford Press, Curdridge, Southampton, Hampshire, 1985.

17. A.J. Barnes, 'Commercial Superplastic Aluminum Alloys – Opportunities and Challenges,' in *Superplasticity in Aerospace-Aluminum*, pp. 424–447, ed. R. Pearce and L. Kelly, Ashford Press, Curdridge, Southampton, Hampshire, 1985.

18. *Superplasticity*, edited by B. Baudelet and M. Suery, Centre National de la Recherche Scientifique, Paris, France, 1985.

19. A.J. Barnes, 'Advances in Superplastic Aluminum Forming,' in *Superplasticity in Aerospace*, pp. 301–313, ed. C. Heikkenen and T.R. McNelly, TMS-AIME, Warrendale, PA, 1988.

20. J. Wadsworth, A.R. Pelton, and R.E. Lewis, 'Superplastic Al-Cu-Li-Mg-Zr Alloys,' *Metall. Trans.*, **16A** (1985), pp. 2319–2332.

21. R. Grimes, 'Superplasticity in Aluminum-Lithium Based Alloys', in *Superplasticity*, pp. 13.1–13.12, ed. B. Baudelet and M. Suery, Centre National de la Recherche Scientifique, Paris, Grenoble, France, 1985.

22. R. Grimes, B.J. Dunwoody, and J.A. Jones, 'Commercial Production, Superplastic Performance and Property Determination of 8090 SPF Alloy,' in *Superplasticity in Metals, Ceramics, and Intermetallics, MRS Proceeding No.196*, pp. 167–172, ed. M.J. Mayo, J. Wadsworth, and M. Kobayashi, Materials Research Society, Pittsburgh, PA, 1990.

23. J.A. Wert, N.E. Paton, C.H. Hamilton, and M.W. Mahoney, 'Grain Refinement in 7075 Aluminum by Thermomechanical Processing,' *Metall. Trans.*, **12A** (1981), pp. 1267–1276.

24. J. Wadsworth, I.G. Palmer, and D.D. Crooks, 'Superplasticity in Al-Li Based Alloys,' *Scr. Metall.*, **17** (1983), pp. 347–352.

25. B. Derby and M.F. Ashby, 'On Dynamic Recrystallization,' *Scr. Metall.*, **21** (1987), pp. 879–884.

26. J. Wadsworth and A.R. Pelton, 'Superplastic Behavior of a Powder-Source Aluminum-Lithium Based Alloy,' *Scr. Metall.*, **18** (1984), pp. 387–392.

27. C.H. Hamilton and A.K. Ghosh, 'Superplastic Sheet Forming,' in *Metal Handbook Vol.14*, pp. 852–863, ed. S.L. Semiatin, ASM International, Metals Park, OH, 1988.

28. J. Wadsworth, C.A. Henshall, and T.G. Nieh, 'Superplastic Aluminum-Lithium Alloys,' in *Aluminum-Lithium Alloys III*, pp. 199–212, ed. C. Baker, P.J. Gregson, S.J. Harris, and C.J. Peel, The Institute of Metals, London, 1986.

29. G. Beghi, R. Matra, and G. Piatti, 'Superplastic Behavior of a Splat Cooled Al–17wt% Cu Alloy,' *J. Mater. Sci.*, **5** (1970), pp. 820–822.

30. J. Wadsworth, 'The Development of Superplasticity in Aluminum-Lithium Based Alloys,' in *Proc. of WESTEC Conference on Superplasticity, ASM Tech. Publication #8401*, pp. 43–57, ed. S. Agrawal, ASM, Metals Park, OH, 1985.

31. J. Wadsworth, C.A. Henshall, A.R. Pelton, and B. Ward, 'Superplastic Properties of an Al-Cu-Li-Zr Alloy,' *J. Mater. Sci. Lett.*, **4** (1985), pp. 674–678.

32. Z. Cui, W. Zhong, and Q. Wei, 'Superplastic Behavior at High Strain Rate of Rapidly Solidified Powder Metallurgy Al-Li Alloy,' *Scr. Metall. Mater.*, **30** (1994), pp. 123–128.

33. N. Furushiro and S. Hori, 'Significance of High Rate Superplasticity in Metallic Materials,' in *Superplasticity in Metals, Ceramics, and Intermetallics, MRS Proceeding No. 196*, pp. 385–390, ed. M.J. Mayo, J. Wadsworth, and M. Kobayashi, Materials Research Society, Pittsburgh, PA, 1990.

34. T.G. Nieh and J. Wadsworth, 'Effects of Zr on the High Strain Rate Superplasticity of 2124 Al,' *Scr. Metall. Mater.*, **28** (1993), pp. 1119–1124.

35. J. Davis, 'Magnesium in Drive Train Components,' in *International Congress and Exposition, SAE Technical Paper #850419*, SAE International, 1985.

36. J. Davis, 'The Potential for Vehicle Weight Reduction Using Magnesium,' in *International Congress and Exposition, SAE Technical Paper #910551*, SAE International, 1991.

37. A. Joshi and R. Lewis, 'Role of RSP on Microstructure and Propreties of Magnesium Alloys,' in *Advance in Magnesium Alloys and Composites*, pp. 89–103, ed. H.G. Paris and W.H. Hunt, TMS, Warrendale, PA, 1988.

38. W.A. Backofen, G.S. Murty, and S.W. Zehr, 'Evidence for Diffusional Creep with Low Strain Rate Sensitivity,' *Trans Metall. Soc.-AIME*, **242** (1968), pp. 329–331.

39. A. Karim, D.L. Holt, and W.A. Backofen, 'Diffusional Creep and Superplasticity in a Mg–6Zn–0.5Zr Alloy,' *Trans. Metall. Soc. -AIME*, **245** (1969 May), pp. 1131–1132.

40. A. Karim, D.L. Holt, and W.A. Backofen, 'Diffusional Flow in a Hydrided Mg–0.5wt pct Zr Alloy,' *Trans. Metall. Soc. -AIME*, **245** (1969 May), pp. 2421–2424.

41. A. Karim and W.A. Backofen, 'Grain Size Dependence of Strain-Rate Hardening Behavior in a Mg-Zn-Zr Alloy,' *Mater. Sci. Eng.*, **3** (1968–69), pp. 306–307.

42. G.S. Murty, 'Internal Stresses in a Superplastic Mg Alloy,' *Scr. Metall.*, **6** (1972), pp. 663–666.

43. M. Mabuchi, K. Kubota, and K. Higashi, 'High Strength and High Strain Rate Superplasticity in a Mg-Mg$_2$Si Composite,' *Scr. Metall. Mater*, **33** (1995), pp. 331–335.

44. O.D. Sherby and J. Wadsworth, 'Development and Characterization of Fine Grain Superplastic Material,' in *Deformation, Processing and Structure*, pp. 355–389, ed. G. Krauss, ASM, Metal Park, Ohio, 1984.

45. M.M. Tilman and L.A. Neumaier, *Superplasticity in Commercial and Experimental Compositions of Magnesium Alloy Sheet*, RI 8662, Bureau of Mines, U.S. Department of the Interior, 1982.

46. J.K. Solberg, J. Torklep, O. Bauger, and H. Gjestland, 'Superplasticity in Magnesium Alloy AZ91,' *Mater. Sci. Eng.*, **A134** (1991), pp. 1201–1203.

47. K. Higashi, K. Kobota, and G. Neite, 'Mechanical Properties of a Rapidly-Solidified Magnesium Alloy,' in *Magnesium Alloys and Their Applications*, pp. 293–298, ed. B.L. Mordike and F. Hehmann, Oberursel, DGM Informationsgesellschaft m.b.H., 1992.

48. P. Metenier, G. Gonzalez-Doncel, O.R. Ruano, J. Wolfenstine, and O.D. Sherby, 'Superplastic Behavior of a Fine-Grained Two-Phase Mg–9wt.%Li Alloys,' *Mater. Sci. Eng.*, **A125** (1990), pp. 195–202.

49. G. Gonzalez-Doncel, J. Wolfenstine, P. Metenier, O.R. Ruano, and O.D. Sherby, 'The Use of Foil Metallurgy Processing to Achieve Ultrafine Grained Mg–9Li Laminates and Mg–9Li–5B$_4$C Particulate Composites,' *J. Mater. Sci.*, **25** (1990), pp. 4535–4540.

50. M.M. Tilman, R.L. Crosby, and L.A. Neumaier, *Superplasticity in Selected Magnesium-Base Alloys*, RI 8382, Bureau of Mines, U.S. Department of the Interior, 1979.

51. O.A. Kaibyshev, *Superplasticity of Alloys, Intermetallides, and Ceramics*. 1st Edition, Springer-Verlag, New York, 1992.

52. J. Wolfenstine, G. Gonzalez-Doncel, and O.D. Sherby. 'Processing and Elevated Temperature Properties of Mg-Li Laminates,' in *Metal & Ceramic Matrix Composites: Processing, Modeling & Mechanical Behavior*, pp. 263–270, ed. R.B. Bhagat, A.H. Clauer, P. Kumar, and A.M. Ritter, The Minerals, Metals & Materials Society, Warrendale, PA, 1990.

53. K. Higashi, O.D. Sherby, G. Gonzalez-Doncel, and J. Wolfenstine. 'Superplastic Two-Phase Mg-Li Alloys,' in *International Conference on Superplasticity in Advanced Materials (ICSAM-91)*, pp. 491–496, ed. S. Hori, M. Tokizane, and N. Furushiro, The Japan Society for Research on Superplasticity, Osaka, Japan, 1991.

54. E.M. Taleff, O.A. Ruano, J. Wolfenstine, and O.D. Sherby, 'Superplastic Behavior of a Fine-Grained Mg–9Li Material at Low

Homologous Temperature,' *J. Mater. Res.*, 7(8) (1992), pp. 2131–2135.

55. J. Wolfenstine, G. Gonzalez-Doncel, and O.D. Sherby, 'Tension versus Compression Superplastic Behavior of a Mg–9wt%Li–5wt% B₄C Composite,' *Mater. Lett.*, **15** (1992), pp. 305–308.

56. T.G. Nieh and J. Wadsworth, 'Superplasticity in a Powder Metallurgy Magnesium Composite,' *Scr. Metall. Mater.*, **32** (1995), pp. 1133–1138.

57. N. Ridley, 'Superplasticity in Iron Base Alloys,' in *Superplastic Forming of Structural Alloys*, ed. N.E. Paton and C.H. Hamilton, TMS-AIME, Warrendale, Pennsylvania, 1982, pp. 191–224.

58. B. Walser and U. Ritter, 'Superplasticity in Iron-Base Alloys,' in *Superplasticity*, pp. 15.1–15.8, ed. B. Baudelet and M. Suery, Edition du CNRS, Paris, 1985.

59. O.D. Sherby and J. Wadsworth, 'Superplasticity in Iron-Based Alloys,' in *Encyclopedia of Materials Science and Engineering, First Supplement*, pp. 519–522, ed. R.W. Cahn, Pergamon Press, Oxford, 1988.

60. Y. Maehara and T.G. Langdon, 'Superplasticity of Steels and Ferrous Alloys,' *Mater. Sci. Eng.*, **A128** (1990), pp. 1–13.

61. G.R. Yoder and V. Weiss, 'Superplasticity in Eutectoid Steel,' *Metall. Trans.*, **3A** (1972), pp. 675–681.

62. O.D. Sherby, T. Oyama, D.W. Kum, B. Walser, and J. Wadsworth, 'Ultrahigh Carbon Steels,' *J. Metals*, **37**(6) (1985), pp. 50–56.

63. J. Wadsworth and O.D. Sherby, 'Superplastic White Cast Irons,' *Foundry MT&T*, **106** (1978), pp. 59–74.

64. J. Wadsworth, L.E. Eiselstein, and O.D. Sherby, 'The Development of Ultrafine, Superplastic Structures in White Cast Irons,' *Mater. Eng. Applications*, **1** (1979), pp. 143–153.

65. O.A. Ruano, L.E. Eiselstein, and O.D. Sherby, 'Superplasticity in Rapidly Solidified White Cast Irons,' *Metall. Trans.*, **13A** (1982), pp. 1785–1792.

66. H.W. Schadler, 'The Stress-Strain Rate Behavior of a Manganese Steel in the Temperature Range of the Ferrite-Austenite Transformation,' *Trans. AIME*, (242) (1968), pp. 1281–1287.

67. C.I. Smith, B. Norgate, and N. Ridley, 'Superplastic Deformation and Cavitation in a Microduplex Stainless Steel,' *Metal Sci.*, **10** (1976), pp. 182–188.

68. M.J. Stewart, 'Superplasticity in Low-Alloy Steels,' *Metall. Trans.*, **7A** (1976), pp. 399–406.

69. S. Lee, J. Wadsworth, and O.D. Sherby, 'Impact Properties of a Laminated Composite Based on Ultrahigh Carbon Steel and Hadfield Manganese Steel,' *Res. Mechanica*, **31** (1990), pp. 233–248.

70. C.I. Smith and N. Ridley, 'Design of a Superplastic Alloy Steel,' *Metals Technol.*, **1** (1974), pp. 191–198.

71. R.C. Gibson, H.W. Hayden, and J.H. Brophy, 'Properties of Stainless Steels with a Microduplex Structure,' *Trans. ASM*, **61** (1968), pp. 85–93.

72. Y. Maehara and Y. Ohmori, 'Microstructural Change during Superplastic Deformation of δ-Ferrite/Austenite Duplex Stainless Steel,' *Metall. Trans.*, **18A** (1987), pp. 663–672.

73. Y. Maehara, 'Superplastic Deformation Mechanism of δ/γ Duplex Stainless Steels,' *Trans. ISIJ*, **27** (1987), pp. 705–712.

74. Y. Maehara, 'High Strain Rate Superplasticity of a 25wt%Cr–7wt%Ni–3wt%Mo–0.14wt%N Duplex Stainless Steel,' *Metall. Trans.*, **22A** (1991), pp. 1083–1091.

75. K. Mineura and K. Tanaka, 'Superplasticity of 20Cr–10Ni–0.7N (wt%) Ultra-High Nitrogen Austenitic Stainless Steel,' *J. Mater. Sci.*, **24** (1989), pp. 2967–2970.

76. K. Osada, 'Properties of a Microduplex Stainless Steel Superplastically Deformed,' in *Superplasticity and Superplastic Forming*, pp. 429–433, ed. C.H. Hamilton and N.E. Paton, The Minerals, Metals & Materials Society, Warrendale, PA, 1988.

77. K. Osada, 'Enhancement of Superplasticity by Cold Rolling in a Duplex Stainless Steel,' in *MRS Intl. Meeting on Advanced Materials Vol 7 (IMAM-7, Superplasticity)*, pp. 101–106, ed. M. Kobayashi and F. Wakai, Materials Research Society, Pittsburgh, PA, 1989.

78. T.R. Parayil, G.L. Dunlop, and P.R. Howell, 'The Role of Dislocations in the Superplastic Deformation of a Duplex Stainless Steel,' in *Superplasticity in Metals, Ceramics, and Intermetallics, MRS Proceeding No. 196*, pp. 93–98, ed. M.J. Mayo, J. Wadsworth, and M. Kobayashi, Materials Research Society, Pittsburgh, PA, 1990.

79. N. Ridley and L.B. Duffy, 'Superplastic Deformation Behaviour of Two Microduplex

Stainless Steels,' in *Strength of Metals and Alloys, ICSMA 7*, pp. 853–864, ed. H.J. McQueen, J.-P. Bailon, J.I. Dickson, J.J. Jonas, and M.G. Akben, Pergamon Press, Oxford, 1985.

80. K. Tsuzaki, H. Matsuyama, M. Nagao, and T. Maki, 'High Strain Rate Superplasticity and Role of Dynamic Recrystallization in Duplex Stainless Steels,' *J. JIM*, **54** (1990), pp. 878–887.

81. Y. Zhang, F. Dabkowski, and N.J. Grant, 'The Superplastic Response of Three Rapidly Solidified Microduplex Stainless Steels of Varying Ferrite-Austenite Content,' *Mater. Sci. Eng.*, **65** (1984), pp. 265–270.

82. H.W. Hayden, R.C. Gibson, H.P. Merrick, and J.H. Brophy, 'Superplasticity in the Ni-Fe-Cr System,' *Trans. ASM*, **60**(1) (1967), pp. 3–13.

83. H.W. Hayden, S. Floreen, and P.D. Goodall, 'The Deformation Mechanisms of Superplasticity,' *Metall. Trans.*, **3A** (1972), pp. 833–842.

84. Y. Maehara, 'Superplasticity of δ-Ferrite/Austenite Duplex Stainless Steels,' *Trans. ISIJ*, **25** (1985), pp. 69–76.

85. B.C. Snyder, J. Wadsworth, and O.D. Sherby, 'Superplastic Behavior in Ferrous Laminated Composites,' *Acta Metall*, **32** (1984), pp. 919–923.

86. G. Daehn, D.W. Kum, and O.D. Sherby, 'Superplasticity of a Stainless Steel Clad Ultrahigh Carbon Steel,' *Metall. Trans,*, **17A** (1986), pp. 2295–2298.

87. G. Daehn, Ph.D. Thesis, Dept. of Materials Science and Engineering, Stanford University, (1987).

88. H.C. Tsai, K. Higashi, and O.D. Sherby, 'Superplasticity in a UHCS-Aluminum Bronze Laminated Composite,' in *Intl. Conf. Advanced Composites*, pp. 151–157, ed. T. Chandra, TMS, Warrendale, PA, 1991.

89. D. Lee and W.A. Backofen, 'Superplasticity in Some Titanium and Zirconium Alloys,' *Trans. AIME*, **239** (1967), pp. 1034–1040.

90. J.A. Wert and N.E. Paton, 'Enhanced Superplasticity and Strength in Modified Ti–6Al–4V Alloys,' *Metall. Trans.*, **14A** (1983), pp. 2535–2545.

91. C.H. Hamilton, 'Superplastic Forming and Diffusion Bonding of Titanium Alloys,' in *Titanium Science and Technology*, pp. 621–647, ed. R.T. Jaffee and H.M. Burke, Plenum Press, New York, 1973.

92. C. Hammond, 'Superplasticity in Titanium Base Alloys,' in *Superplastic Forming of Structural Alloys*, pp. 131–146, ed. N.E. Paton and C.H. Hamilton, TMS-AIME, Warrendale, PA, 1982.

93. C.H. Hamilton, 'Superplasticity in Titanium Alloys', in *Superlasticity*, pp. 14.1–14.16, ed. B. Baudelet, M. Suery, Centre National de la Recherche Scientifique, Paris, Grenoble, France, 1985.

94. A.K. Ghosh and C.H. Hamilton, 'Mechanical Behavior and Hardening Characteristics of a Superplastic Ti–6Al–4V Alloy,' *Metall. Trans.*, **10A** (1979), pp. 699–706.

95. N.E. Paton and C.H. Hamilton, 'Microstructural Influences on Superplasticity in Ti–6Al–4V,' *Metall. Trans.*, **10A** (1979), pp. 241–250.

96. M.T. Cope, D.R. Evetts, and N. Ridley, 'Superplastic Deformation Characteristics of Two Microduplex Titanium Aloys,' *J. Mater. Sci.*, **21** (1986), pp. 4003–4008.

97. J.R. Leader, D.F. Neal, and C. Hammond, 'The Effect of Alloying Additions on the Superplastic Properties of Ti–6%Al–4%V,' *Metall. Trans.*, **17A** (1986), pp. 93–106.

98. A. Wisbey and P.G. Partrige, 'Superplastic Deformation in a High Temperature Ti-Alloy IMI834,' in *International Conference on Superplasticity in Advanced Materials (ICSAM–91)*, pp. 465–470, ed. S. Hori, M. Tokizane, and N. Furushiro, The Japan Society for Research on Superplasticity, Osaka, Japan, 1991.

99. F.H. Froes, C.F. Yolton, J.C. Chestnutt, C.H. Hamilton, and M.E. Rosenblum, 'Superplastic Forming of Corona 5 (Ti–4.5Al–5Mo–1.5Cr),' in *Titanium '80 Sicence and Technology*, pp. 1025–1032, ed. H. Kimura and O. Izumi, TMS-AIME, Warrendale, PA, 1980.

100. J. Ma, R. Kent, and C. Hammond, 'Superplastic Deformation in Ti–4%Al–4%Mo–2%Sn–0.5%Si (IMI550),' *J. Mater. Sci.*, **21** (1986), pp. 475–487.

101. P. Griffiths and C. Hammond, 'Superplasticity in Large Grained Materials,' *Acta Metall.*, **20** (1972), pp. 953–945.

102. N. Furushiro and S. Hori, 'Strain Rate Sensitivity of Flow Stress and Effective Stress in Superplastic Deformation of Ti–8Mn Alloys,' in *Titanium '80 Sicence and Technology*, pp. 1067–1070, ed. H. Kimura

and O. Izumi, TMS-AIME, Warrendale, PA, 1980.

103. S. Hashimoto, F. Moriwaki, T. Mimaki, and S. Miura, 'Sliding Along the Interphase Boundary in Austenitic/Ferritic Duplex Stainless Steel Bicrystals,' in *International Conference on Superplasticity in Advanced Materials (ICSAM-91)*, pp. 23–32, ed. S. Hori, M. Tokizane, and N. Furushiro, The Japan Society for Research on Superplasticity, Osaka, Japan, 1991.

104. A. Eberhardt and B. Baudelet, 'Interphase Boundary Sliding at High Temperature in Two-Phase (α/β)-Brass Bicrystals,' *Phil. Mag.*, **41**(6) (1980), pp. 843–867.

105. Q. Liu, W. Yang, and G. Chen, 'On Superplasticity of Two Phase α-Titanium-Intermetallics Ti-(Co,Ni)-Al Alloys,' *Acta Metall. Mater.*, **43** (1995), pp. 3571–3582.

106. M.W. Mahoney and R. Crooks, 'Mechanisms of Superplastic Flow in Inconel 718,' in *Superplasticity and Superplastic Forming*, pp. 73–77, ed. C.H. Hamilton and N.E. Paton, The Minerals, Metals, and Materials Society, Warrendale, PA, 1988.

107. M.W. Mahoney and R. Crooks, 'Superplastic Forming of Inconel 718,' in *Superplasticity in Aerospace*, pp. 331–344, ed. C. Heikkenen and T.R. McNelly, TMS-AIME, Warrendale, PA, 1988.

108. G.D. Smith and H.L. Flower, 'Superplastic Forming of Alloy 718,' *Adv. Mater. Proc.*, **145**(4) (1994), pp. 32–34.

109. I. Kuboki, Y. Motohashi, and M. Imabayashi, 'Grain Refinement and Superplasticity in a Hard Nickel-Base Alloy,' in *Superplasticity and Superplastic Forming*, pp. 413–417, ed. C.H. Hamilton and N.E. Paton, The Metallurgical Society of AIME, Warrendale, PA, 1988.

110. S.H. Reichman and J.W. Smythe, 'Superplasticity in P/M IN–100 Alloy,' *Intl. J. Powder Metall.*, **6** (1970), pp. 65–74.

111. R.G. Menzies, J.W. Edington, and G.J. Davies, 'Superplastic Behavior of Powder Consolidated Nickel-Base Superalloy IN–100,' *Metal Sci.*, **15** (1981), pp. 210–216.

112. S. Kikuchi, S. Ando, S. Futami, T. Kitamura, and M. Koiwa, 'Superplastic Deformation and Microstructure in PM IN–100 Superalloy,' *J. Mater. Sci.*, **25** (1990), pp. 4712–4716.

113. S. Kikuchi, 'Superplasticity of PM Nickel-Base Superalloy,' in *International Conference on Superplasticity in Advanced Materials (ICSAM-91)*, pp. 485–490, ed. S. Hori, M. Tokizane, and N. Furushiro, The Japan Society for Research on Superplasticity, Osaka, Japan, 1991.

114. J.K. Gregory, J.C. Gibeling, and W.D. Nix, 'High Temperature Deformation of Ultra-Fine-Grained Oxide Dispersion Strengthened Alloys,' *Metall. Trans.*, **16A** (1985), pp. 777–787.

115. H.F. Merrick, 'Superplasticity in Nickel-base Alloys,' in *Superplastic Forming of Structural Alloys*, ed. N.E. Paton and C.H. Hamilton, The Metallurgical Society of AIME, Warrendale, PA, 1982, pp. 209–224.

116. J.B. Moore, J. Tequesta, and R.L. Athey, *U.S. Patent 3,519,503*, 1970.

117. R.C. Gibson and J.H. Brophy, 'Microduplex Nickel-Iron-Chromium Alloys,' in *Ultrafine-Grain Metals*, pp. 377–394, ed. J.J. Burke and V. Weiss, Syracuse University Press, 1969.

118. E.E. Brown, R.C. Boettner, and D.L. Ruckle. 'Minigrain Processing of Nickel-Base Alloys,' in *Superalloys – Processing, Proc. 2nd Intl. Conf.*, pp. L–1 to L–12, AIME, New York, NY, 1972.

119. L. Ceschini, G.P. Cammarota, G.L. Garagnani, F. Persiani, and A. Afrikantov, 'Superplastic Behavior of Fine-Grained IN–718 Superalloy,' in *Superplasticity in Advanced Materials – ICSAM-94*, pp. 351–356, ed. T.G. Langdon, Trans Tech Publications Ltd, Switzerland, 1994.

Chapter 6

Fine-structure superplastic ceramics

Despite extensive studies of superplastic behavior in metallic systems since the 1960s, work on superplasticity in ceramics and ceramic composites is of very recent origin. This is primarily because ceramics normally fracture inter-granularly at low strain values, as a result of a weak grain-boundary cohesive strength. The low grain-boundary cohesive strength is a result of inherent high grain-boundary energy [1]. Research in superplastic ceramics began actively only in the late 1980s but has expanded very rapidly since then.

The ceramics and ceramic composites made superplastic to date are essentially based on the principles developed for metallic alloys. Existing data indicate that for polycrystalline ceramics, however, a grain size of less than 1 μm is necessary for superplastic behavior. This is in contrast to superplastic metals, in which grain sizes are typically only required to be less than 10 μm. To highlight the dominant effect of grain size on the deformation behavior of ceramics, Figure 6.1 shows the modulus-compensated flow stresses measured from a number of studies on tetragonal zirconia as a function of diffusivity-compensated strain rate [2–6]. It is evident that for a given stress, the strain rate increases dramatically as grain size decreases. (Or, conversely, that for a given imposed strain rate the stress required decreases dramatically as grain size decreases.) Figure 6.1 illustrates the importance of grain-boundary-sliding (GBS) mechanisms in the deformation of fine-grained ceramics.

The first observation of fine-structure superplasticity in ceramics, in a yttria-stabilized tetragonal zirconia (YTZP), is generally attributed to Wakai in 1986 [7], although an elongation to failure of ~100% in polycrystalline MgO was reported in 1965 by Day and Stokes [8]. The historical result obtained by Wakai is summarized in Figure 6.2. Many other claims have been made of superplastic behavior in ceramics, but nearly all are based on tests carried out only in compression [9–21]. It is important to point out that superplasticity refers to high ductility in *tension*; therefore, enhanced plasticity data from compression tests cannot necessarily be considered as convincing evidence of superplasticity because ceramics often exhibit high values of the strain-rate-sensitivity exponent (*m* is often equal to 1) [22] in compression tests but, nonetheless, show very limited tensile ductility. Therefore, ceramics that are successfully deformable in compression may fail prematurely in tension from grain-boundary separation or other related tensile fracture effects, e.g., grain-boundary cavitation and cracking. Tensile ductility at elevated temperatures is limited primarily by grain-boundary cavitation, which is initiated by tensile stresses but usually suppressed by compressive stresses. Furthermore, there is also increasing evidence suggesting that the stress–strain-rate relationships, and therefore, the mechanisms of plastic flow may be different in compression and tension [23–25]. Nevertheless, compression tests still provide information on the mechanisms of plastic flow.

Since 1986, a number of fine-grained polycrystalline ceramics have been demonstrated to be superplastic in tension. These include YTZP [6, 7, 26–29], Y- or MgO-doped Al_2O_3 [23, 30], hydroxyapatite [31], β-spodumene glass ceramics [25, 32], Al_2O_3-reinforced YTZP (Al_2O_3/YTZP) [33–37], SiC-reinforced Si_3N_4 (SiC/Si_3N_4) [38–41], and iron/iron carbide (Fe/Fe_3C) [42] composites. The area of superplastic ceramics has been the subject of considerable interest, and some

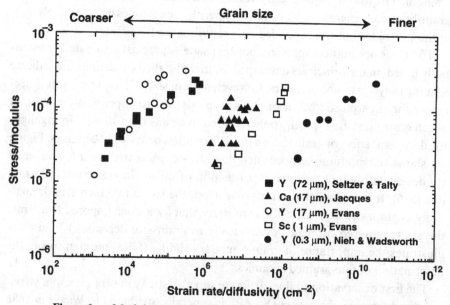

Figure 6.1   Modulus-compensated, plastic flow stress as a function of diffusion-compensated strain rate for tetragonal zirconia. It is evident that flow stress decreases as grain size decreases (data from Refs. [2–6]).

review papers are now available [43–47]. In the following sections, these super-plastic materials are divided into monolithic ceramics and ceramic composites and are discussed accordingly.

6.1 Monolithic ceramics

6.1.1 Yttria-stabilized tetragonal zirconia polycrystal

6.1.1.1 Microstructure

The first true polycrystalline ceramic shown to be superplastic was a 3-mol % YTZP (3YTZP), which is now considered to be the model ceramic system. It has been the most widely studied, yet the experimental results are also probably the most controversial. The material is relatively pure: chemical compositions of the superplastic 3YTZP are typically (by weight) 6.2% Y_2O_3, 0.005% Al_2O_3, 0.002% SiO_2, 0.002% Fe_2O_3, 0.003% Na_2O, and balance ZrO_2. An electron micrograph from the material (Figure 6.3) reveals equilibrated, hexagonal-shaped grains with sharp apexes and a mean grain size of about 0.3 μm. Results from conventional X-ray diffraction show that 3YTZP consists primarily of the tetragonal phase with only a limited amount of the monoclinic phase (<1%).

The microstructure of 3YTZP is thermally unstable at high temperatures and undergoes concurrent grain growth during superplastic deformation [48–50].

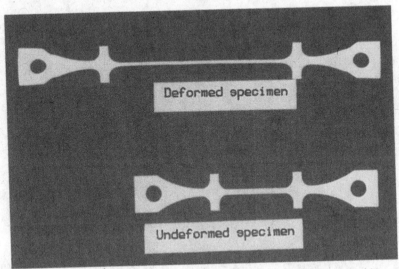

Figure 6.2 Historical discovery of superplasticity in fine-grained, 3-mol % YTZP. The sample was deformed over 200% at 1450 °C in tension (from Ref. [7]).

Several mathematical expressions have been used to describe concurrent grain growth in superplastic metals [51–54] and ceramics [18, 50, 55]. The grain-growth behavior of 3YTZP has been investigated [48, 56], and experimental results show both static and dynamic grain growth in 3YTZP with the following equation:

$$kt = D^\ell - D_o^\ell, \tag{6.1}$$

where k is a kinetic constant that depends primarily on temperature and grain-boundary energy, t is time, $\ell = 3$, and D and D_o represent the instantaneous and initial grain sizes, respectively. Values for ℓ are the same for both static and dynamic grain growth, suggesting that both processes are driven by a similar mechanism. In fact, the result that $\ell = 3$ indicates that grain growth in 3YTZP is controlled by solute or impurity drag [57, 58], which presumably is a result of strong yttrium segregation to the grain boundaries [43, 59–62].

Figure 6.3   Transmission electron micrograph of as-received 3YTZP showing equilibrated, hexagonal-shaped grains.

6.1.1.2 Stress–strain curve

Typical true stress vs true strain curves for a superplastic 3YTZP sample are shown in Figure 6.4. The $\sigma - \varepsilon$ curves show a significant degree of strain hardening, resulting primarily from grain coarsening during deformation. The steady-state region that is often observed in superplastic metal alloys does not exist in Figure 6.4. The lack of a steady-state region is also a result of concurrent grain growth taking place in fine-grained superplastic ceramics. Also evident in Figure 6.4 is the fact that macroscopic necking does not occur until near the final fracture of the test sample.

Tensile elongations of YTZP are only 5 and 60% at 1000 and 1200 °C, respectively, but elongations of greater than 100% can be routinely obtained at temperatures above 1350 °C. A maximum value of elongation to failure of 800% has been recorded in a sample tested at 1550 °C and at a strain rate of $8.3 \times 10^{-5}\,\text{s}^{-1}$. A direct comparison of this superplastically deformed sample with an untested sample is shown in Figure 6.5. (A maximum tensile elongation of over 1000% has been recorded in a YTZP containing 5 wt% SiO_2 (grain size=0.26 μm) deformed at 1400 °C [29].)

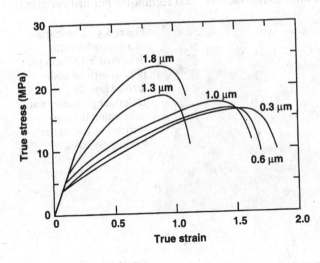

Figure 6.4 True stress–strain curves for YTZP tested at 1550 °C and a strain rate of $2.7 \times 10^{-5}\,\text{s}^{-1}$ with various grain sizes. Tensile elongation decreases and flow stress increases with an increase in grain size.

6.1.1.3 Grain size

Grain size strongly affects the superplastic properties of 3YTZP [49, 63]. As indicated in Figure 6.4, the elongation to failure decreases and the flow strength increases as grain size increases. The dependence of flow stress on grain size has been investigated by Nieh and Wadsworth [63]. The results, summarized in Figure 6.6, indicates that the flow stress approximately depends on the grain size raised to a second power:

$$\sigma \propto d^2, \tag{6.2}$$

where σ is the flow stress and d is the instantaneous grain size.

6.1.1.4 Strain-rate-sensitivity exponent

The strain-rate-sensitivity exponent m is often used as a useful guide to mechanisms of plastic flow. For 3YTZP, there exists a great disparity in the values of m that have been reported in the literature; these vary from 0.3 [27, 64, 65] to 0.5 [7, 11, 14, 28, 49]. At least three major factors – test technique, impurity content,

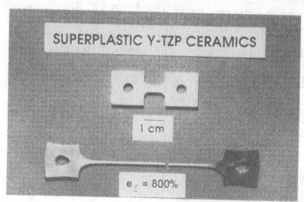

Figure 6.5 Comparison of a superplastically deformed YTZP sample (800%) with an undeformed sample. Deformation occurs not only within the gage section but also at the loading pinholes.

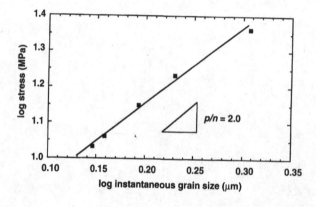

Figure 6.6 Flow stress as a function of the instantaneous grain size for YTZP tested at 1550 °C and a strain rate of 2.7×10^{-5} s^{-1}.

and microstructural evolution during deformation – can contribute to the different *m* values that have been measured. For example, the *m* value for 3YTZP has been shown to depend on the testing technique, specifically whether a strain-rate-increase, strain-rate-decrease, or individual test was used [6]. Because of the microstructural evolution during deformation, these various testing techniques produce different *m* values.

Another viewpoint has been proposed by Ma and Langdon [65] who suggest that the measured *m* value is dependent on the impurity level in 3YTZP (Figure 6.7). The data in Figure 6.7 suggest that the higher the impurity content, and hence the higher the glass content at the grain boundaries, the higher the *m* value. Carry and coworkers [64, 66] conducted experiments with YTZP containing different glass-forming impurities and also showed that the presence of a grain-boundary glassy phase not only increases the deformation rate, but also changes the *m* value. However, these results were challenged by Gust *et al.* [67] who have shown that the presence of a grain-boundary glassy phase (BaS) enhances the deformation rate of YTZP, but it does not change the *m* value.

Because of the strong grain-size dependence of the flow stress, microstructural evolution during deformation and concurrent grain growth, in particular, are

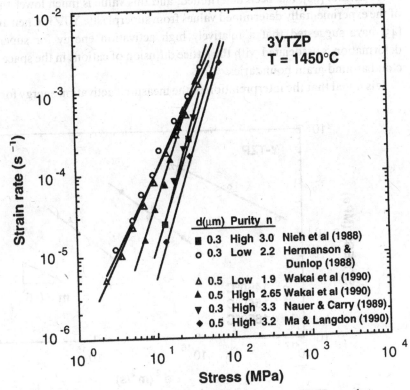

Figure 6.7 Strain rate vs stress for low- and high-purity YTZP tested at 1450 °C (from Ref. [65]).

expected to result in specific artificial effects on the determination of m values. In fact, the measured m value from a conventional $\log \dot{\varepsilon}$ vs $\log \sigma$ plot would represent only an *apparent* value. To evaluate the true m value (i.e., measured under constant structure conditions), one must be cautious to correct for the effect of concurrent grain growth. By normalizing the stress with the square of the final grain size [63], the *true m* value for 3YTZP is shown to be about 0.67 (Figure 6.8).

6.1.1.5   Activation energy

The activation energy for superplastic deformation of 3YTZP under constant structure conditions has been determined to be about 510 kJ/mol [6]. In contrast, a higher value of 590 kJ/mol is obtained if concurrent grain growth is neglected [7]. By comparison, an activation energy of 540 kJ/mol has been reported for the creep of YTZP [5]. Even in the absence of concurrent grain growth, the interpretation of the observed activation energies in YTZP is difficult. Theoretically, diffusion of the Zr^{+4} cation is the slower diffusing species and should be the rate-controlling one. However, no data for cation diffusion in tetragonal zirconia exist. Only the activation energy for the diffusion of Zr^{+4} in cubic zirconia, which is 423 kJ/mol [68], has been determined, and this value is much lower than any of the experimentally determined values from superplastic YTZP. Chen and Xue [43] have suggested that a relatively high activation energy for superplastic deformation is associated with the lattice diffusion of cations in the space charge cloud around grain boundaries.

It is noted that the interpretation of the measured activation energy for super-

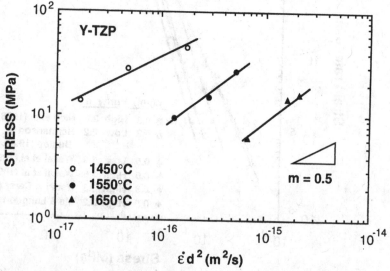

Figure 6.8   Flow stress as a function of grain-size normalized strain rate. The strain-rate-sensitivity exponents are greater than 0.5 for all temperatures.

plastic flow is generally not straightforward. In principle, it is possible to relate a high-temperature deformation mechanism to a certain, specific, diffusional process once the activation energy for the deformation process is accurately determined. Unfortunately, basic diffusion data of this type are usually not available for complex ceramics. Even if they are available, their values, depending on the exact measuring methods, can vary quite significantly. More often than not, diffusion coefficients are reported that vary over several orders of magnitude. This poses a serious problem in selecting precise diffusion data and, subsequently, in identifying the deformation mechanisms. Furthermore, ceramics usually contain a significant number of defects and impurities, and diffusion processes in ceramics are, therefore, even more complicated. Diffusion processes in these materials depend upon the stoichiometric composition, charge state of the diffusing species, and exact test environment. Data obtained in a ceramic from one particular experiment may not be useful to another experiment conducted on the same material. Therefore, from a practical point of view, to assess precise superplastic deformation mechanisms by comparison of the measured activation energy with diffusion data is not a very promising approach. By contrast, it is probably more useful to assess other experimental parameters, such as the stress and grain-size exponents.

6.1.2   Alumina

Alumina is probably one of the most widely used structural ceramics, but superplasticity is difficult to obtain in alumina, as a result of rapid anisotropic grain growth during high-temperature deformation [69, 70]. Nonetheless, several studies on superplasticity in doped, fine-grained Al_2O_3 have been performed. These were carried out by Carry and colleagues [23, 55, 64, 69, 71, 72] and Yoshizawa and Sakuma [30]. For example, Gruffel et al. [23] demonstrated that the grain size of Al_2O_3 containing 500-ppm MgO can be further refined by adding various dopants, such as Cr_2O_3, Y_2O_3, and TiO_2. A grain size of about 0.66 μm was obtained in a 500-ppm Y_2O_3-doped Al_2O_3. As a result of this fine grain size, the Al_2O_3 exhibits a rupture elongation of 65% at 1450 °C under an applied stress of 20 MPa. Although a tensile elongation of 65% is not as high as those observed in superplastic metals, it is a large plastic tensile deformation and implies at least the potential for superplasticity in Al_2O_3.

The flow behavior of the Y_2O_3-doped Al_2O_3 was not fully characterized, and therefore, the deformation mechanisms could not be further assessed. However, it was indicated that the tensile elongation, as well as the flow stresses, strongly depended on the grain size in Al_2O_3. Specifically, the tensile elongation systematically increases with a decrease in grain size. Also, the flow stress is proportional to the grain size raised to a 1.5 power, i.e., $\sigma \propto d^{1.5}$ [23, 63]. These results were proposed to be indicative of a GBS deformation mechanism.

In a similar manner to YTZP, severe dynamic grain growth was also observed in fine-grained Al_2O_3 within the superplastic temperature range (1350–1500 °C) [69]. The grain growth was observed to follow Eq. (6.1) with $\ell = 4$. It is suggested that the $t^{1/4}$ kinetic law results from the viscous drag of intergranular precipitates (presumably, Al_2O_3–Y_2O_3 garnet) with material transport governed by interfacial diffusion [73]. Some experimental evidence from fine-grained Al_2O_3 [23] indicates that the tension and compression behavior of the material might be different.

6.1.3   Hydroxyapatite

Hydroxyapatite $[Ca_{10}(PO_4)_6(OH)_2]$ is the human body's own ceramic and a primary constituent in enamel, dentine, and bone tissue. Its similarity to bone in composition and structure provides a good biological compatibility for implantation in bone tissue. Dense, equiaxed, and fine grained (~0.64 μm), hydroxyapatite can be produced through advanced powder metallurgy processes [74]. The fine-grained material has been demonstrated to be superplastic [31]; a maximum elongation of greater than 150% was recorded at 1050 °C at a strain rate of 1.4×10^{-4} s^{-1}. Substantial strain hardening, caused by grain growth during superplastic deformation was also observed. The yield stress of hydroxyapaptite was found to be linearly dependent on the grain size, i.e., $\sigma \propto d$, and the *apparent* stress exponent of the yield stress was determined to be greater than $n = 3$ ($m = 0.33$), which is relatively high compared to those observed in other superplastic ceramics. The high n value is attributed to a rapid grain growth during superplastic testing. Based on measurements that the apparent intragranular strain was less than 20% of the total strain, Wakai *et al.* [31] concluded that GBS was the dominant mode of deformation.

Cavitation occurred during the superplastic deformation of hydroxyapatite. In fact, because of the cavitation and concurrent grain growth, the hydroxyapatite, which was initially optically transparent, became translucent after superplastic deformation. The demonstration of superplasticity in hydroxyapatite was technologically significant because it was the first functional ceramic demonstrated to be superplastic.

6.1.4 β-Spodumene glass ceramics

Wang and Raj [32] studied the deformation mechanisms of two fine-grained (0.91–2 μm) lithium aluminosilicate β-spodumene glass ceramics (containing 4–9 vol % liquid phase) and observed superplasticity in both materials. Tensile elongations of 135% and 400%, respectively, were recorded from the two materials tested at a strain rate of 10^{-4} s^{-1} and at temperatures between 1150 and 1200 °C, which are noted to be above the solidus temperatures of the test materials. Experimental results showed that the stress exponent was unity and $\dot{\varepsilon} \propto d^{-3}$, from which it was concluded that the deformation was limited by the diffusion of atoms through the liquid phase, a process which was modeled by Raj [75] and Pharr and Ashby [76]. Because a large volume of liquid phase was present, the work of Wang and Raj may represent a rather special case that is not representative of a true polycrystalline solid. Nevertheless, their results clearly illustrate that the presence of glassy phases can significantly enhance mass flow and change the plastic flow properties in a ceramic.

6.2 Ceramic composites

6.2.1 Zirconia-based composites

Two zirconia-based composite systems have been explored: Al_2O_3-reinforced YTZP (Al_2O_3/YTZP) and mullite-reinforced YTZP. Alumina–zirconia composite is considered as a model material system because there is no intermediate phase existing between alumina and zirconia; Al_2O_3 and ZrO_2 form a eutectic at about 1850 °C. Among the Al_2O_3/YTZP duplex composites, the 20 wt% Al_2O_3/3YTZP has been the most extensively studied. The typical microstructure of the composite after thermal etching at 1450 °C is presented in Figure 6.9. The material consists of equiaxed ZrO_2 (bright) and Al_2O_3 (dark) grains, both of which are about 0.5 μm. Wakai and Kato [33] first showed that this material was superplastic and reported a maximum tensile elongation of over 200%. Nieh and Wadsworth [35] further recorded a maximum elongation of 625%, obtained from testing at 1650 °C and at a strain rate of 4.0×10^{-4} s^{-1} (Figure 6.10). The composite generally exhibited a superplastic elongation of over 200% at temperatures above 1450 °C at strain rates from 2.7×10^{-5} to 1.7×10^{-3} s^{-1}.

Deformation properties of the 20 wt% Al_2O_3/YTZP are summarized in Figure 6.11 for temperatures of 1450, 1550, and 1650 °C. Some of the data in Figure 6.11 are noted to be obtained from testing in air [33] and some in vacuum [35]. Nonetheless, all the data are in good agreement. A behavioral difference between YTZP and the 20 wt% Al_2O_3/YTZP is the fact that the m values obtained in the 20 wt% Al_2O_3/YTZP, using either strain-rate change or individual tests, were virtually the same. The m value for the composite was about 0.5 and remained constant throughout the temperature range of 1250 to 1650 °C, suggesting that a single superplastic deformation mechanism, specifically GBS, prevails.

Concurrent grain growth also took place during the superplastic deformation of 20 wt% Al_2O_3/YTZP [33, 36, 77]. The growth rate of the ZrO_2 grains was found to be faster than that of the Al_2O_3 grains in the 20 wt% Al_2O_3/YTZP composite [36]. This is because ZrO_2 is the major phase, and during coarsening (which is essentially a diffusional process) the diffusion path is shorter for ZrO_2 than it is for Al_2O_3. The growth rate of the ZrO_2 grains in the composite, however, is similar to

Figure 6.9 Microstructure of the 20 wt% Al_2O_3/YTZP composite showing equiaxed grains. The dark and light phases are Al_2O_3 and ZrO_2, respectively.

that in the unreinforced 3YTZP [77]. These results indicate that the Al_2O_3 phase has an insignificant effect on inhibiting the growth of ZrO_2 grains. By contrast, the presence of ZrO_2 grains can greatly inhibit grain growth in Al_2O_3 [78].

The effect of grain size on the superplastic flow in 20 wt% Al_2O_3/YTZP has been studied, and the experimental results indicated that the elongation to failure of the material decreases and the flow strength increases with increasing grain size. The flow stress at a given strain rate and at a given strain was determined to be proportional to the grain size of the major phase ZrO_2 raised to a 0.75 power [79]:

$$\sigma \propto d^{0.75}. \tag{6.3}$$

This result, together with the observation that $n=2$ ($m=0.5$), lead to the constitutive equation for superplastic flow in the 20 wt% Al_2O_3/YTZP, as follows:

$$\dot{\varepsilon} = A d^{-1.5} \sigma^2 \exp\left(-\frac{Q}{RT}\right), \tag{6.4}$$

where $Q=590$ kJ/mol is the activation energy for superplastic flow [33, 35].

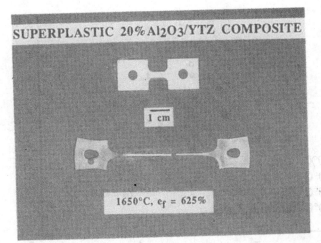

Because of a weaker grain-size dependence of the superplastic flow stress ($\sigma \propto d^2$ for YTZP vs $\sigma \propto d^{0.75}$ for 20 wt% Al_2O_3/YTZP), strain hardening, which is primarily caused by concurrent grain growth, is less pronounced in the composite than that in YTZP. This offers an explanation as to why, in contrast to YTZP, the determination of strain rate sensitivity in the composite is insensitive to certain test techniques.

In addition to the 20 wt% Al_2O_3/YTZP, Wakai et al. [34] have systematically studied the superplastic properties of ZrO_2/Al_2O_3 duplex composites with various amounts of Al_2O_3, ranging from 20 to 80 wt%. Tensile elongation to failure of these composites decreases with increasing Al_2O_3 content but is still over 120% in the 80 wt% Al_2O_3 composite tested at 1550 °C with a strain rate of 2.8×10^{-4} s^{-1}. Strain-rate–stress behavior for these duplex composites is summarized in Figure 6.12. The m value is approximately 0.5 for all ZrO_2/Al_2O_3 composites. The value was lowest (~0.4) for the 50–50 composite and increased along either side of the equiatomic composition. The apparent activation energy was 550 kJ/mol for the unreinforced matrix, i.e., the monolithic YTZP, and increased with Al_2O_3 content. The activation energy became essentially a constant (700 kJ/mol) for composites that contained more than 50% Al_2O_3. Assuming that alumina is much harder than zirconia, Wakai et al. suggested that the superplastic flow behavior of the composite can be qualitatively described by a rheological equation [80]:

$$\dot{\varepsilon} = (1-V)^q A \sigma^n, \tag{6.5}$$

where V is the volume fraction of alumina and q is a constant, depending upon the n value and nature of the second phase. It would be misleading to claim from Figure 6.12 that composites with more Al_2O_3 are stronger; this may be a result of different grain sizes. In fact, grain sizes are coarser in the materials containing more alumina.

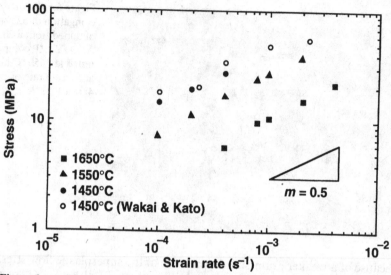

Figure 6.11   Flow stress of Al$_2$O$_3$/YTZP as a function of strain rate, determined from strain-rate increase tests, at different temperatures. The strain-rate-sensitivity value is about 0.5 and remains virtually constant from 1450 to 1650 °C (data from Refs. [35] and [33]).

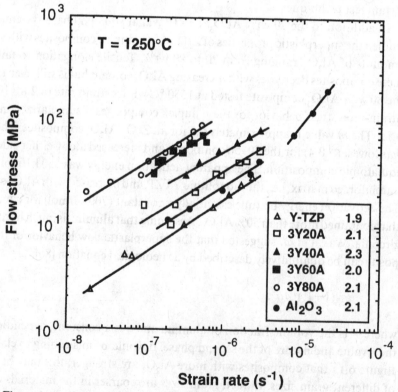

Figure 6.12   Flow stress of Al$_2$O$_3$/YTZP duplex composites as a function of strain rate at 1250 °C (data from Ref. [34]).

The superplastic flow of mullite/zirconia (2YTZP) composites with various amounts of mullite, ranging from 10 to 50 vol %, has been characterized by Yoon and Chen [21]. In these composites, mullite exhibits a fibrous morphology. The flow stress of these composites was found, phenomenologically, to be proportional to the grain size of ZrO_2 raised to a third power:

$$\sigma \propto d^3. \tag{6.6}$$

The strength of these composites increases with mullite content. Nevertheless, all composites exhibit a similar flow behavior, namely an m value of about 0.67. Yoon and Chen [21] extended the rheological equation, Equation (6.5), to incorporate the morphological feature of the hard phase (mullite) and develop the following constitutive equation to express the superplastic deformation of the mullite/YTZP composites:

$$\dot{\varepsilon} = (1 - V_{mullite})^{2 + n/2} A \left(\frac{\sigma^n}{d^3} \right) \exp\left(-\frac{Q}{RT} \right). \tag{6.7}$$

This equation is useful for describing the superplastic behavior of a composite, provided the deformation of a monolithic material and its composite can both be described by a similar power-law mathematical form.

6.2.2 Alumina-based composites

Lange and Hirlinger [81] first found that the addition of certain oxides, and particularly ZrO_2, can greatly stabilize the grain structure of Al_2O_3. Using this result, Xue et al. [78] successfully prepared a fine-grained (~ 0.5 μm) Al_2O_3 composite containing 10 vol % ZrO_2 and further demonstrated that the composite can be superplastically stretched under biaxial tension at 1400 °C. Without the ZrO_2 addition, alumina grains would grow exceedingly fast and result in a flow stress that is about 2.5 times higher than that for the composite (Figure 6.13).

Nagano et al. [82], however, performed tests with an Al_2O_3 composite containing 30 wt% TiC, which has superior wear and corrosion resistance and is a potential cutting tool material. The composite deformed at a faster strain rate than pure alumina with a similar grain size. The composite exhibits an m value of approximately 0.26, which is relatively low for superplasticity. Nonetheless, a maximum tensile elongation of 66% was obtained at 1550 °C and at a strain rate of 1.2×10^{-4} s^{-1}. This relatively low tensile elongation was primarily attributed to the coarse initial grain size (~ 1.2 μm). Also, there existed a chemical reaction between Al_2O_3 and TiC at elevated temperatures, which resulted in cavitation and, thus, a reduction of tensile elongation.

6.2.3   Silicon nitride-based composites

Silicon nitride is an attractive structural ceramic because of its high-temperature strength and oxidation resistance. Chen and colleagues [83, 84] first observed superplasticity in this material system. The superplastic material was a glass-phase-rich, yttria-doped silicon nitride (β'-SiAlON). It was superplastic with 230% elongation at 1550 °C and a strain rate of $10^{-4}\,s^{-1}$. The material exhibited unusual strain hardening resulting from the concurrent development of elongated grains and texture during deformation. The strengthening effect, as a result of the formation of fibrous grains, causes a reduction of tensile elongation of the material.

Silicon carbide–silicon nitride (SiC/Si_3N_4) composite is another promising material for high-temperature structures. Although both SiC and Si_3N_4 are extremely difficult to form plastically, even at elevated temperatures [85], Wakai and coworkers [38–40] demonstrated that a fine-grained SiC/Si_3N_4 composite can be synthesized and behaves superplastically. The composite was prepared initially from an amorphous Si–C–N powder that was produced by vapor-phase reaction of $[Si(CH_3)_3]_2NH/NH_3/N_2$ at 1000 °C [86, 87]. Bulk materials were produced by mixing the amorphous powder with 6 wt% Y_2O_3 and 2 wt% Al_2O_3 as sintering aids and further hot-pressing at 1650 °C and 34 MPa in N_2.

The above processes resulted in a composite containing approximately 20 wt% SiC in a β-Si_3N_4 matrix. The microstructure of the composite consisted of both spherical grains (<0.2 μm) and elongated grains (<0.5 μm). The material exhibited a superplastic elongation of over 150% at 1600 °C at an initial strain rate of $4\times10^{-5}\,s^{-1}$. The elongation value increased with decreasing SiC content and probably attributed to a decrease in strength. In a manner similar to β'-SiAlON, a strong texture was developed in SiC/Si_3N_4 during deformation, which was expected to degrade the superplastic properties of the composite. It should

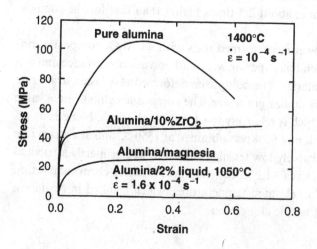

Figure 6.13 Comparison of stress–strain curves for pure alumina and an alumina composite containing 10 vol % ZrO_2 (data from Ref. [43]).

be noted that as a result of the reactions among Y_2O_3, Al_2O_3, and Si_3N_4 at the superplastic temperature (1600 °C), a substantial amount of intergranular liquid phase was present in the composite. In theory, a covalent polycrystal containing a large amount of intergranular liquid phase is expected to behave like a Newtonian fluid, i.e., with a stress exponent of unity [75]. However, a stress exponent of approximately 2 rather than 1 was measured in this SiC/Si_3N_4 composite. The exact deformation mechanisms are difficult to specify because of the microstructural dynamics, i.e. a continuous evolution of the microstructure.

Silicon carbide–silicon nitride composite was the first covalent compound demonstrated to be superplastic. Recently, in an attempt to reduce concurrent grain growth in a Si_3N_4 composite during superplastic deformation, Rouxel *et al.* [41] developed a composite containing 30 wt% SiC, 2 wt% Al_2O_3, and 6 wt% Y_2O_3. No grain growth was observed during superplastic deformation of the composite, but the superplastic properties (a maximum elongation of 114%) were inferior to the 20 wt% SiC composite. The lower tensile elongation may be related to the higher flow stress in the 30 wt% SiC composite.

6.2.4 Iron carbide-based composites

Kim *et al.* [24, 42] demonstrated superplasticity in a fine-grained iron–iron carbide (Fe/FeC$_3$) composite. The material was prepared from rapidly solidified powders of a 6.25%C–1.5%Cr hypereutectic iron using two different powder metallurgy procedures: (1) hot isostatic pressing followed by uniaxial pressing and (2) hot extrusion followed by uniaxial pressing. The microstructure consists of 80 vol % fine-equiaxed grains (~4 μm) of Fe_3C and 20 vol % iron-based second phase. Both of the above procedures yield materials that are superplastic at elevated temperatures (700–1000 °C) with low values of stress exponent ($n=1$ to 2) and tensile elongations as high as 600%. The grain shape and size after deformation remains equiaxed and fine. The strain rate in the $n=2$ region is inversely proportional to approximately the cube of the grain size with an activation energy for superplastic flow of between 200 and 240 kJ/mol. Superplastic flow in the iron carbide material in the $n=2$ region is proposed to be governed by GBS accommodated by slip, i.e., it is controlled by the diffusion of iron along iron carbide grain boundaries.

One of the interesting findings was that the flow stress in compression is about 2 times higher than that in tension in the region where GBS is the rate-controlling process. A direct comparison between these two sets of data is shown in Figure 6.14. Interestingly, a difference in strength between a tension and compression test was also observed during superplastic deformation in a Mg–9 wt% Li–5 wt% B_4C composite [88]. In contrast to these cases, Owen and Chokshi [89] have shown that the tensile and compression behavior of a 20 wt% Al_2O_3/YTZP composite are quite similar. This result is depicted in Figure 6.15.

Kim *et al.* [24, 42] argued that the difference in flow stress between tension and compression is a result of the greater ease of GBS in tension than in compression. Wang and Raj [32], however, argued that friction between a test sample and the platform in compression may give rise to the difference. Nevertheless, it is still a subject of dispute.

Tensile elongations of the Fe/Fe$_3$C composite were observed to increase with a decrease in flow stress and a decrease in grain size. To understand the parameters controlling the optimum superplastic elongation in this Fe/Fe$_3$C composite, as well as YTZP, Al$_2$O$_3$/YTZP, and Al$_2$O$_3$, Kim *et al.* [90] have correlated the maximum tensile elongation as a function of temperature-compensated strain rate, i.e., $\dot{\varepsilon} \cdot \exp(Q/RT)$, for a number of superplastic ceramics; this result is given in Figure 6.16. Assuming that the plastic flow of superplastic ceramics can be described by the Dorn equation, Figure 6.16 essentially indicates that tensile elongation is determined by flow stress. A similar conclusion has also been pro-

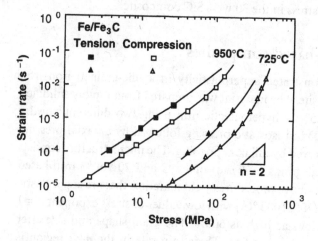

Figure 6.14 Comparison of the tension and compression strain-rate-change tests of Fe/Fe$_3$C at 725 and 950 °C (data from Ref. [42]).

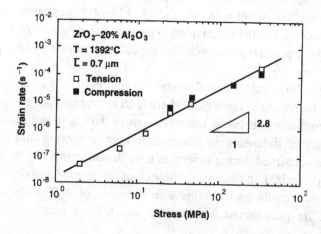

Figure 6.15 Variation in steady state creep rate with imposed stress for the 20 wt% Al$_2$O$_3$/YTZP composite tested at 1392 °C in tension and compression (data from Ref. [89]).

posed by Chen and Xue [43]. They illustrated that the tensile ductility of super-plastic ceramics is limited by grain-boundary decohesion, which is in turn determined by flow stress.

6.3 Constitutive equations and microstructures

6.3.1 Constitutive equations

Generally, superplastic flow in ceramics is a diffusion-controlled process and the strain rate $\dot{\varepsilon}$ can be expressed as

$$\dot{\varepsilon} = Ad^{-p}\sigma^n\exp\left(-\frac{Q}{RT}\right), \tag{6.7}$$

where A, p, and n are constants, d is the grain size, σ is the flow stress, Q is the activation energy for flow, and R is the gas constant, T is the absolute tempera-

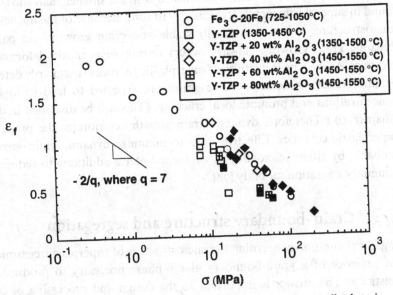

Figure 6.16 Elongation to failure plotted as a function of normalized strain rate for various superplastic ceramics and ceramic composites (data from Ref. [90]).

ture. The values for p, n, and Q vary according to the microstructure, specific flow/diffusion law, and sometimes the impurity content in a material. In principle, these parameters can be associated with various deformation mechanisms, e.g., GBS, Newtonian flow, Coble creep, Nabarro–Herring creep, interfacial-reaction-controlled diffusional creep, and liquid-phase, transport-controlled diffusional creep.

A summary of the microstructure and properties of some of the superplastic ceramics and ceramic composites is given in Table 6.1. As shown in this table, the specific functional relationship among the material variables are dependent on the material and strain rate and temperature ranges. The n value is noted, however, to be less than 2 for all materials except hydroxyapatite. Also, superplastic flow in these materials is noted to be always dependent on grain size. The grain-size exponent p ranges from 1 to 3, which is within the range of various diffusion-controlled flow models [75, 91–93]. It is noted once again that the grain sizes in all the superplastic ceramics with tensile elongations greater than 100%, except Fe/Fe$_3$C, are less than 1 μm, which is much smaller than those usually found in superplastic metals (typically, <10 μm). Because of the fineness of the microstructure, it is not thermally stable and grain growth – in particular, dynamic grain growth – always occurs during superplastic deformation of ceramics. As the tensile ductility of superplastic ceramics is primarily determined by intergranular cracking, coarse grains are expected to lead to high stress concentrations and promote local cracking. (This will be discussed in detail in Chapter 10.) Therefore, dynamic grain growth deteriorates the properties of superplastic ceramics. Effective ways to minimize dynamic grain growth are probably by either microalloying or second-phase additions to reduce grain-boundary migration mobility [78].

6.3.2   Grain-boundary structure and segregation

A general question concerning the microstructure of superplastic ceramics is: Is the presence of a grain-boundary glassy phase necessary to produce superplasticity? The answer is important to the design and processing of ceramic materials. The presence of a glassy phase in a ceramic is often a result of the addition of some sintering aids in order to improve densification. An equilibrium, thin (~2 nm) glassy phase at grain boundaries may have little influence on the mechanical properties of a ceramic [94]. But excessive grain-boundary glassy phases can degrade both the mechanical properties and environmental resistance of a ceramic [67]. If the presence of a glassy phase at grain boundaries is critical for the superplastic forming of a ceramic, then the amount and distribution of this glassy phase must be carefully controlled so that the structural performance of the ceramic will not be compromised.

Microstructural studies on grain boundaries have mainly been carried out in

Table 6.1 Property data of superplastic ceramics and ceramic composites in tension

Material	Microstructure	Testing parameters	Maximum elongation (%)	Material variables	Ref.
Monolithics					
3YTZP	$d=0.3\ \mu m$		200	$n=2, p=2,$ $Q=590$ kJ/mol	[7]
3YTZP	$d=0.30\ \mu m$ no glassy phase	1550 °C $8.3\times10^{-5}\ s^{-1}$	800	$n=1.5, p=3,$ $Q=510$ kJ/mol	[6]
3YTZP	$d=0.3\ \mu m$	1450 °C $4.8\times10^{-5}\ s^{-1}$	246	$n=2, p=NA,$ $Q=580$ kJ/mol	[28]
Al_2O_3/Y_2O_3 (500 ppm)	$d=0.66\ \mu m$	1450 °C $\sim10^{-4}\ s^{-1}$	>65	$n=NA, p/n=1.5,$ $Q=NA$	[23]
Hydroxyapatite	$d=0.64\ \mu m$	1050 °C $1.4\times10^{-4}\ s^{-1}$	>150	$n>3, p/n=1,$ $Q=NA$	[31]
β-spodumene glass	$d=0.91\text{–}2.0\ \mu m$ >4 vol % glassy phase	1200 °C $10^{-4}\ s^{-1}$	>400	$n=1, p=3.1,$ $Q=707$ kJ/mol	[25]
Composites					
20 wt% $Al_2O_3/YTZP$	$d=0.50\ \mu m$ no glassy phase	1650 °C $4\times10^{-4}\ s^{-1}$	620	$n=2, p=1.5,$ $Q=380$ kJ/mol	[35]
20 wt% $Al_2O_3/YTZP$	$d=0.50\ \mu m$	1450 °C $10^{-4}\ s^{-1}$	200	$n=2, p=NA,$ $Q=600$ kJ/mol	[33]
10 vol % ZrO_2/Al_2O_3	$d=0.5\ \mu m$	1400 °C $10^{-4}\ s^{-1}$	>100	$n=NA, p=NA$ $Q=NA$	[78]
20 wt% $YTZP/Al_2O_3$	$d(ZrO_2)=0.47\ \mu m$ $d(Al_2O_3)=1.0\ \mu m$	1550 °C $2.8\times10^{-4}\ s^{-1}$	110	$n=2, p=NA,$ $Q=700$ kJ/mol	[34]
30 wt% TiC/Al_2O_3	$d=1.2\ \mu m$	1550 °C $1.2\times10^{-4}\ s^{-1}$	66	$n=4, p=NA,$ $Q=853$ kJ/mol	[82]
β'-SiAlON	$d=0.4\ \mu m$ with glassy phase	1550 °C $10^{-4}\ s^{-1}$	230	$n=1.5, p=NA,$ $Q=NA$	[84]
20 wt% SiC/Si_3N_4	$d=0.2\text{–}0.5\ \mu m$ with glassy phase	1600 °C $4\times10^{-5}\ s^{-1}$	>150	$n=2, p=NA,$ $Q=649\text{–}698$ kJ/mol	[39]
20 vol % Fe/Fe_3C	$d=3.4\ \mu m$	1000 °C $10^{-4}\ s^{-1}$	600	$n=1.6, p=2.9,$ $Q=200\text{–}240$ kJ/mol	[42]

3YTZP. The findings are, however, quite controversial. For example, Nieh *et al.* [61] have presented several pieces of experimental evidence to show that there is no grain-boundary glassy phase in their superplastic 3YTZP and 20 wt% Al_2O_3/YTZP samples. These experimental results include high-resolution lattice images of grain-boundary triple junctions in 3YTZP. An example is shown in Figure 6.17, in which lattice fringes from adjoining grains can be followed to their intersections at both the grain-boundary interface and the triple junction, indicating the absence of any second phase. Also, both Auger electron spectroscopy and X-ray photoelectron spectroscopy (XPS) from the intergranular fracture surfaces of superplastically deformed specimens, shown in Figure 6.18, show the absence of low-melting-point glassy phases in 3YTZP and 20 wt% Al_2O_3/YTZP. These results suggest that the presence of a grain-boundary glassy phase is unnecessary for superplasticity in fine-grained ceramics.

Conversely, the existence of glassy phases at grain boundaries in superplastic 3YTZP has also been reported. For example, Duclos *et al.* [11] claimed the presence of a glassy phase at triple points in 3YTZP. Because ceramics usually contain glassy phases at grain boundaries, the observations of superplasticity and the GBS mechanisms in YTZP might sometimes have led to the arbitrary assumption that a glassy phase is present at grain boundaries. This assumption may not always be true. Some advanced ceramics are indeed free of grain-boundary glassy phases. It is very important to point out that the microstructural characteristics of a material strongly depend on the exact processing techniques. For example, Duclos *et al.* [11] observed a grain-boundary glassy phase in bulk 3YTZP, but Nieh *et al.* [61] did not, despite the fact that the same starting

Figure 6.17 High-resolution lattice image of a grain-boundary triple junction in fine-grained superplastic YTZP, indicating the absence of any second phase.

ceramic powders were used; this is probably because the bulk samples were processed differently. As a result, the impurity contents may also be different in these bulk samples.

Several pieces of experimental evidence exist to indicate that a strong segregation of yttrium occurs at the free surfaces and grain boundaries of YTZP [60, 62, 95, 96]. Yttrium concentrations can sometimes be as high as 30 at. % at a free surface (~2 nm), as compared to 4 at.% in the bulk [61]. Therefore, the crystal structure of zirconia near the grain-boundary region may be cubic rather than tetragonal. This would produce a phase contrast in transmission electron imaging. High-resolution, lattice-image techniques can minimize such artifacts. Hwang and Chen [96] have proposed that the grain boundaries of zirconia are positively charged. As a result, any divalent and trivalent cations are expected to be enriched at the grain-boundary and to thereby decrease the grain-boundary mobility.

Although the presence of a liquid phase at grain boundaries may not be a necessary prerequisite for superplasticity in ceramics, its presence can definitely affect some of the kinetic processes, such as sintering and GBS [67, 76, 97]. For example, Wakai et al. [98] demonstrated that the temperature for superplasticity in 3YTZP can be reduced by doping manganese oxide (~3 mol %) into the material to create a grain-boundary glassy phase. Also, Hwang and Chen [96] and Xue [99] examined the effect of various dopants on the sintering and superplastic forming temperatures of YTZP and found that a minor addition (0.3 mol %) of CuO to YTZP and Al_2O_3/YTZP could lower the sintering temperature without significantly increasing the grain size. The CuO-doped YTZP

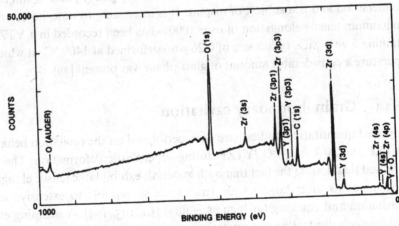

Figure 6.18 XPS spectrum from the fracture surface of a superplastic YTZP sample, indicating the absence of any second phase.

can be even formed superplastically at 1150 °C. Chen and Xue [43] further demonstrated that the presence of grain-boundary glassy phases can, in general, reduce the superplastic forming temperatures for various ceramics, including YTZP and Si_3N_4. Evidently, the creation of glassy phases via oxide doping can greatly enhance the diffusion processes in, as well as the deformation rate of, superplastic ceramics.

In terms of the nature of glassy phase, Yashizawa and Sakuma [30, 100, 101] found empirically that lithium silicate is the most effective in reducing the superplastic temperature of 3YTZP. The optimum amount of the volume fraction of the liquid phase, as pointed out by Hermansson et al. [102], is not important once it is beyond a critical value. To evaluate the influence of glassy grain-boundary phases, Gust et al. [67] performed experiments on fine-grained YTZP deliberately added with different glassy phases, such as barium silicate (BaS) and borosilicate (BS). Several observations were made. The glass-containing materials deformed about 10 times faster than the pure YTZP, as shown in Figure 6.19, but the deformation mechanisms, indicated by the stress exponent values, appear to be similar. The strain-rate enhancement was found to be highest in samples containing the grain-boundary phase with the highest solubility for Y_2O_3 and ZrO_2, but the strain rate did not scale inversely with the viscosity of the silicate phase. In addition, the degree of strain rate enhancement is related to the volume fraction (0.1–5 wt%) of the glassy phases, as shown in Figure 6.20. Also, as a result of the glass additions, the activation energy for superplasticity reduces from 640 kJ/mol in the pure material to 510 kJ/mol in the glass-containing materials.

The influence of glassy phase on the superplastic elongation was not investigated. Therefore, the exact role of grain-boundary glassy phase is still not fully understood and is a subject of dispute. It is interesting to note, however, that a maximum tensile elongation of over 1000% has been recorded in a YTZP containing 5 wt% SiO_2 (grain size of 0.26 μm) deformed at 1400 °C, at which temperature a considerable amount of glass phase was present [29].

6.3.3 Grain-boundary cavitation

Limited quantitative studies have been performed on the cavitation behavior of YTZP and 20 wt% Al_2O_3/YTZP during superplastic deformation. The results showed that despite the fact that both materials exhibit large tensile elongations, the materials after large tensile strains are susceptible to extensive internal cavitation and cracking (as high as ~30%) [50, 103, 104]. A scanning electron micrograph illustrating the cavitation in a superplastically deformed YTZP is given in Figure 6.21. Cavitation in composite materials (e.g., Al_2O_3/YTZP) is more extensive than that in the monolithic ceramics (e.g., YTZP). Also, the level of cavitation in superplastic ceramics was observed to increase with increasing strain rate, and the elongation to failure is inversely proportional to the level of cavitation. These observations suggest that the tensile ductility of superplastic ceramics is limited by cavity interlinkage in a direction perpendicular to the tensile axis during superplastic deformation. Interestingly, this conclusion – that the fracture of superplastic ceramics is dictated by intergranular cracking – has also been proposed by Chen and Xue [43] and Kim *et al.* [24]. Cavitation is discussed in detail in Chapter 10.

6.4   Ingot processing route for superplastic ceramics

All fine-grained ceramics were prepared by the traditional powder processing approaches. An alternative route to achieve fine structures involving thermo-mechanical working of an ingot material has recently been explored [105]. One of

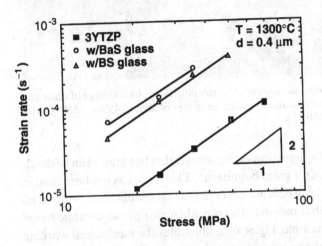

Figure 6.19 Steady-state strain rate vs true stress for 3YTZP both with and without 1 wt% glass (data from Ref. [67]).

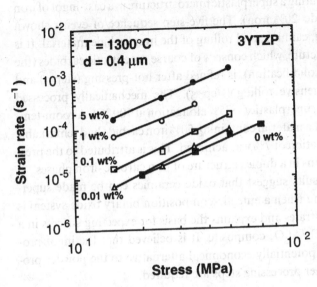

Figure 6.20 Steady-state strain rate vs true stress for 3YTZP with various amounts of BaS glass (data from Ref. [67]).

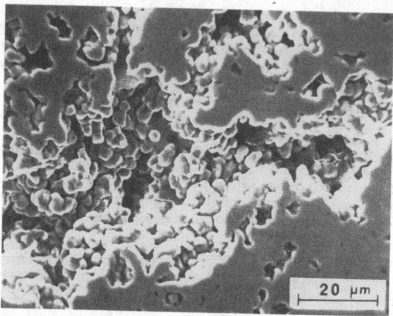

Figure 6.21   Scanning electron micrograph illustrating interlinkage of cracks in a YTZP sample tested to an elongation to failure of about 245% at 1550 °C with a strain rate of 8.3×10^{-5} s^{-1}.

the major advantages of the ingot processing route is that fine grains can be developed containing impurity-free grain boundaries. This route has not been used in traditional ceramic processing because it is believed that ceramics are brittle at all temperatures. However, that may not be true when two-phase, ceramic-based materials are used. For example, Figure 6.22 illustrates the mechanical working steps developed for generating a superplastic microstructure in a cast-ingot of iron carbide (80% iron carbide–20% iron). The five-step sequence of events shown involves the hot-pressing, canning, and rolling of the iron carbide material. It is noted that the initial structure, which consists of coarse proeutectic carbides (the first ones to form upon solidification), is refined after hot-pressing (step 3) and further refined after extensive rolling (step 5). The mechanically processed material was shown to be superplastic (~75% elongation at 700 °C). A room-temperature test revealed that a material containing 50% iron carbide had remarkable tensile ductility: an elongation of 7% was achieved. This is attributed to the presence of fine-equiaxed grains in a duplex structure of iron carbide–iron phases.

These encouraging results suggest that oxide ceramics can be made superplastic by ingot processing when a eutectic-composition binary oxide system is selected. Figure 6.23 illustrates and explains the basis for expecting success in a eutectic composition ZrO_2/Al_2O_3 composite. It is believed that the ingot processing route will offer a potentially economical alternative to the powder processing route because fewer processing steps are required.

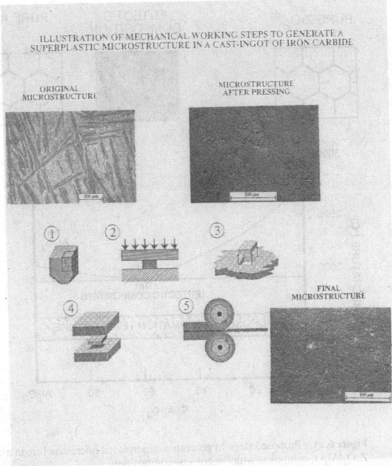

Figure 6.22 Mechanical working steps to generate a superplastic microstructure in a cast iron carbide–iron composite.

6.5 Superplasticity in geological materials

Claims of both internal-stress and fine-structure superplasticity have been made in several geological materials. The movement of the earth's upper mantle $[(Mg_{0.91}Fe_{0.09})SiO_4]$ is considered to be an example of internal-stress superplasticity. Rocks and minerals claimed to be superplastic include limestone [106], $MgGeO_4$ spinel [107], mylonites (olivine-rich) [108], and even ice [109]. Because extremely low strain rates of geological flow ($\sim 10^{-14}\,s^{-1}$) are generally involved, it is difficult to obtain mechanical data in the laboratory to support the prevalence of superplasticity. Therefore, evidence for superplasticity is primarily based on the extrapolation of experimental data.

As a result, most of the claims of superplasticity in geological materials were

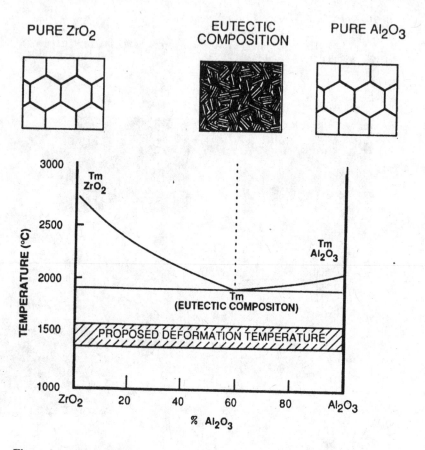

Figure 6.23 Proposed steps to generate a superplastic microstructure in a cast ZrO$_2$/Al$_2$O$_3$ composite with the eutectic composition.

made from either theoretical considerations [110–113] or microstructural evidence [114–122] and are indirect. Incorporating superplasticity into deformation mechanism maps for geological materials is difficult because of the lack of understanding of the precise superplastic flow law [106]. Nevertheless, it should be noted that at an extremely low strain rate of about 10^{-14} s^{-1}, superplastic mechanisms may operate at coarse grain sizes by comparison with conventional superplastic materials. For example, assuming $p=2$ in Eq. (6.7), superplasticity is predicted to occur at a strain rate of 10^{-14} s^{-1} in a ceramic with grain size greater than 1 m! For a review on superplasticity in geological materials readers are referred to Ref. [123].

References

1. J.J. Gilman, 'Mechanical Behavior of Crystalline Solids,' *National Bureau of Standards Monograph*, **59** (1963), pp. 79–104.

2. R.G. St-Jacques and R. Angers, 'Creep of CaO-Stabilized ZrO_2,' *J. Am. Ceram. Soc.*, **55**(11) (1972), pp. 571–574.

3. P.E. Evans, 'Creep in Yttria- and Scandia-Stabilized Zirconia,' *J. Am. Ceram. Soc.*, **53** (1970), pp. 365–369.

4. R.G. St-Jacques and R. Angers, 'The Effect of CaO-Concentration on the Creep of CaO-Stabilized ZrO_2,' *Trans. Br. Ceram. Soc.*, **72** (1972), pp. 285–289.

5. M.S. Seltzer and P.K. Talty, 'High-Temperature Creep of Y_2O_3-Stabilized ZrO_2,' *J. Amer. Ceram. Soc.*, **58**(3–4) (1975), pp. 124–130.

6. T.G. Nieh and J. Wadsworth, 'Superplastic Behavior of a Fine-Grained, Yttria-Stabilized, Tetragonal Zirconia Polycrystal (Y-TZP),' *Acta Metall. Mater.*, **38** (1990), pp. 1121–1133.

7. F. Wakai, S. Sakaguchi, and Y. Matsuno, 'Superplasticity of Yttria-Stabilized Tetragonal ZrO_2 Polycrystals,' *Adv. Ceram. Mater.*, **1** (1986), pp. 259–263.

8. R.B. Day and R.J. Stokes, 'Mechanical Behavior of Polycrystalline Magnesium Oxide at High Temperatures,' *J. Am. Ceram. Soc.*, **49**(7) (1966), pp. 345–354.

9. C. Carry and A. Mocellin, 'Superplastic Creep of Fine-Grained $BaTiO_3$ in a Reducing Environment,' *J. Am. Ceram. Soc.*, **69**(9) (1986), pp. C–215–C–216.

10. C. Carry and A. Mocellin, 'Structural Superplasticity in Single Phase Crystalline Ceramics,' *Ceramics International*, **13** (1987), pp. 89–98.

11. R. Duclos, J. Crampon, and B. Amana, 'Structural and Topological Study of Superplasticity in Zirconia Polycrystals,' *Acta Metall.*, **70** (1987), pp. 877–883.

12. J.L. Hart and A.C.D. Chaklader, 'Superplasticity in Pure ZrO_2,' *Mater. Res. Bull.*, **2** (1967), pp. 521–526.

13. P.E.D. Morgan, 'Superplasticity in Ceramics,' in *Ultrafine-Grain Ceramics*, pp. 251–271, ed. J.J. Burke, N.L. Reed, and V. Weiss, Syracuse University Press, New York, 1970.

14. Y. Okamoto, J. Ieuji, Y. Yamada, K. Hayashi, and T. Nishikawa, 'Creep Deformation of Yttria-Stabilized Tetragonal Zirconia (Y-TZP),' in *Advances in Ceramics, Vol. 24*, pp. 565–571, ed. S. Somiya, N. Yamamoto, and H. Yanagida, American Ceramic Society, Westerville, Ohio, 1988.

15. P.C. Panda, R. Raj, and P.E.D. Morgan, 'Superplastic Deformation in Fine-Grained $MgO•2Al_2O_3$ Spinel,' *J. Am. Ceram. Soc.*, **68** (1985), pp. 522–529.

16. J.R. Smythe, R.C. Bradt, and J.H. Hoke, 'Isothermal Deformation in Bi_2O_3-Sm_2O_3,' *J. Mater. Sci.*, **12** (1977), pp. 1495–1502.

17. K.R. Venkatachari and R. Raj, 'Superplastic Flow in Fine-Grained Alumina,' *J. Am. Ceram. Soc.*, **69**(2) (1986), pp. 135–138.

18. L.A. Xue and R. Raj, 'Superplastic Deformation of Zinc Sulfide near Transformation Temperature (1020°C),' *J. Am. Ceram. Soc*, **72** (1989), pp. 1792–1796.

19. L.A. Xue and R. Raj, 'Grain Growth in Superplastically Deformed Zinc Sulfide/Diamond Composites,' *J. Am. Ceram. Soc*, **74**(7) (1989), pp. 1729–1731.

20. L.A. Xue and R. Raj, 'Effect of Diamond Dispersion on the Superplastic Rheology of Zinc Sulfide,' *J. Am. Ceram. Soc*, **73**(8) (1989), pp. 2213–2216.

21. C.K. Yoon and I.W. Chen, 'Superplastic Flow of Two-Phase Ceramics Containing Rigid Inclusion-Zirconia/Mullite Composites,' *J. Amer. Ceram. Soc*, **73**(6) (1990), pp. 1555–1565.

22. W.R. Cannon and T.G. Langdon, 'Review – Creep of Ceramics,' *J. Mater. Sci.*, **23** (1988), pp. 1–20.

23. P. Gruffel, P. Carry, and A. Mocellin, 'Effects of Testing Conditions on Superplastic Creep of Alumina Doped with Ti and Y,' in *Science of Ceramics, Volume 14*, pp. 587–592, ed. D. Taylor, The Institute of Ceramics, Shelton, Stoke-on-Trent, UK, 1987.

24. W.J. Kim, J. Wolfenstine, O.S. Ruano, G. Frommeyer, and O.D. Sherby, 'Processing and Superplastic Properties of Fine-Grained Iron Carbide,' *Metall. Trans.*, **23A** (1992), pp. 527–535.

25. J.-G. Wang and R. Raj, 'Mechanism of Superplastic Flow in a Fine-Grained Ceramic Containing Some Liquid Phase,' *J. Am. Ceram. Soc.*, **67**(6) (1984), pp. 399–409.

26. F. Wakai, Y. Kodama, and T. Nagano, 'Superplasticity of ZrO$_2$ Polycrystals,' in *Japan J. Appl. Phy. Series 2, Lattice Defects in Ceramics*, Japan Journal of Applied Physics, Tokyo, Japan, 1989, pp. 57–67.

27. T.G. Nieh, C.M. McNally, and J. Wadsworth, 'Superplastic Behavior of a Yttria-Stabilized Tetragonal Zirconia Polycrystal,' *Scr. Metall.*, **22** (1988), pp. 1297–1300.

28. T. Hermanson, K.P.D. Lagerlof, and G.L. Dunlop, 'Superplastic Deformation of YTZP Zirconia,' in *Superplasticity and Superplastic Forming*, pp. 631–635, ed. C.H. Hamilton and N.E. Paton, TMS, Warrendale, PA, 1988.

29. K. Kajihara, Y. Yoshizawa, and T. Sakuma, 'Superplasticity in SiO$_2$-Containing Tetragonal Zirconia Polycrystal,' *Scr. Metall. Mater.*, **28** (1993), pp. 559–562.

30. Y. Yoshizaka and T. Sakuma, 'Improvement of Tensile Ductility in High-Purity Alumina due to Magnesia Addition,' *Acta Metall. Mater.*, **40** (1992), pp. 2943–2950.

31. F. Wakai, Y. Kodama, S. Sakaguchi, and T. Nonami, 'Superplasticity of Hot Isostatically Pressed Hydroxyapatite,' *J. Am. Ceram. Soc*, **73**(2) (1990), pp. 257–260.

32. J.-G. Wang and R. Raj, 'Influence of Hydrostatic Pressure and Humidity on Superplastic Ductility of two β-Spodumene Glass-Ceramics,' *J. Am. Ceram. Soc.*, **67**(6) (1984), pp. 385–390.

33. F. Wakai and H. Kato, 'Superplasticity of TZP/Al$_2$O$_3$ Composite,' *Adv. Ceram. Mater.*, **3**(1) (1988), pp. 71–76.

34. F. Wakai, Y. Kodama, S. Sakaguchi, N. Murayama, H. Kato, and T. Nagano, 'Superplastic Deformation of Al$_2$O$_3$/ZrO$_2$ Duplex Composites,' in *MRS Intl. Meeting on Advanced Materials Vol 7 (IMAM-7, Superplasticity)*, pp. 259–266, ed. M. Doyama, S. Somiya, and R.P.H. Chang, Materials Research Society, Pittsburgh, PA, 1989.

35. T.G. Nieh and J. Wadsworth, 'Superplasticity in Fine-Grained 20%Al$_2$O$_3$/YTZ Composite,' *Acta Metall. Mater.*, **39** (1991), pp. 3037–3045.

36. T.G. Nieh, C.M. McNally, and J. Wadsworth, 'Superplastic Behavior of a 20% Al$_2$O$_3$/YTZ Ceramic Composite,' *Scr. Metall.*, **23** (1989), pp. 457–460.

37. T. Kuroishi, K. Uno, and F. Wakai, 'Characterization of Superplastic ZrO$_2$-Toughened Al$_2$O$_3$ Prepared by Slip Casting,' in *MRS Intl. Meeting on Advanced Materials Vol 7 (IMAM-7), Superplasticity*, pp. 267–274, ed. M. Kobayashi and F. Wakai, Materials Research Society, Pittsburgh, PA, 1989.

38. F. Wakai, 'Superplasticity of Non-Oxide Ceramics,' in *Superplasticity in Metals, Ceramics, and Intermetallics, MRS Proceeding No. 196R*, pp. 349–358, ed. M.J. Mayo, J. Wadsworth, and M. Kobayashi, Materials Research Society, Pittsburgh, PA, 1990.

39. F. Wakai, Y. Kodama, S. Sakaguchi, N. Murayama, K. Izaki, and K. Niihara, 'A Superplastic Covalent Crystal Composite,' *Nature (London)*, **334**(3) (1990), pp. 421–423.

40. T. Rouxel, F. Wakai, K. Izaki, and K. Niihara, 'Superplasticity of Si$_3$N$_4$/SiC Composites,' in *Proc. 1st Intl. Symp. on Science of Engineering Ceramics*, pp. 437–442, ed. S. Kimura and K. Niihara, Ceramic Society of Japan, Tokyo, Japan, 1991.

41. T. Rouxel, F. Wakai, and K. Izaki, 'Tensile Ductility of Superplastic Al$_2$O$_3$-Y$_2$O$_3$-Si$_3$N$_4$/SiC Composites,' *J. Am. Ceram. Soc.*, **75**(9) (1992), pp. 2363–2372.

42. W.J. Kim, G. Frommeyer, O.A. Ruano, J.B. Wolfenstine, and O.D. Sherby, 'Superplastic Behavior of Iron Carbide,' *Scr. Metall.*, **23** (1989), pp. 1515–1520.

43. I.-W. Chen and L.A. Xue, 'Development of Superplastic Structural Ceramics,' *J. Am. Ceram. Soc.*, **73**(9) (1990), pp. 2585–2609.

44. F. Wakai, 'A Review of Superplasticity in ZrO$_2$-Toughened Ceramics,' *Brit. Ceram. Trans. J.*, **88** (1989), pp. 205–208.

45. Y. Maehara and T.G. Langdon, 'Review: Superplasticity in Ceramics,' *J. Mater. Sci.*, **25** (1990), pp. 2275–2286.

46. T.G. Nieh and J. Wadsworth, 'Superplastic Ceramics,' *Ann. Rev. – Mater. Sci.*, **20** (1990), pp. 117–140.

47. T.G. Nieh, J. Wadsworth, and F. Wakai, 'Recent Advances in Superplastic Ceramics and Ceramic Composites,' *Inter. Mater. Rev.*, **36**(4) (1991), pp. 146–161.

48. T.G. Nieh and J. Wadsworth, 'Dynamic Grain Growth in Yttria-Stabilized Tetragonal Zirconia during Superplastic Deformation,' *J.*

Am. Ceram. Soc., **72**(8) (1989), pp. 1469–1472.

49. F. Wakai, S. Sakaguchi, and H. Kato, 'Compressive Deformation Properties and Microstructures in the Superplastic Y-TZP,' *J. Ceram. Soc. Japan (In Japanese)*, **94** (1986), pp. 72–75.

50. D.J. Schissler, A.H. Chokshi, T.G. Nieh, and J. Wadsworth, 'Microstructural Aspects of Superplastic Tensile Deformation and Cavitation Failure in a Fine-Grained, Yttria Stabilized, Tetragonal Zirconia,' *Acta Metall. Mater.*, **39**(12) (1991), pp. 3227–3236.

51. A.K. Ghosh and C.H. Hamilton, 'Mechanical Behavior and Hardening Characteristics of a Superplastic Ti–6Al–4V Alloy,' *Metall. Trans.*, **10A** (1979), pp. 699–706.

52. C.H. Caceres and D.S. Wilkinson, 'Large Strain Behavior of a Superplastic Copper Alloys-II. Cavitation and Fracture,' *Acta Metall.*, **32** (1984), pp. 423–434.

53. D.S. Wilkinson and C.H. Caceres, 'On the Mechanism of Strain-Enhanced Grain Growth during Superplastic Deformation,' *Acta Metall.*, **32** (1984), pp. 1335–1345.

54. O.N. Senkov and M.M. Myshlyaev, 'Grain Growth in a Superplastic Zn–22%Al Alloy,' *Acta Metall.*, **34** (1986), pp. 97–106.

55. C. Carry and A. Mocellin, 'Superplastic Forming of Alumina,' *Proc. Brit. Ceram. Soc*, **33** (1983), pp. 101–115.

56. B.J. Kellett, C. Carry, and A. Mocellin, 'High-Temperature Extrusion Behavior of a Superplastic Zirconia-Based Ceramic,' *J. Amer. Ceram. Soc*, **74**(7) (1990), pp. 1922–1927.

57. J.W. Cahn, 'The Impurity-Drag Effect in Grain Boundary Motion,' *J. Am. Ceram. Soc.*, **10** (1962), pp. 789–798.

58. R.J. Brook, 'The Impurity-Drag Effect and Grain Growth Kinetics,' *Scripta Metall.*, **2** (1968), pp. 375–378.

59. P.J. Whalen, F. Reidinger, S.T. Correale, and J. Marti, 'Yttria Migration in Y-TZP during High-Temperature Annealing,' *J. Mater. Sci.*, **22** (1987), pp. 4465–4469.

60. G.S.A.M. Theunissen, A.J.A. Winnubst, and A.J. Burggraaf, 'Segregation Aspects in the ZrO_2-Y_2O_3 Ceramic System,' *J. Mater. Sci. Lett.*, **8** (1989), pp. 55–57.

61. T.G. Nieh, D.L. Yaney, and J. Wadsworth, 'Analysis of Grain Boundaries in a Fine-Grained, Superplastic, Yttria-Containing, Tetragonal Zirconia,' *Scr. Metall.*, **23** (1989), pp. 2007–2012.

62. A.E. Hughes and B.A. Sexton, 'XPS Study of an Intergranular Phase in Yttria-Zirconia,' *J. Mater. Sci.*, **24** (1989), pp. 1057–1061.

63. T.G. Nieh and J. Wadsworth, 'Effect of Grain Size on Superplastic Behavior of Y-TZP,' *Scr. Metall. Mater.*, **24** (1990), pp. 763–766.

64. C. Carry, 'High Ductilities, Superplastic Behaviors and Associated Mechanisms in Fine Grained Ceramics,' in *MRS Intl. Meeting on Advanced Materials Vol.7 (IMAM-7, Superplasticity)*, pp. 251–258, ed. M. Doyama, S. Somiya, and R.P.H. Chang, Materials Research Soc., Pittsburgh, PA, 1989.

65. Y. Ma and T.G. Langdon, 'An Investigation of the Mechanical Behavior of a Superplastic Yttria-Stabilized Zirconia,' in *Superplasticity in Metals, Ceramics, and Intermetallics, MRS Proceeding No. 196*, pp. 325–330, ed. M.J. Mayo, J. Wadsworth, and M. Kobayashi, Materials Research Society, Pittsburgh, PA, 1990.

66. A. Lakki, R. Schaller, M. Nauer, and C. Carry, 'High Temperature Superplastic Creep and International Friction of Yttria Doped Zirconia Polycrystals,' *Acta Metall. Mater.*, **41** (1993), pp. 2845–2853.

67. M. Gust, G. Goo, J. Wolfenstine, and M. Mecartney, 'Influence of Amorphous Grain Boundary Phases on the Superplastic Behavior of 3-mol%-Yttria-Stabilized Tetragonal Zirconia Polycrystals (3Y-TZP),' *J. Am. Ceram. Soc.*, **76**(7) (1993), pp. 1681–1690.

68. Y. Oishi, K. Ando, and Y. Sakka, 'Lattice and Grain-Boundary Diffusion Coefficients of Cations in Stabilized Zirconias,' in *Advances in Ceramics Vol.7 – Additives and Interfaces in Electronic Ceramics*, ed. M.F. Yan and A.H. Heuer, American Ceramic Society, Columbus, Ohio, 1983, pp. 208–219.

69. J.D. Fridez, C. Carry, and A. Mocellin, 'Effects of Temperature and Stress on Grain-Boundary Behavior in Fine-Grained Alumina,' in *Structure and Properties of MgO and Al_2O_3 Ceramics*, pp. 720–740, ed. W.D. Kingery, American Ceramic Society, Columbus, Ohio, 1984.

70. L.A. Xue and I.W. Chen, 'Deformation and Grain Growth of Low-Temperature-Sintered High-Purity Alumina,' *J. Am. Ceram. Soc.*, **73**(11) (1990), pp. 3518–3521.

71. C. Carry, 'Microstructures, Grain Boundaries and Superplasticity in Fine Grained Ceramics,' in *Superplasticity in Metals,*

Ceramics, and Intermetallics, MRS Proceeding No.196, pp. 313–324, ed. M.J. Mayo, J. Wadsworth, and M. Kobayashi, Materials Research Society, Pittsburgh, PA, 1990.

72. C. Carry and A. Mocellin, 'Example of Superplastic Forming Fine-Grained Al_2O_3 and ZrO_2 Ceramics,' in *High Tech Ceramics*, pp. 1043–1052, ed. P. Vincenzini, Elsevier Science Publishers, Amsterdam, 1987.

73. F.A. Nichols, 'On the Diffusional Mobilities of Particles, Pores and Loops,' *Acta Metall.*, **20** (1972), pp. 207–214.

74. M. Jarco, C.H. Bolen, M.B. Thomas, J. Bobick, J.F. Kay, and R.H. Doremus, 'Hydroxyapatite Synthesis and Characterization in Dense Polycrystalline Form,' *J. Mater. Sci.*, **11** (1976), pp. 2027–2035.

75. R. Raj, 'Creep in Polycrystalline Aggregates by Matter Transport Through a Liquid Phase,' *J. Geophys. Res.*, **87**(B6) (1982), pp. 4731–4739.

76. G.M. Pharr and M.F. Ashby, 'On Creep Enhanced by a Liquid Phase,' *Acta Metall.*, **31** (1983), pp. 129–138.

77. T.G. Nieh, C.M. Tomasello, and J. Wadsworth, 'Dynamic Grain Growth in Superplastic Ceramics and Ceramic Composite,' in *Superplasticity in Metals, Ceramics, and Intermetallics, MRS Proceeding No. 196*, pp. 343–348, ed. M.J. Mayo, J. Wadsworth, and M. Kobayashi, Materials Research Society, Pittsburgh, PA, 1990.

78. L.A. Xue, X. Wu, and I.W. Chen, 'Superplastic Alumina Ceramics with Grain Growth Inhibitors,' *J. Am. Ceram. Soc.*, **74**(4) (1991), pp. 842–845.

79. T.G. Nieh and J. Wadsworth, 'Effect of Grain Size on Superplastic Behavior of Al_2O_3/YTZ,' *J. Mater. Res.*, **5**(11) (1990), pp. 2613–2615.

80. I.W. Chen, 'Superplastic Flow of Two-Phase Alloys,' in *Superplasticity*, pp. 5.1–5.20, ed. B. Baudelet and M. Suery, Centre National de la Recherche Scientifique, Paris, Grenoble, France, 1985.

81. F.F. Lange and M.M. Hirlinger, 'Hindrance of Grain Growth in Al_2O_3 by ZrO_2 Inclusions,' *J. Amer. Ceram. Soc.*, **67**(3) (1984), pp. 164–168.

82. T. Nagano, H. Kato, and F. Wakai, 'Deformation of Alumina/Titanium Carbide Composite at Elevated Temperatures,' *J. Am. Ceram. Soc.*, **74**(9) (1991), pp. 2258–2262.

83. I.W. Chen and S.-J. Hwang, 'Shear Thickening Creep of Superplastic Silicon Nitrides,' *J. Am. Ceram. Soc.*, **75**(5) (1992), pp. 1073–1079.

84. X. Wu and I.-W. Chen, 'Exaggerated Texture and Grain Growth in a Superplastic SiAlON,' *J. Am. Ceram. Soc.*, **75**(10) (1992), pp. 2733–2741.

85. C. Carry and A. Mocellin, '*High Temperature Creep of Dense Fine Grained Silicon Carbides*', in *Deformation of Ceramic Materials II*, pp. 391–403, ed. R.E. Tressler and R.C. Bradt, Plenum Publishing Corp., New York, 1984.

86. K. Izaki, K. Hakkei, K. Ando, and K. Niihara, 'Fabrication and Mechanical Properties of Si_3N_4-SiC Composites from Fine, Amorphous Si-C-N Powder Precursors,' in *Ultrastructure Processing of Advanced Ceramics*, pp. 891–900, ed. J.D. Mackenzie and D.R. Ulrich, John Wiley & Sons, New York, 1989.

87. K. Niihara, K. Suganuma, A. Nakahira, and K. Izaki, 'Interfaces in Si_3N_4-SiC Nano-Composite,' *J. Mater. Sci. Lett.*, **9** (1990), pp. 598–599.

88. J. Wolfenstine, G. Gonzalez-Doncel, and O.D. Sherby, 'Tension versus Compression Superplastic Behavior of a Mg–9wt%Li–5wt% B_4C Composite,' *Mater. Lett.*, **15** (1992), pp. 305–308.

89. D.M. Owen and A.H. Chokshi, 'A Comparison of the Tensile and Compressive Creep Behavior of a Superplastic Yttria-Stabilized Tetragonal Zirconia–20 wt% Alumina Composite,' in *International Conference on Superplasticity in Advanced Materials (ICSAM–91)*, pp. 215–220, ed. S. Hori, M. Tokizane, and N. Furushiro, The Japan Society for Research on Superplasticity, Osaka, Japan, 1991.

90. W.J. Kim, J. Wolfenstine, and O.D. Sherby, 'Tensile Ductility of Superplastic Ceramics and Metallic Alloys,' *Acta Metall. Mater.*, **39**(2) (1991), pp. 199–208.

91. R.L. Coble, 'A Model for Boundary Diffusion Controlled Creep in Polycrystalline Materials,' *J. Appl. Phys.*, **34** (1964), pp. 1679–1682.

92. O.D. Sherby and J. Wadsworth, 'Development and Characterization of Fine Grain Superplastic Material,' in *Deformation, Processing and Structure*, pp. 355–389, ed. G. Krauss, ASM, Metal Park, Ohio, 1984.

93. C. Herring, 'Diffusion Viscosity of a Polycrystalline Solid,' *J. Appl. Phys.*, **21** (1951), pp. 437–445.

94. D.R. Clarke and G. Thomas, 'Grain Boundary Phases in a Hot-Pressed MgO Fluxed Silicon Nitride,' *J. Am. Ceram. Soc.*, **60**(11–12) (1977), pp. 491–495.

95. T.G. Nieh and J. Wadsworth, 'Characterization of Superplastic Y-TZP Using a Hot Indentation Technique,' *Scr. Metall.*, **23** (1989), pp. 1261–1264.

96. C.-M.J. Hwang and I.W. Chen, 'Effect of a Liquid Phase on Superplasticity of 2 mol%Y_2O_3-Stabilized Tetragonal Zirconia Polycrystals,' *J. Amer. Ceram. Soc.*, **73** (1990), pp. 1623–1632.

97. N. Kimura, S. Abe, J. Morishita, and H. Okamura, 'Preparation of Low-Y_2O_3 TZP by Low-Temperature Sintering,' in *Proc. Int'l Institute for the Science of Sintering (IISS) Symp.*, pp. 1142–1148, ed. S. Somiya, M. Shimada, M. Yoshimura, and R. Watanabe, Elsevier Applied Science, London, 1988.

98. F. Wakai, H. Okamura, N. Kimura, and P.G.E. Descamps, 'Superplasticity of Transition Metal Oxide-Doped Y-TZP at Low Temperatures,' in *Proc. 1st Japan Int'l SAMPE Symp.*, pp. 267–271, ed. N. Igata, K. Kimpara, T. Kishi, E. Nakata, A. Okura, and T. Uryu, Society for the Advancement of Materials and Process Engineering, Corvida, CA, 1989.

99. L.A. Xue, 'Enhanced Superplastic Deformation of 2 mol% Yttria-Stabilized Tetragonal Zirconia Polycrystals-Alumina Composite by Liquid-Forming Additives,' *J. Mater. Sci. Lett.*, **10** (1991), pp. 1291–1292.

100. Y. Yoshizaka and T. Sakuma, 'Effect of Grain-Boundary Glassy Phase on Superplastic Deformation in ZrO_2–2.5mol% Y_2O_3,' in *Proc. 1st Japan Int'l SAMPE Symp.*, pp. 272–277, ed. N. Igata, K. Kimpara, T. Kishi, E. Nakata, A. Okura, and T. Uryu, Society for the Advancement of Materials and Process Engineering, Corvida, CA, 1989.

101. Y. Yoshizaka and T. Sakuma, 'Role of Grain-Boundary Glass Phase on the Superplastic Deformation of Tetragonal Zirconia Polycrystal,' *J. Am. Ceram. Soc.*, **73**(10) (1990), pp. 3069–3073.

102. T. Hermansson, H. Swan, and G. Dunlop, 'The Role of the Intergranular Glassy Phase in the Superplastic Deformation of Y-TZP Zirconia,' in *Euro-Ceramics*, pp. 3.329–3.333, ed. G. de With, R.A. Terstra, and R. Metselaar, Elsevier Applied Science, London, 1989.

103. A.H. Chokshi, T.G. Nieh, and J. Wadsworth, 'Role of Concurrent Cavitation in the Fracture of a Superplastic Zirconia-Alumina Composite,' *J. Am. Ceram., Soc*, **74** (1991), pp. 869–873.

104. A. Chokshi, D.J. Schissler, T.G. Nieh, and J. Wadsworth, 'A Comparative Study of Superplastic Deformation and Cavitation Failure in a Yttria Stabilized Zirconia and a Zirconia Alumina Composite,' in *Superplasticity in Metals, Ceramics, and Intermetallics, MRS Proceeding No. 196*, pp. 379–384, ed. M.J. Mayo, J. Wadsworth, and M. Kobayashi, Materials Research Society, Pittsburgh, PA, 1990.

105. W.J. Kim, O. D. Sherby, T.G. Nieh, and G. Frommeyer, 'Ingot Processed Eutectic Iron Carbide (Fe–4.3 wt% C),' *Scr. Mater.*, **34**, (1996).

106. P.J. Vanghan and R.S. Coe, 'Creep Mechanisms in Mg_2GeO_4: Effects of a Phase Transition,' *J. Geophys. Res.*, **86** (1981), pp. 389–404.

107. A.M. Boullier and Y. Gueguen, 'SP-Mylonites: Origin of Some Mylonites by Superplastic Flow,' *Contrib. Mineral. Petrol.*, **50** (1975), pp. 93–104.

108. P. Duval, M.F. Ashby, and I. Anderson, 'Rate Controlling Processes in the Creep of Polycrystalline Ice,' *J. Phys. Chem.*, **87** (1983), pp. 4066–4074.

109. L. Lliboutry and P. Duval, 'Various Isotropic and Anisotropic Ices Found in Glaciers and Polar Ice Caps and Their Corresponding Rheologies,' *Ann. Geophys.*, **3** (1985), pp. 207–224.

110. J.P. Poirier, 'On Transformation Plasticity,' *J. Geophys. Res.*, **87** (1982), pp. 6791–6797.

111. E.M. Parmentier, 'A Possible Mantle Instability due to Superplastic Deformation Associated with Phase Transformation,' *Geophys. Res. Lett*, **8** (1981), pp. 143–146.

112. S.M. Schmid, 'Rheological Evidence for Changes in the Deformation Mechanism of Solenhofen Limestone Towards Low Stresses,' *Tectonophys.*, **31** (1976), pp. T21–T78.

113. S.H. Kirby and C.B. Raleigh, 'Mechanisms of High-Temperature, Solid-State Flow in Minerals and Ceramics and Their Bearing on the Creep Behavior of the Mantle,' *Tectonophys.*, **19** (1973), pp. 165–194.

114. L. Ruff and H. Kanamori, 'Seismic Coupling and Uncoupling at Subduction Zones,' *Tectonophys*, **99** (1983), pp. 99–117.

115. S. White, 'The Effects of Strain on the Microstructures, Fabrics and Deformation Mechanisms in Quartzites,' *Phil. Trans. R. Soc. Lond. Ser. A*, **283** (1976), pp. 69–86.

116. A.M. Boullier and A. Nicholas, 'Classification of Textures and Fabrics of Peridotite Xenoliths from South African Kimberlites,' in *Physics and Chemistry of the Earth, Vol. 9*, ed. L.H. Ahrens, Pergamon, Oxford, 1975, pp. 97–105.

117. Y. Gueguen and A.M. Boullier, 'Evidence of Superplasticity in Mantle Peridolites,' in *Proc. NATO Petrophysics Meeting*, p. 19, ed. R.G.J. Sterns, Wiley/Academic Press, New York, 1976.

118. S. White, 'Geological Significance of Recovery and Recrystallization Processes in Quartz,' *Tectonophys*, **39** (1977), pp. 143–170.

119. I. Allison, R.L. Barnett, and R. Kerrich, 'Superplastic Flow and Changes in Crystal Chemistry of Feldspars,' *Tectonophys*, **53** (1979), pp. T41-T46.

120. M.A. Etherridge and J.C. Wilkie, 'Grain Size Reduction, Grain Boundary Sliding and the Flow Strength of Mylonites,' *Tectonophys*, **58** (1979), pp. 159–178.

121. J.H. Behrmann, 'Crystal Plasticity and Superplasticity in Quartzite; a Natural Example,' *Tectonophys*, **115** (1985), pp. 101–129.

122. J.H. Behrmann and D. Mainprice, 'Deformation Mechanisms in a High-Temperature Quartz-Feldspar Mylonite: Evidence for Superlastic Flow in the Lower Continental Crust,' *Tectonophys*, **140** (1987), pp. 297–305.

123. M.S. Paterson, 'Superplasticity in Geological Materials,' in *Superplasticity in Metals, Ceramics, and Intermetallics, MRS Proceeding No. 196*, pp. 303–312, ed. M.J. Mayo, J. Wadsworth, and M. Kobayashi, Materials Research Society, Pittsburgh, PA, 1990.

Chapter 7

Fine-structure superplastic intermetallics

Ordered intermetallic alloys and their composites generally have good high-temperature strength, low density, and environmental resistance and are, therefore, potential materials for high-temperature structures [1–5]. However, ordered intermetallic alloys are also known to be brittle, have low toughness because of their ordered structure, and show a propensity for grain-boundary embrittlement. As a result, intermetallic alloys often either have poor fabricability and machinability or require a fabrication process that is complicated and tedious. The generic brittleness problem in intermetallics, particularly aluminides, has been studied extensively in recent years and some breakthroughs have been made [6–8]. For example, polycrystalline Ni_3Al that has an $L1_2$ structure exhibits almost no ductility, but Ni_3Al containing a small amount (0.2 wt%) of boron exhibits room-temperature tensile ductility of up to 40% [7]. Because of these technological breakthroughs, there is great interest in using these materials for engineering structures.

Superplasticity in intermetallics has only been recently demonstrated. Although large tensile elongations (~100%) for an intermetallic (also known as Sendust, Fe–9.6 wt% Si–5.4 wt% Al) were indicated as early as 1981 [9], true superplastic intermetallics were not observed until 1987 [10]. At present, several intermetallics of the $L1_2$ structure (e.g., Ni_3Al [10–16] and Ni_3Si [17–21]), iron aluminide [22, 23], titanium aluminide (TiAl [24–33]), and trititanium aluminides (Ti_3Al [34–38]) have demonstrated superplasticity. These intermetallics are being investigated for their structural applications.

Similar to superplastic metals, superplasticity in intermetallics has been recorded in both quasi-single-phase and two-phase materials. The strain-rate-sensitivity values (m) range typically from 0.32 to 1, suggesting that several independent deformation mechanisms operate, and that they all result in extended ductility in these alloys. One interesting observation is that some intermetallics exhibit superplastic properties, even in the relatively coarse-grained (>10 μm) conditions [9, 17, 22, 23]. For example, nickel silicides having a grain size of greater than 15 μm still behave superplastically [18, 21]. This grain size is noted to be coarser than that required in superplastic metals, which is typically less than 10 μm. In the following sections, we will discuss the results obtained from nickel-, titanium-, and iron-based intermetallic compounds.

7.1   Nickel-based intermetallic compounds

7.1.1   Nickel silicide (Ni$_3$Si)

Nickel silicide (Ni$_3$Si), which has an L1$_2$ structure, is the major constituent of Hastelloy D (composition: Ni–9 wt% Si–3 wt% Cu), a superalloy that resists attack by sulfuric acid solutions [39]. However, Hastelloy D is a cast alloy that not only has low-room-temperature ductility but is also difficult to fabricate. Major developments on the physical metallurgy of Ni$_3$Si-based alloys are largely the result of work in the late 1960s by Williams [40].

The superplastic behavior of Ni$_3$Si-based intermetallics has been studied by several investigators [17–19, 41]. All of these Ni$_3$Si alloys were produced by casting techniques. Superplasticity was observed only in materials that had been thermomechanically processed and recrystallized into a fine, equiaxed grain structure. The typical microstructure of a processed Ni$_3$Si (nominal composition: Ni–9 wt% Si–3.1 wt% V–2 wt% Mo) is shown in Figure 7.1. It is a duplex structure consisting of grains that are primarily ordered L1$_2$ cubic β phase and grains that are composed of a disordered cubic β phase dispersion in an α-Ni solid solution matrix. The grains have a mean linear intercept grain size of approximately 15 μm, and the volume fraction of these two types of grains is approximately 50/50. Transmission electron micrograph microstructures indicate that, in addition to the α and β phases, a third phase (hexagonal MoNiSi), with a size of about 1 μm, was also present along grain boundaries. This third phase may act as a grain-growth inhibitor. The microstructure of duplex Ni$_3$Si alloys is unusually stable. For example, the grain size (~18 μm) in a sample tested at 1080 °C (>0.9T_m) is only slightly larger than the grain size of an untested sample (~15 μm grain size).

In a study of Ni–Si–V–Mo intermetallics (both V and Mo additions to Ni$_3$Si are made to stabilize the ordered L1$_2$ phase), superplasticity was observed over a limited temperature range from 1000 to 1100 °C but over a wide strain-rate range from 6×10^{-4} to 1 s^{-1} [17, 18]. A tensile elongation of over 200% can generally be obtained under all the above test conditions, and a maximum elongation to failure of 710% has been recorded at 1080 °C with a strain rate of $8 \times 10^{-3} \, \text{s}^{-1}$. A typical true stress–true strain curve for the superplastic Ni_3Si, under a constant crosshead speed, is shown in Figure 7.2. The curve exhibits a high initial strain-hardening rate, then the flow stress decreases rapidly with increasing

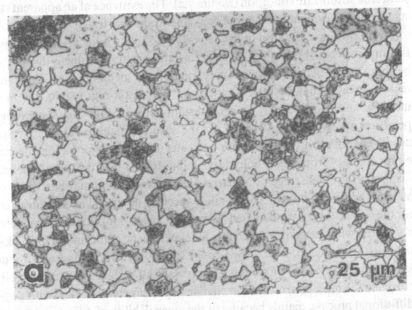

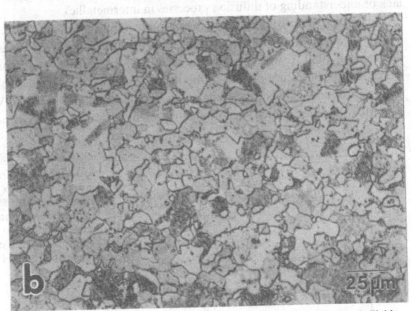

Figure 7.1 Microstructures of (a) as hot rolled and (b) annealed nickel silicide. The light phase is the $L1_2$ cubic β structure and the dark phase is the $\alpha + \beta$ phase (from Ref. [17]).

strain. The rapid reduction in flow stress after the peak stress with increasing strain is not a result of dynamic recrystallization *per se*; rather it is an indication of the strong strain-rate dependence on the flow stress. This is illustrated by directly comparing the corresponding true stress–true strain curve under a constant true strain-rate condition (Figure 7.2). The existence of an apparent steady-state region is evident for the true constant strain-rate test.

Strain rate as a function of flow stress is plotted in a $\log \dot{\varepsilon} - \log \sigma$ format for the duplex Ni_3Si (Figure 7.3). The strain-rate-sensitivity value is determined to be about 0.5, which is typical of many superplastic metals, in the high strain-rate region (greater than 4×10^{-3} s^{-1}). In the low strain-rate region, the m value increases and approaches 1, probably resulting from the fact that the testing temperatures are over $0.9T_m$, where T_m is the absolute melting point of Ni_3Si. This near-Newtonian viscous behavior indicates that the deformation mechanism either changes from grain-boundary sliding (GBS) to a diffusion-type mechanism, such as Coble creep or Nabarro–Herring creep, or it may be a result of slip accommodation by a dislocation–glide mechanism [42]. The activation energy for the $m=0.5$ region is calculated to be about 555 kJ/mol, which is relatively high compared with the activation energy for superplastic deformation of 267 kJ/mol measured in a fine-grained, nickel-based superalloy MA754 [43] and about 350 kJ/mol for some of γ'-containing, nickel-based alloys [44, 45]. It is impossible to correlate the measured activation energies from superplastic flow with a specific diffusional process, mainly because of the unavailability of diffusion data and the lack of understanding of diffusion processes in intermetallics.

The microstructure and phases present in the duplex Ni_3Si are quite complicated. A transmission electron micrograph of the duplex Ni_3Si superplastically deformed at 1080 °C with a strain rate of 3.3×10^{-3} s^{-1} is shown in Figure 7.4. Three major phases are present in the microstructure: β, α-Ni solid solution, and grain-boundary γ-$Ni_{31}Si_{12}$ (containing a limited amount of Mo and V) [46]. The MoNiSi phase, which was originally present at grain boundaries prior to

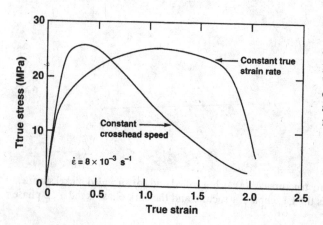

Figure 7.2
True stress–true strain curves for a duplex superplastic Ni_3Si at 1080 °C under the conditions of constant crosshead speed and constant true strain rate (data from Ref. [17]).

superplastic deformation, dissociated into the alloy matrix during superplastic tests. The presence of γ-Ni$_{31}$Si$_{12}$ in the microstructure is a result of the decomposition of β'-Ni$_3$Si (i.e., the high-temperature L1$_2$ phase) during cool down, rather than an indication of the presence of that phase at superplastic test temperatures.

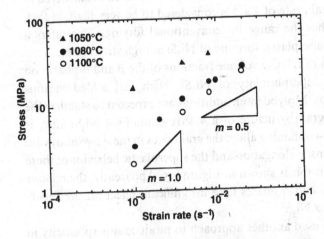

Figure 7.3 Strain rate as a function of flow stress for superplastic duplex Ni$_3$Si, where m is approximately 0.5 and approaches 1 in the low strain-rate region (data from Ref. [17]).

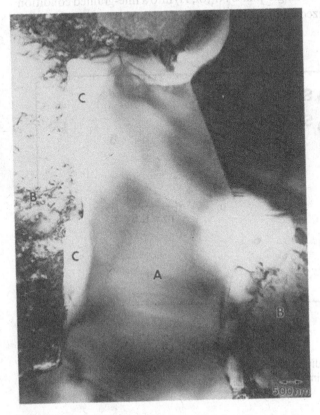

Figure 7.4 Transmission electron micrograph of Ni$_3$Si superplastically deformed 600% at 1080 °C, showing the presence and distribution of the β (A), $\sigma + \beta$ (B), and γ-Ni$_{31}$Si$_{12}$ (C) phases (from Ref. [17]).

One of the interesting characteristics of the superplastic duplex Ni_3Si is that the temperature range is quite limited (1000–1100 °C) [17, 18]; Ni_3Si is single phase below 1000 °C; however, superplasticity can exist over a relatively wide strain-rate range. For example, at 1080 °C a change in strain rate over three orders of magnitude, from 10^{-3} to $1\ s^{-1}$, only results in a decrease in tensile elongation from 650 to 300%. An elongation value of 300% is still considered to be superplastic, and a strain rate of $1\ s^{-1}$ is considered to be very high; in fact, such a strain rate is within the range for conventional forging. This offers a technological benefit of superplastic forming of Ni_3Si at high strain rates.

To determine the effect of relative volume fractions of the β and α phases on the microstructure and superplasticity of Ni_3Si, Nieh [21] added various amounts of Mo to the alloy. Molybdenum additions are expected to stabilize the ordered $L1_2$ phase. The grain structure of a 4%Mo-containing Ni_3Si alloy is similar to that of a 2%Mo-containing alloy; the grain sizes in these two alloys are both about 15 μm. The tensile elongation and the superplastic behavior of these two alloys are also similar; this is shown in Figure 7.5. Apparently, the relative volume fractions of the β and α phases have insignificant effect on the super-plastic properties of duplex Ni_3Si.

Takasugi *et al.* [19, 47] used another approach to produce superplasticity in Ni_3Si alloys. They processed single-phase $Ni_3(Si,Ti)$ into a fine-grained condition (~4 μm) and characterized the superplastic behavior of the alloys. The fine-grained material exhibits superplasticity at temperatures between 800 and

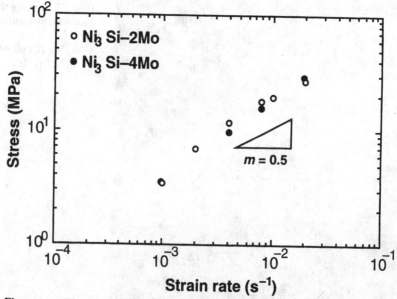

Figure 7.5 The addition of Mo to Ni_3Si at 1080 °C to increase the amount of $L1_2$ phase has little influence on the superplastic properties of Ni_3Si alloys (data from Ref. [17]).

900 °C and at strain rates between 6×10^{-5} and 10^{-3} s^{-1}. These deformation temperatures are only about 0.8 T_m, where T_m is the absolute melting point of Ni$_3$(Si,Ti), whereas superplasticity is found at about 0.85 to 0.94T_m for duplex Ni$_3$Si alloys [17]. This is primarily attributed to the fact that grains in the single-phase Ni$_3$Si coarsen quickly at T_m >850 °C, while the microstructures of the duplex alloys are more thermally stable.

A maximum elongation of only 180% was recorded from single-phase Ni$_3$(Si,Ti). This elongation value is much lower than the value of 710% obtained from the duplex Ni$_3$Si alloys. The strain-rate-sensitivity value of the Ni$_3$(Si,Ti) is about 0.43. From mechanical properties and microstructural observations, Takasugi *et al.* [19] suggested that a GBS-based mechanism is responsible for superplastic deformation in single-phase Ni$_3$(Ti,Si) alloys. Takasugi *et al.* also studied the effect of microalloying on superplastic properties of Ni$_3$(Ti,Si). They found that although adding boron (36 ppm) to the alloy can reduce the grain size of Ni$_3$(Ti,Si), it also retards the sliding mobility of grain boundaries. In addition, the activation energy for superplasticity increases from 273 to 318 kJ/mol. Moreover, boron addition is detrimental to superplastic elongation.

In contrast to duplex Ni$_3$Si alloys, which still exhibit high elongation at high strain rates (~1 s^{-1}), the superplastic elongation of single-phase Ni$_3$(Ti,Si) was found to decrease rapidly with increasing strain rate. In fact, the elongation values of single-phase Ni$_3$(Ti,Si) never exceeded 60% at strain rates higher than 2×10^{-4} s^{-1}. The rapid reduction in elongation at high strain rates may be related to the nature of the grain boundary. Specifically, single-phase Ni$_3$(Ti,Si) contains only homogeneous grain boundaries, while duplex Ni$_3$Si contains heterogeneous grain boundaries. The sliding mobility of a heterogeneous grain boundary is much faster than the sliding mobility of a homogeneous grain boundary, as discussed in Section 3.1.3. In theory, a dopant that stabilizes grain size but does not reduces the sliding mobility of grain boundaries is expected to yield the optimum superplasticity in single-phase Ni$_3$(Si,Ti). The superplastic elongation of Ni$_3$(Si,Ti) was also noted to be very sensitive to the grain size, as shown in Figure 7.6. As indicated in the figure, elongation values decrease from 130% at a 3.7-μm grain size to only about 10% at an 11-μm grain size.

7.1.2   Nickel aluminide (Ni$_3$Al)

Superplasticity has been produced in both single-phase [12] and duplex [13, 15, 16] nickel aluminide (Ni$_3$Al). In addition, both powder-metallurgy [10, 13–15] and ingot-metallurgy [12, 48] Ni$_3$Al alloys have been reported to be superplastic. In the case of powder-metallurgy materials, only an alloy with a composition of Ni–18 at. % Al–8 at. % Cr–1 at. % Zr–0.15 at. % B (or Ni–8.5%Al–7.8%Cr–0.8% Zr–0.02%B by weight, designated as IC–218) has been extensively studied for superplasticity. For this alloy, Cr is used to overcome the dynamic embrittlement associated with environmental moisture [49] and Zr is used to provide solid-solution strengthening at elevated temperatures [50].) Sikka *et al.* [10] first reported superplasticity in this alloy, and the work was further expanded by Wright and Sikka [11], Choudhury *et al.* [13], and Mukhopadhyay *et al.* [14, 15].

The superplastic IC–218 alloy had a duplex microstructure, containing about 10 to 15% of disordered γ phase in an ordered γ' phase matrix with a grain size of about 6 μm. It exhibited superplasticity at temperatures between 950 and 1100 °C and strain rates from 10^{-5} to 10^{-2} s^{-1}; a maximum elongation value of 640% was obtained at 1100 °C with at a strain rate of 8.3×10^{-4} s^{-1}. The strain rate–stress relationship in IC–218 is shown in Figure 7.7. The deformation can apparently be divided into three regions according to the strain-rate-sensitivity values. In the high-strain-rate region (greater than 7×10^{-2} s^{-1}), m is about 0.32. It increases to 0.75 over the strain-rate range from 7×10^{-2} to 4×10^{-4} s^{-1}, below which m appears to increase to an even higher value (i.e., it approaches 1). This change of the m values is somewhat similar to that observed in Ni$_3$Si (Figure 7.3), except that Ni$_3$Si exhibits only two regions. Also, Ni$_3$Si shows a lower m value (0.5) in the high strain-rate region than does Ni$_3$Al (0.75). This difference may be the result of the difference in their grain sizes (6 μm for Ni$_3$Al and 15 μm for Ni$_3$Si).

In contrast to duplex Ni$_3$Si, in which the superplastic strain to failure is weakly dependent upon strain rate, the superplastic strain of duplex Ni$_3$Al

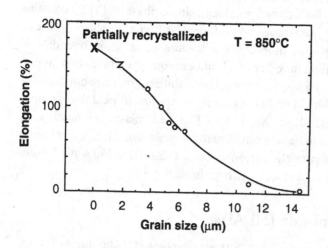

Figure 7.6 Superplastic elongation of Ni$_3$(Si,Ti) is strongly dependent upon grain size (from Ref. [19]).

depends strongly on the strain rate. For example, the tensile elongation of Ni_3Al reduces from 440% to only 100% when the strain rate increases by one order of magnitude (8×10^{-4} to 8×10^{-3} s^{-1}) at 1050 °C [13]. However, the temperature range for superplasticity of Ni_3Al is similar to that for Ni_3Si, i.e., it is narrow (1000–1100 °C). For both materials, the result can be rationalized by the fact that this is the temperature range for the existence of a two-phase microstructure. The activation energies for Ni_3Al in regions where the m values are 0.75 and 0.32 were measured to be 290 and 383 kJ/mol, respectively. Both activation energies are noted to be relatively low compared with that observed in Ni_3Si. Nevertheless, a value of 290 kJ/mol compares well with the activation energies for superplastic deformation measured in a nickel-based superalloy MA 754 [43].

Both duplex Ni_3Al and Ni_3Si are noted to be susceptible to cavitation during superplastic deformation. Most cavities are either along interphase boundaries between the two types of grains or associated with grain-boundary triple points. No systematic study has been performed to evaluate quantitatively the cavitation behavior during superplastic deformation of either Ni_3Si or Ni_3Al.

Superplasticity in single-phase, boron-doped (0.24 at. %) Ni_3Al has been characterized by Kim et al. [12]. The material is of ultrafine grain size ($\sim$1.6 μm). A maximum elongation of 160% was recorded at 700 °C with a strain rate of less than 10^{-4} s^{-1}. Superplasticity in this single-phase Ni_3Al takes place via dynamic recrystallization, suggesting the thermal and dynamic instability of the fine-

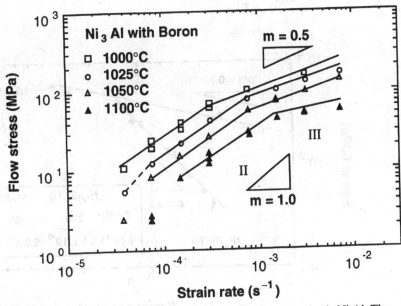

Figure 7.7 Strain rate as a function of flow stress for superplastic Ni_3Al. The strain-rate-sensitivity value is about 0.5 and approaches 1 in the low-strain-rate region (data from Ref. [14]).

grained microstructure. Although the single-phase Ni_3Al has a finer grain size than the duplex Ni_3Al (1.6 μm vs 6 μm), superplastic elongation of the single-phase alloy is inferior to that of the duplex alloy. Similar to Ni_3Si, a lower superplastic elongation in the single-phase Ni_3Al is believed to be related to the fact that sliding between homogeneous grain boundaries in single-phase alloys is, in general, more difficult than in the heterogeneous grain boundaries that exist in duplex alloys [51, 52]. Also, fine grains in the single-phase alloy are expected to be less stable under high-temperature and dynamic conditions.

For $L1_2$ intermetallics, although it is known that grain size can affect superplastic flow, a quantitative relationship between grain size and plastic flow stress has not yet been reported. A plot showing such a relationship for both single-phase Ni_3Al and $Ni_3(Si,Ti)$ is shown in Figure 7.8. The p values (0.33 for Ni_3Al and 2 for $Ni_3(Si,Ti)$) are noted to be quite different for the two alloys, despite the fact that they have a similar $L1_2$ crystal structure. The explanation for this discrepancy is not clear. The difference may be caused by the different grain sizes, and thus microstructural stability during the superplastic deformation, of the two alloys.

A summary of the superplastic properties of $L1_2$ intermetallics is given in Table 7.1. Despite the ordered crystal structure of $L1_2$ intermetallics, the superplastic behavior of $L1_2$ intermetallics and conventional metal alloys are quite similar. Specifically, the superplastic deformation of Ni_3Si and Ni_3Al can be well described by a classical equation, used to describe superplastic metals, of the form [42]

$$\dot{\varepsilon} = A\sigma^n d^{-p}\exp\left(-\frac{Q}{RT}\right). \tag{7.1}$$

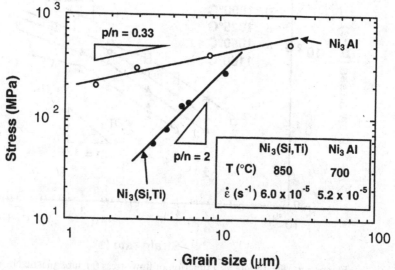

Figure 7.8 Superplastic flow stress as a function of grain size for single-phase $Ni_3(Si,Ti)$ and Ni_3Al doped with boron (data from Refs. [12, 19]).

Exact values for n, p, Q, and the material constant A for each material are determined by the temperature and strain rate regions in which superplastic properties of the material are characterized.

7.2   Titanium-based intermetallic compounds

The titanium aluminides TiAl (γ) and Ti_3Al (α_2) have considerable potential for use in advanced structures because of their attractive elevated temperature strength and low density [53]. Both intermetallic compounds, however, normally have low ductility at room temperature, which thus far has restricted their application. Therefore, methods to improve their ductility, while retaining their strength, are major challenges [53, 54]. The poor intermediate and room-temperature ductility causes the conventional manufacturing operations such as rolling, forging, or drawing to be difficult for titanium aluminides. As a result, the development of superplastic titanium aluminides is expected to be technologically useful.

7.2.1 α_2-Titanium aluminides (Ti_3Al)

The basic Ti_3Al structure is DO_{19}, an ordered hexagonal phase (α_2). Various elements (e.g., Nb, V, W, and Mo) are often added to the base Ti_3Al alloy to stabilize the β phase (disordered body-centered cubic) to improve room-temperature ductility. As a result, these alloys usually have a duplex β and α_2 microstructure [55]. Superplasticity has only been observed in duplex Ti_3Al alloys

Table 7.1 Summary of data for single-phase and duplex superplastic L12 intermetallics

Composition (wt%)	d (μm)	Temperature (°C)	$\dot{\varepsilon}$ (s^{-1})	m	Elongation (%)	Ref.
Ni–9%Si–3.1%V–2%Mo (duplex)	15	1080	8×10^{-3}	0.5	710	[17]
Ni–9%Si–3.1%V–4%Mo (duplex)	>10	1070	1×10^{-3}	0.43	560	[17, 18]
Ni_3(Si,Ti) (single-phase)	4	850	6×10^{-5}	0.43	180	[19]
0.24 at. % B–doped Ni_3Al (single-phase)	1.6	700	5×10^{-5}	0.42	160	[12]
Ni–8.5%Al–7.8%Cr–0.8%Zr–0.02%B (duplex)	6	1100	8.3×10^{-4}	0.75	640	[13]

Ti–25Al–10Nb–3V–1Mo (super α_2) [35–38] and Ti–24Al–11Nb (α_2) [34, 35, 37, 56,57]. These alloys were thermomechanically processed to fine-grained ($\sim$5 μm) conditions.

Superplasticity exists in a narrow temperature range (950–1020 °C); within this temperature range, the volume fraction of β phase is about 40 to 60% [57]. Maximum tensile elongation values of about 500 and 1350% were recorded in the α_2 and super α_2 alloys, respectively, at a strain rate of approximately $10^{-5}\,\mathrm{s}^{-1}$. The measured strain-rate-sensitivity values of both alloys are normally above 0.5 under superplastic conditions, and it was believed that a GBS mechanism is responsible for the observed superplasticity. Ridley *et al.* [57] pointed out that super α_2 remained free from cavitation after large superplastic tensile elongation ($>$1000%), although dynamic grain growth was observed. This suggests the alloy is suitable for stretch forming without the need to impose back pressure to inhibit cavitation. It is interesting to note that Wittenauer *et al.* [36], using the above superplastic properties, have fabricated super α_2 foil from sheet material by vacuum pack rolling. The foils can be used as facing sheet for further fabrication of intermetallic composites.

7.2.2 γ-Titanium aluminides (TiAl)

Studies of superplastic TiAl are of the most recent origin. A summary of published work on superplastic TiAl is given in Table 7.2. All materials, except PM Ti–47Al, were produced by ingot-casting techniques.

Superplasticity has only been observed in fine-grained, two-phase ($\gamma + \alpha_2$)

Table 7.2   Summary of data from superplastic TiAl

Composition (at.%)	d (μm)	Temperature (°C)	$\dot{\varepsilon}$ (s^{-1})	m	Elongation (%)	Ref.
Ti–43Al	5	1000–1100	10^{-5} to 2×10^{-2}	0.5	275	[24]
Ti–50Al	<5	900–1050	2×10^{-4} to 8.3×10^{-3}	$\sim$0.4	250	[27, 33]
Ti–47Al–2Nb– 1.6Cr–0.5Si–0.4Mn	20	1180–1310	2×10^{-5} to 2×10^{-3}	0.65	470	[30]
Ti–47.4Al	8	927	1×10^{-4}	NA	$\sim$400	[29]
Ti–43Al–13V	NA	800–1143	3×10^{-4} to 10^{-1}	NA	580	[32]
PM Ti–47Al	2	950	10^{-4} to 10^{-3}	0.3	NA	[58]

TiAl alloys, with the exception of the Ti–43Al–13V alloy, which is ($\gamma+\beta$). The fine microstructure is achieved if a composition is selected in the two-phase region where approximately 50 vol % of each of the two aluminides TiAl and Ti$_3$Al coexist. An examination of the Ti–Al phase diagram, shown in Figure 7.9, reveals that such an alloy composition has many characteristics of a hyper-eutectoid steel. That is, the alloy can be heated into a single-phase region at high temperature where the aluminum is completely in solution in the hexagonal close-packed structure of titanium; upon cooling, proeutectoid TiAl will precipitate followed by a eutectoid reaction wherein a lamellar structure of TiAl and Ti$_3$Al is obtained (similar to pearlite in steel). If the alloy is mechanically worked below the eutectoid temperature, the lamellar structure will break up to form two equiaxed, fine-grained phases (equivalent to strain-enhanced spheroidization in eutectoid steels).

The grain structure has a significant effect on the deformation behavior of TiAl. The influence of microstructure on the true stress–true strain curve in compression for a Ti(57 at. %)–Al(43 at. %) alloy at 1000 °C is shown in Figure 7.10. In this figure, the flow stress is plotted as a function of the true plastic strain at an engineering strain rate of $3\times10^{-4}\,\text{s}^{-1}$ for samples having initially either a lamellar or a spheroidized microstructure. As can be seen, there is a large decrease in the flow stress with strain for the lamellar structure Ti–Al alloy, and eventually the flow stress remains virtually constant with strain. The large decrease in flow stress is a result of a change in microstructure from a lamellar structure to fine-grained equiaxed structure.

Based upon Figure 7.10, Cheng et al. [24] deformed a Ti–43Al alloy in compression at 1050 °C, perpendicular to the extrusion direction, and with a 3.5:1 reduction to obtain a fine-grained, two-phase microstructure. The microstructure of the resultant alloy after pressing indeed reveals that the original

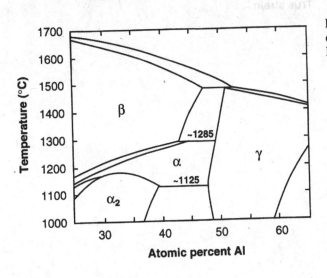

Figure 7.9 Ti–Al phase diagram (data from Ref. [59]).

lamellar structure had been broken up to form a fine-grained, equiaxed, two-phase structure. The volume fraction at the temperature for superplastic testing (1000–1100 °C) was determined to be 52% TiAl and 48% Ti₃Al. The grain size after pressing, including both phases, is 5 μm.

The strain rate as a function of flow stress for the fine-grained, equiaxed material is presented in Figure 7.11. The curves in Figure 7.11 divide into two stress-exponent regions as a function of the flow stress. At high values of the flow stress, a stress exponent (n) approaching 4 is found. In this region, the rate-controlling deformation mechanism was suggested to be a diffusion-controlled dislocation process. At low values of the flow stress, a stress exponent of about 2 (or $m=0.5$) is observed at all temperatures. At the highest temperature of testing (1100 °C), the $n=2$ region extends to a strain rate of as high as 10^{-3} s^{-1}. In this region, the material is superplastic and the rate-controlling deformation process is attributed to a GBS mechanism. An elongation-to-failure value of 275% was obtained at an initial strain rate of 2.5×10^{-4} s^{-1} at 1050 °C in air. Because

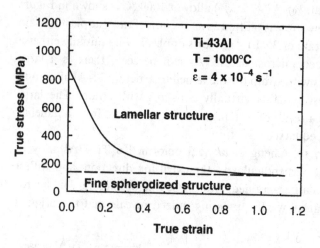

Figure 7.10   Compression true stress–true strain curves at 1000 °C for the two-phase Ti(57 at. %)–Al(43 at. %) alloy in the lamellar and fine spheroidized conditions (data from Ref. [24]).

considerable surface oxidation occurred, an elongation-to-failure value of greater than 275% was expected if tested in an inert atmosphere. In fact, a higher maximum elongation of 470% was recorded from a two-phase TiAl tested in vacuum (3×10^{-3} Pa) by Lee *et al.* [30].

It is noted that the grain size used by Lee *et al.* was about 20 μm, which is quite coarse. The microstructure of the alloy is noted to consist of equiaxed γ grains and lamellar $\alpha_{2+}\gamma$. The effect of grain size on the superplastic behavior of TiAl-based alloys, particularly the elongation and flow stress, has not yet been systematically investigated.

The apparent activation energy in the $n=2$ region was determined to be 390 kJ/mol. Again, this activation energy could not be compared to activation energies for diffusion because no data are available for self-diffusion of titanium and aluminum either in TiAl or in Ti$_3$Al. Nonetheless, the value is close to that measured from the creep of a fine-grained Ti(53 at. %)–Al(47 at. %) by Kampe *et al.* [60]. Based upon all these results, the rate-controlling deformation mechanism is

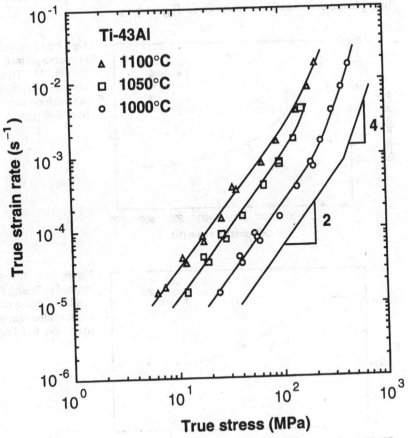

Figure 7.11   Strain rate–stress relations for the fine-grained, two-phase Ti (57 at. %)–Al (43 at. %) alloy at 1000, 1050, and 1100 °C (data from Ref. [24]).

proposed to be GBS accommodated by slip controlled by lattice diffusion in TiAl. This mechanism has also been suggested by Lee *et al.* [30] and Imayev *et al.* [27].

7.3   Iron-based intermetallic compounds

Only limited studies have been carried out on iron aluminides, i.e. Fe_3Al and FeAl [22, 23]. For example, a Fe_3Al alloy (Fe–28Al–2Ti) showed superplasticity at strain rates from 2×10^{-4} to 4×10^{-3} at temperatures ranging from 700 to 900 °C. The maximum elongation was over 620% at 850 °C at a strain rate of $10^{-3}\,s^{-1}$. Interestingly, superplasticity in iron aluminides was observed in coarse-grained conditions; grain sizes are 100 and 350 μm in superplastic Fe_3Al and FeAl, respectively. Li *et al.* [23] argued that a continuous recovery and recrystallization process was responsible for the observed superplasticity. This is supported by the observation that grain size was continously decreased during superplastic deformation, as shown in Figure 7.12.

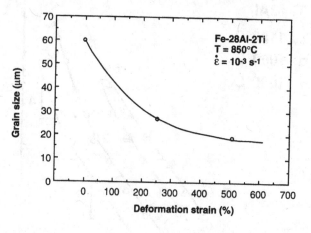

Figure 7.12   Grain size of coarse-grained FeAl (Fe–36.5Al–2Ti) as a function of superplastic deformation strain (data from Ref. [23]).

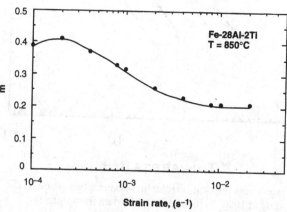

Figure 7.13   Coarse-grained (100 μm) Fe_3Al exhibits a strain-rate-sensitivity value near 0.33 (data from Ref. [22]).

It is noted, however, that coarse-grained Fe_3Al and FeAl both have a strain-rate-sensitivity value near 0.33 (Figure 7.13). They have also shown all the important microstructural features for Class I solid solution [61], e.g., high dislocation density, absence of boundary sliding, no grain-boundary cavity formation, and samples that neck to a point. Yang et al. [62] have argued that, when disorder is introduced into an ordered intermetallic solid by the glide of a dislocation, the steady-state velocity is limited by the rate at which chemical diffusion can reinstate order behind the glide dislocation. In such cases, the deformation is controlled by a viscous glide process.

There is yet another possible explanation for viscous glide in ordered intermetallics. Intermetallics often contain excessive solutes as a result of non-stoichiometry or an appreciable amount of interstitial impurities, e.g., oxygen, nitrogen, and carbon. The presence of these excessive solutes or impurities may contribute to the drag processes. In addition, Yaney and Nix [63] have suggested that the addition of solute atoms is not the only way to cause dislocations to move in a viscous manner. For example, lattice friction effects may also reduce glide mobility, particularly in covalently bonded solids (such as Ge and Si) and ordered intermetallic compounds. In this latter group, strong repulsive forces exist between atoms of like character. Thus, the deformation behavior of ordered intermetallics, in which lattice friction effects limit glide mobility, is expected to be quite similar to that of Class I solid solutions. Thus, the exact superplastic deformation mechanisms in coarse-grained iron aluminides are still unclear but are both scientifically and technologically interesting and important.

References

1. *High-Temperature Ordered Intermetallic Alloys, Vol. 39*, ed. C.C. Koch, N.S. Stoloff, and C.T. Liu, Materials Research Society, Pittsburgh, PA, 1985.

2. *High Temperature Ordered Intermetallic Alloys II, MRS Sym. Vol.81*, ed. N.S. Stoloff, C.C. Koch, C.T. Liu, and O. Izumi, Materials Research Society, Pittsburgh, PA, 1987.

3. *High Temperature Ordered Intermetallic Alloys III, MRS Sym. Vol.133*, ed. C.T. Liu, A.I. Taub, N.S. Stoloff, and C.C. Koch, Materials Research Society, Pittsburgh, PA, 1989.

4. *High Temperature Ordered Intermetallic Alloys IV, MRS Sym. Vol. 213*, ed. L.A. Johnson, D. Pope, and J. Stiegler, Materials Research Society, Pittsburgh, PA, 1991.

5. *High Temperature Ordered Intermetallic Alloys V, MRS Sym. Vol. 288*, ed. I. Baker, R. Darolia, J.D. Whittenberger, and M.H. Yoo,

Materials Research Society, Pittsburgh, PA, 1993.

6. A. Aoki and O. Izumi, 'Improvement in Room Temperature Ductility of the $L1_2$ Type Intermetallic Compound Ni_3Al by Boron Addition,' *J. JIM*, **43** (1979), pp. 1190–1195.

7. C.T. Liu, C.L. White, and J.A. Horton, 'Effect of Boron on Grain Boundaries in Ni_3Al,' *Acta Metall.*, **33**(2) (1985), pp. 213–229.

8. C.T. Liu and J.O. Stiegler, 'Ductile Ordered Intermetallic Alloys,' *Science*, **2–6** (1989), pp. 636.

9. S. Hanada, T. Sato, S. Watanabe, and O. Izumi, 'Deformability of Sendust Polycrystals,' *J. JIM*, **45**(12) (1981), pp. 1293–1299.

10. V.K. Sikka, C.T. Liu, and E.A. Loria, 'Processing and Properties of Powder Metallurgy Ni_3Al-Cr-Zr-B for Use in

Oxidizing Environments,' in *Processing of Structural Metals by Rapid Solidification*, pp. 417–427, ed. F.H. Froes and S.J. Savage, Am. Soc. Metals, Metals Park, OH, 1987.

11. R.N. Wright and V.K. Sikka, 'Elevated Temperature Tensile Properties of Powder Metallurgy Ni_3Al Alloyed with Chromium and Zirconium,' *J. Mater. Sci.*, **23** (1988), pp. 4315–4318.

12. M.S. Kim, S. Hanada, S. Wantanabe, and O. Izumi, 'Superplasticity in a Recrystallized Ni_3Al Polycrystal Doped with Boron,' *Mater. Trans. JIM*, **30**(1) (1989), pp. 77–85.

13. A. Choudhury, A.K. Muhkerjee, and V.K. Sikka, 'Superplasticity in an Ni_3Al Base Alloy with 8wt%Cr,' *J. Mater. Sci.*, **25** (1990), pp. 3142–3148.

14. J. Mukhopadhyay, G.C. Kaschner, and A.K. Muhkerjee, 'Superplasticity and Cavitation in Boron Doped Ni_3Al,' in *Superplasticity in Aerospace II*, pp. 33–46, ed. T.R. McNelly and C. Heikkenen, The Minerals, Metals & Materials Society, Warrendale, PA, 1990.

15. J. Mukhopadhyay, G. Kaschner, and A.K. Muhkerjee, 'Superplasticity in Boron Doped Ni_3Al,' *Scr. Metall.*, **24** (1990), pp. 857–862.

16. S. Ochiai, Y. Doi, I. Yamada, and Y. Kojima, 'Superplasticity in Boron-Doped $(\beta + \gamma')$ Two-Phase Nickel Aluminide,' *J. JIM*, **57**(2) (1993), pp. 214–219.

17. T.G. Nieh and W.C. Oliver, 'Superplasticity of a Nickel Silicide,' *Scr. Metall.*, **23** (1989), pp. 851–854.

18. S.L. Stoner and A.K. Muhkerjee, 'Superplasticity in Fine Grain Nickel Silicide,' in *International Conference on Superplasticity in Advanced Materials (ICSAM–91)*, pp. 323–328, ed. S. Hori, M. Tokizane, and N. Furushiro, The Japan Society for Research on Superplasticity, Osaka, Japan, 1991.

19. T. Takasugi, S. Rikukawa, and S. Hanada, 'Superplastic Deformation in $Ni_3(Si,Ti)$ Alloys,' *Acta Metall. Mater.*, **40** (1992), pp. 1895–1906.

20. T. Takasugi, S. Rikukawa, and S. Hanada, 'Superplasticity in $L1_2$ Type $Ni_3(Si,Ti)$ Intermetallics,' in *International Conference on Superplasticity in Advanced Materials (ICSAM–91)*, pp. 329–338, ed. S. Hori, M. Tokizane, and N. Furushiro, The Japan Society for Research on Superplasticity, Osaka, Japan, 1991.

21. T.G. Nieh, 'Superplasticity in $L1_2$ Intermetallic Alloys,' in *Superplasticity in Metals, Ceramics, and Intermetallics, MRS Proceeding No.196*, pp. 343–348, ed. M.J. Mayo, J. Wadsworth, and M. Kobayashi, Materials Research Society, Pittsburgh, PA, 1990.

22. D. Lin, A. Shan, and D. Li, 'Superplasticity in Fe_3Al-Ti Alloy with Large Grains,' *Scr. Metall. Mater.*, **31**(11) (1994), pp. 1455–1460.

23. D. Li, A. Shan, Y. Liu, and D. Lin, 'Study of Superplastic Deformation in an FeAl Based Alloy with Large Grains,' *Scr. Metall. Mater.*, **33** (1995), pp. 681–685.

24. S.C. Cheng, J. Wolfenstine, and O.D. Sherby, 'Superplastic Behavior of Two-Phase Titanium Aluminides,' *Metall. Trans.*, **23A** (1992), pp. 1509–1513.

25. R.M. Imayev and V.M. Imayev, 'Mechanical Behavior of TiAl Submicrocrystalline Intermetallic Compound at Elevated Temperatures,' *Scr. Metall. Mater.*, **25** (1991), pp. 2041–2046.

26. R.M. Imayev, V.M. Imayev, and G.A. Salishchev, 'Formation of Submicrocrystalline Structure in TiAl Intermetallic Compound,' *J. Mater. Sci.*, **27** (1992), pp. 4465–4471.

27. R.M. Imayev, O.A. Kaibyshev, and G.A. Salishchev, 'Mechanical Behavior of Fine Grained TiAl Intermetallic Compound – I. Superplasticity,' *Acta Metall. Mater.*, **40**(3) (1992), pp. 581–587.

28. T. Maeda, M. Okada, and Y. Shida, 'Superplasticity in Ti-Rich TiAl,' in *International Conference on Superplasticity in Advanced Materials (ICSAM–91)*, pp. 311–316, ed. S. Hori, M. Tokizane, and N. Furushiro, The Japan Society for Research on Superplasticity, Osaka, Japan, 1991.

29. T. Tsujimoto, K. Hashimoto, and M. Nobuki, 'Alloy Design for Improvement of Ductility and Workability of Alloys Based on Intermetallic Compound TiAl,' *Mater. Trans., JIM*, **33**(11) (1992), pp. 989–1003.

30. W.B. Lee, H.S. Yang, Y.-W. Kim, and A.K. Muhkerjee, 'Superplastic Behavior in a Two-Phase TiAl Alloy,' *Scr. Metall. Mater.*, **29** (1993), pp. 1403–1408.

31. T. Wajata, K. Isonishi, K. Ameyama, and M. Tokizone, 'Superplasticity in TiAl and Ti_3Al Two Phase Material Made from PREPed Powder,' *ISIJ Inter.*, **33**(8) (1993), pp. 884–888.

32. D. Vanderschueren, M. Nobuki, and M. Nakamura, 'Superplasticity in a Vanadium Alloyed Gamma Plus Beta Phased Ti-Al

Intermetallics,' *Scr. Metall. Mater.*, **28** (1993), pp. 605–610.

33. R. Imayev, V. Imayev, and G. Salishchev, 'Effect of Grain Size on Ductility and Anomalous Yield Strength of Micro- and Submicrocrystalline TiAl,' *Scr. Metall. Mater.*, **29** (1993), pp. 713–718.

34. A. Dutta and D. Banerjee, 'Superplastic Behavior in a Ti₃Al-Nb Alloy,' *Scr. Metall. Mater.*, **24** (1990), pp. 1319–1322.

35. A.K. Ghosh and C.-H. Cheng, 'Superplastic Deformation in Titanium Aluminides and Modeling of Transient Deformation,' in *International Conference on Superplasticity in Advanced Materials (ICSAM–91)*, pp. 299–310, ed. S. Hori, M. Tokizane, and N. Furushiro, The Japan Society for Research on Superplasticity, Osaka, Japan, 1991.

36. J.P. Wittenauer, C. Bassi, and B. Walser, 'Hot Deformation Characteristics of Nb-Modified Ti₃Al,' *Scr. Metall.*, **23** (1989), pp. 1381–1386.

37. H.S. Yang, P. Jin, E. Dalder, and A.K. Muhkerjee, 'Superplasticity in a Ti₃Al-base Alloy Stabilized by Nb, V, and Mo,' *Scr. Metall. Mater.*, **25** (1991), pp. 1223–1228.

38. H.S. Yang, P. Jin, and A.K. Muhkerjee, 'Superplastic Properties of Ti₃Al,' *Mater. Sci. Eng.*, **A153** (1992), pp. 457–464.

39. P. Kumar, 'Influence of Microstructure on the Corrosion Behavior of a Ni-Si Alloy,' in *High-Temperature Ordered Intermetallic Alloys*, pp. 537–554, ed. C.C. Kock, C.T. Liu, and N.S. Stoloff, Materials Res. Soc., Pittsburgh, PA, 1985.

40. K.J. Williams, 'The Microstructure and Tensile Properties of Nickel-Rich Nickel-Silicon and Nickel-Silicon-Titanium Alloys,' *J. Inst. Metals*, **97** (1969), pp. 112–118.

41. S.L. Stoner and A.K. Muhkerjee, 'Superplasticity in a Nickel Silicide Alloy-Microstructure and Mechanical Correlation,' *Mater. Sci. Eng.*, **A153** (1992), pp. 465–469.

42. O.D. Sherby and J. Wadsworth, 'Development and Characterization of Fine Grain Superplastic Material,' in *Deformation, Processing and Structure*, pp. 355–389, ed. G. Krauss, ASM, Metal Park, Ohio, 1984.

43. J.K. Gregory, J.C. Gibeling, and W.D. Nix, 'High Temperature Deformation of Ultra-Fine-Grained Oxide Dispersion Strengthened Alloys,' *Metall. Trans.*, **16A** (1985), pp. 777–787.

44. R.G. Menzies, J.W. Edington, and G.J. Davies, 'Superplastic Behavior of Powder Consolidated Nickel-Base Superalloy IN–100,' *Metal Sci.*, **15** (1981), pp. 210–216.

45. J.-P.A. Immarigeon and P.H. Floyd, 'Microstructural Instabilities during Superplastic Forging of a Nickel-base Superalloy Compact,' *Metall. Trans.*, **12A** (1981), pp. 1177–1186.

46. E.B. Badtiev, O.S. Petrushkova, and L.A. Vestin, 'Identification and Crystallography of Ni₃₁Si₁₂ Compound,' *Mosk. Univ. Khim.*, **15**(3) (1974), pp. 367–368 (in Russian).

47. T. Takasugi, S. Rikukawa, and S. Hanada, 'The Boron Effect on the Superplastic Deformation of Ni₃(Si,Ti) Alloys,' *Scr. Metall. Mater.*, **25** (1991), pp. 889–894.

48. S. Hanada, 'Deformation Microstructure in Superplastic Intermetallics,' in *International Conference on Superplasticity in Advanced Materials (ICSAM–91)*, pp. 289–298, ed. S. Hori, M. Tokizane, and N. Furushiro, The Japan Society for Research on Superplasticity, Osaka, Japan, 1991.

49. C.T. Liu and C.L. White, 'Dynamic Embrittlement of Boron-Doped Ni₃Al Alloys at 600°C,' *Acta Metall.*, **35**(3) (1987), pp. 643–649.

50. C.T. Liu and C.L. White, 'Design of Ductile Polycrystalline Ni₃Al Alloys,' in *High Temperature Ordered Intermetallic Alloys, MRS Sym. no. 39*, pp. 365–380, ed. C.C. Koch, C.T. Liu, and N.S. Stoloff, Material Research Society, Pittsburgh, PA, 1985.

51. S. Hashimoto, F. Moriwaki, T. Mimaki, and S. Miura, 'Sliding Along the Interphase Boundary in Austenitic/Ferritic Duplex Stainless Steel Bicrystals,' in *International Conference on Superplasticity in Advanced Materials (ICSAM–91)*, pp. 23–32, ed. S. Hori, M. Tokizane, and N. Furushiro, The Japan Society for Research on Superplasticity, Osaka, Japan, 1991.

52. A. Eberhardt and B. Baudelet, 'Interphase Boundary Sliding at High Temperature in Two-Phase α/β-Brass Bicrystals,' *Phil. Mag.*, **41**(6) (1980), pp. 843–867.

53. Y.W. Kim and D.M. Dimiduk, 'Progress in the Understanding of Gamma Titanium Aluminides,' *JOM*, **43**(8) (1991), pp. 40–47.

54. M. Yamaguchi, 'High Temperature Intermetallics – with Particular Emphasis on TiAl,' *Mater. Sci. Technol.*, **8** (1992), pp. 299–307.

55. J. Wadsworth and F.H. Froes, 'Developments in Metallic Materials for Aerospace Applications,' *JOM*, **41**(5) (1989), pp. 12–19.

56. O.A. Kaibyshev, *Superplaticity of Alloys, Intermetallides, and Ceramics.* 1st Edition, 257, Springer-Verlag, New York, 1992.

57. N. Ridley, M.F. Islam, and J. Pilling, 'Superplasticity and Diffusion Bonding of Microduplex Super Alpha 2,' in *Structural Intermetallics*, pp. 63–68, ed. R. Darolia, J.J. Lewandowski, C.T. Liu, P.L. Martin, D.B. Miracle, and M.V. Nathal, The Minerals, Metals & Materials Society, Warrendale, PA, 1993.

58. M. Tokizane, T. Fukami, and T. Inaba, 'Structure and Mechanical Properties of the Hot Pressed Compact of Ti-Rich TiAl Powder Produced by the Plasma Rotating Electrode Process,' *ISIJ Inter.*, **10** (1991), pp. 1088–1092.

59. J.J. Valencia, C. McCullough, C.G. Levi, and R. Mehrabian, 'Microstructure Evolution during Conventional and Rapid Solidification of a Ti–50at%Al Alloy,' *Scr. Metall.*, **21** (1987), pp. 1341–1346.

60. S.L. Kampe, J.D. Bryant, and L. Christodoulou, 'Creep Deformation of TiB_2-Reinforced Near-γ Titanium Aluminides,' *Metall. Trans.*, **22A** (1991), pp. 447–454.

61. O.D. Sherby and J. Wadsworth, 'Superplasticity-Recent Advances and Future Directions,' *Prog. Mater. Sci.*, **33** (1989), pp. 166–221.

62. H.S. Yang, W.B. Lee, and A.K. Muhkerjee, 'Superplastic Characteristics of Two Titanium Aluminides: γ-TiAl and α_2-Ti_3Al,' in *Structural Intermetallics*, pp. 69–76, ed. R. Darolia, J.J. Lewandowski, C.T. Liu, P.L. Martin, D.B. Miracle, and M.V. Nathal, The Minerals, Metals & Materials Society, Warrendale, PA, 1993.

63. D.L. Yaney and W.D. Nix, 'Mechanisms of Elevated-Temperature Deformation in the B2 Aluminides NiAl and CoAl,' *J. Mater. Sci.*, **23** (1988), pp. 3088–3098.

Chapter 8

Fine-structure superplastic metal composites and laminates

Many metal-matrix composites (MMCs) and laminates have been developed in recent years for advanced structures [1–5]. Both materials are attractive for many structural applications because they exhibit unusual combinations of structural, physical, and thermal properties including high modulus and strength, good wear resistance [6, 7], good dimensional stability and low thermal expansion [8], and low density. Many studies have shown that discontinuously reinforced MMCs can behave superplastically [9–17]. These composites are mainly aluminum-based, but some magnesium-based [18–23] and zinc-based [24–26] composites have also been studied.

8.1 Aluminum-based metal-matrix composites

Comparative superplasticity data from some of the representative Al-based composites are summarized in Table 8.1. All of these composites are reinforced by SiC, either in whisker or particulate form. Up to the present time, superplasticity has not only been observed in MMCs produced by powder metallurgy (PM) methods but also in MMCs produced by ingot metallurgy (IM) methods [27–29]. The composites listed in Table 8.1 were made by conventional PM techniques, except for the 15 vol % SiC_w–7475Al, which was manufactured using SiC_w layered between specially prepared foils of superplastic 7475Al alloy [12]. The IM composite, 10 vol % SiC_p–2024Al [28], was produced by stir-casting. For discussion purposes, the composites listed in Table 8.1 are hereafter abbreviated as reinforcement–matrix alloy, e.g., 20 vol % SiC_w–2124Al becomes SiC_w/2124.

As shown in Table 8.1, the reported strain rate sensitivities (m) vary significantly. An exceptionally large elongation of 1400% was found in one of the thermally cycled materials (SiC$_w$/6061). The matrix grain sizes for the SiC$_w$/7475, SiC$_p$/64, and SiC$_p$/2024 alloys were about 6 μm, whereas the grain sizes for the other composites were about 1 μm.

The results in Table 8.1 can be classified into two groups according to the test method. The first group, in which test samples are thermally cycled between two temperatures, is termed *thermal-cycling superplasticity*; this phenomenon has been mentioned in Chapter 3 and will be further discussed in detail in Chapter 11. The other group, which is the most commonly known, is based on isothermal tests.

Table 8.1   Superplastic properties of SiC reinforced Al composites

Material composition[a]	Material code	T (°C)	Maximum elongation (%)	m	Strain rate s^{-1}	Stress (MPa)
20 vol % SiC$_w$–2124Al [9]	SiC$_w$/2124	475–550	~300	~0.33	3.3×10^{-1}	~10
20 vol % SiC$_w$–2024Al [10]	SiC$_w$/2024	100↔450[b]	~300	1.0	5×10^{-4}	~15
20 vol % SiC$_w$–6061Al [11]	SiC$_w$/6061	100↔450[b]	~1400	1.0	1×10^{-5}	~7
15 vol % SiC$_w$–7475Al [12]	SiC$_w$/7475	~520	350	≥0.5	2×10^{-4}	~7
10 vol % SiC$_p$–PM 64Al[c] [12]	SiC$_p$/64	~516	~250	≥0.5	2×10^{-4}	~1.4
15 vol % SiC$_p$–2014Al[d] [30]	SiC$_p$/2014	480	395	0.4	4×10^{-4}	
15 vol % SiC$_p$–7475Al[d] [30]	SiC$_p$/7475	515	442	0.38	2×10^{-4}	
10 vol % SiC$_p$–2024Al [28]	SiC$_p$/2024	515	685	0.4	5×10^{-4}	~5

[a] w=whisker; p=particulate.

[b] Thermal cycling.

[c] Back pressure of 4.14 MPa applied.

[d] Back pressure of 5.25 MPa applied.

8.1.1   Thermal-cycling superplasticity

In Table 8.1, for the cases of SiC$_w$/2024 and SiC$_w$/6061, superplasticity was observed under nonisothermal test conditions. In these cases [10, 11], the concept of internal-stress superplasticity (an example of which is phase-transformation superplasticity [31, 32]) was used to demonstrate superplasticity under conditions of thermal cycling between 100 and 450 °C (at 100 s per cycle). Under the thermal-cycling conditions, internal stresses are generated from the thermal coefficient mismatch between the reinforcement (i.e., SiC whisker) and aluminum matrix. With a small externally applied stress, the composites behave like Newtonian viscous fluids ($m=1$). This results in superplastic behavior of about 300% in the SiC$_w$/2024 and an exceptionally large elongation of 1400% in the SiC$_w$/6061. In contrast, the composites exhibit a low m value (~0.1) with only 10% elongation under isothermal creep deformation conditions. For both the thermal-cycling and isothermal tests, the average strain rate is about 10^{-4} to 10^{-5} s^{-1}, i.e. a range of values similar to those used for conventional superplastic forming of aluminum alloys. A direct comparison of the results from SiC$_w$/2024 between the two different test conditions is shown in Figure 8.1. The mechanistic description of thermal-cycling superplasticity is given in Chapter 11.

8.1.2   Isothermal superplasticity

In the case of isothermal superplasticity, the results can be further categorized into two subgroups according to the strain rate at which superplasticity occurs.

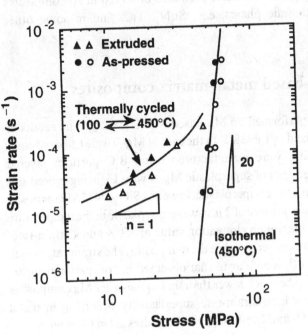

Figure 8.1 Comparison of the plastic properties of 20 vol % SiC$_w$–reinforced 2024Al deformed under isothermal and thermal-cycling conditions (data from Ref. [10]).

As shown in Table 8.1, the strain rates at which superplasticity occurs in all the composites except the $SiC_w/2124$ composite are approximately 10^{-4} to $10^{-5} s^{-1}$, which results in elongation values of about 300 to 400%. These strain rates are of the same orders of magnitude as for conventional superplasticity. The m values for these composites are about 0.4.

As mentioned previously, $SiC_w/7475$ was manufactured using SiC_w layered between specially prepared foils of superplastic 7475Al alloy. As a result, the superplastic behavior of the $SiC_w/7475$ composite is similar to that of the monolithic 7475Al alloy. The three PM SiC particulate-reinforced composites are noted to show superplasticity only under conditions in which a back pressure was applied. In the absence of back pressure, the composites behave like conventional alloys. The applied back pressure is expected to reduce cavitation and thus delay fracture in the materials. The fact that the materials do not behave superplastically without back pressure indicates severe cavitation in these materials at strain rates of approximately $10^{-4} s^{-1}$. This is consistent with Figure 8.1, i.e., under isothermal conditions, it is difficult to make Al-based composites superplastic at the strain rates usually observed in conventional superplastic Al alloys. In contrast, the IM composite $SiC_p/2024Al$ is superplastic without the application of back pressure. This may be attributed to a relatively homogeneous microstructure in the stir-cast composite. To obtain superplasticity in PM composites without applying back pressure, it is necessary to carry out experiments at an extraordinarily high strain rate (greater than $10^{-2} s^{-1}$). This high-strain-rate phenomenon, denoted as high-strain-rate superplasticity (HSRS), will be treated separately in the next chapter. HSRS has been also observed in Al composites reinforced with other ceramic phases, e.g., Si_3N_4, TiC, and in some other advanced Al alloys.

8.2   Magnesium-based metal-matrix composites

Several studies has been performed on Mg-based MMC [18–23]. The results of these studies are summarized in Table 8.2. In the case of Mg–9 wt% Li-based composite [20], the composites were manufactured using B_4C particles layered between specially prepared foils of superplastic Mg–9 wt% Li, using procedures similar to the preparation of the superplastic 15 vol % SiC_w–7475Al composite [12]. The composite, with a grain size of 2 μm, was superplastic in the temperature range of 150 to 200 °C; a maximum elongation value of 350% and a strain-rate-sensitivity exponent of 0.5 were recorded in the composite. The strain rate, which is approximately $3\times10^{-4} s^{-1}$, is similar to that observed in conventional superplastic alloys but is noted to be much lower than that in the other Mg composites ($\sim10^{-1} s^{-1}$). The relatively high strain-rate superlasticity occurring in metal matrix composites (both Al- and Mg-based) will be addressed in Chapter 9.

8.3   Zinc-based metal–matrix composite

A 30 vol % Al_2O_3–reinforced Zn composite has been illustrated to behave super-plastically under thermal-cycling conditions ($100\leftrightarrow300$ °C and $150\leftrightarrow250$ °C) [24]. Similar to SiC–Al composites, this composite behaves like conventional oxide-dispersion strengthened materials under isothermal conditions, having low elongation, deforming by dislocation slip (low strain-rate sensitivity), and exhibiting an apparent threshold stress. Under thermal-cycling conditions, however, the composite behaves superplastically (with $m\approx1$) and can be deformed at a stress level that is less than the apparent threshold stress. The latter observation suggests that the threshold stress may not be associated with the dis-location creep mechanism but rather with a grain-boundary sliding mechanism. A mathematical description of the plastic flow behavior in an MMC under thermal-cycling conditions is presented in Chapter 11.

Zinc–22%Al alloy-based MMCs have also been explored. Eutectoid Zn–22%Al alloy was the first commercial superplastic alloy developed [33]. The alloy can be readily formed superplastically with injection molds used in pro-ducing plastics. To enhance further its high-temperature resistance, the alloy has been reinforced with SiC whiskers using squeeze casting techniques [25]. Efforts have also been made by various research groups [25, 26] to make the composite superplastic but with only limited success.

8.4   Metal laminates

In the area of superplastic laminates, studies have been carried out mainly in ultrahigh carbon (UHC) steel-based laminates. In fact, several UHC steel lami-nates consisting of superplastic and nonsuperplastic components have been

Table 8.2 Properties of superplastic Mg-based composites

Material composition	T (°C)	Maximum elongation (%)	m	Strain rate (s^{-1})	Stress (MPa)
5 wt % B_4C–Mg–9 wt%Li [20]	200	~350	0.5	3×10^{-4}	~3
17 vol % SiC_p–ZK60 [22]	450	~320	~0.5	1.3×10^{-1}	~7
20 vol % TiC–Mg–5%Zn [21]	475	300	0.3	7×10^{-2}	~10
20 vol% Mg_2Si-Mg [23]	~500	~350	0.3	1×10^{-1}	~14

shown to exhibit superplasticity [34, 35]. For example, UHC steel–mild steel laminated composites exhibit strain-rate-sensitivity exponents of over 0.30 and
elongations to fracture of over 400% [34]. The strain rate–stress results show
good agreement with constitutive equations for creep that have been developed
based on an isostrain creep deformation model.

In an isostrain model, the strain-rate-sensitivity exponent of a two-compo-
nent laminated composite may be related to the properties of its components by
[35]

$$m_c = (m_{sp} f_{sp} \sigma_{sp} + m_{nsp} f_{nsp} \sigma_{nsp})/\sigma_c, \qquad (8.1)$$

where subscripts c, sp, and nsp are the composite, superplastic component, and
nonsuperplastic component, respectively. The equation predicts that a laminated
composite containing a nonsuperplastic material may exhibit nearly ideal fine-
structure superplasticity (i.e., $m_c = 0.5$) provided the following conditions are met:
(1) the load necessary to deform the superplastic component is significantly
greater than that for the nonsuperplastic component at the appropriate temperature and strain rate, (2) the nonsuperplastic material is inherently ductile (i.e., the
reduction of area at failure is high), (3) interdiffusion at the interface does not
produce any harmful products that may impair superplasticity, and (4) cohesion
of the laminate layers is sound. In essence, during deformation, the nonsuperplastic layers are constrained by interfacial bonding to deform in conformity with
the superplastic layers. Under such constraints, the development of necks in the
nonsuperplastic layers is delayed, and the nonsuperplastic layers can be stretched
plastically to an appreciable degree of strain without fracturing.

The above model provides useful material guidelines for achieving ideal
superplasticity ($m = 0.5$) in laminated composites. From these guidelines, a laminated composite consisting of a ferritic stainless-steel and UHC steel (containing 3 wt% Si) was predicted and, subsequently, demonstrated to exhibit
superplasticity ($m = 0.5$ and elongation to failure of 850%) at 815 °C; in contrast,
the ferritic stainless steel exhibited only 250% elongation when tested under the
same conditions. This combination of components leads to the unexpected result
that coarse-grained stainless steels can be made superplastic. This superplastic
capability has been further demonstrated in gas-pressure, blow-forming experiments on cones of UHC–stainless-steel laminates. For example, Daehn [36] has
performed gas-pressure, blow-forming experiments with stainless steel and
stainless-steel-clad–UHC steel to assess die-filling capabilities. The experimental
arrangement and results are shown in the top section of Figure 8.2.

Figure 8.2 (lower left) shows a stainless-steel plate formed by gas pressure at
775 °C. The sample ruptured after 5 min without properly filling the die; the
radius of curvature at the apex is 16 mm. Figure 8.2 (lower right) illustrates a
stainless-steel–UHC steel laminate part formed by the same process and under

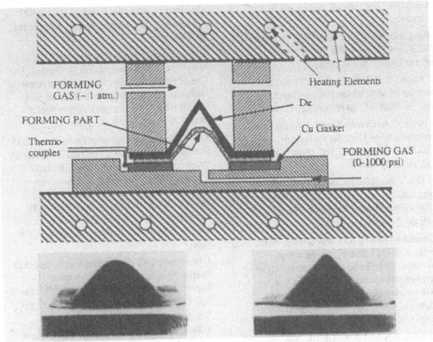

Figure 8.2 Experimental arrangement for gas-pressure, blow-forming on cones of UHC–stainless laminates to assess die-filling capabilities (data from Ref. [35]).

the same test conditions as the stainless steel. In this case, the sample did not rupture and filled the die satisfactorily. An advantage in the manufacture of such stainless-steel–UHC-steel-clad composites is that they will have extraordinarily high strengths at ambient temperature because of the presence of the strong and ductile matrix UHC steel. The above lamination technique has also been successfully used for manufacturing tough UHC steel–brass (70Cu–30Zn) laminates [37, 38]. In addition to UHC-steel-based laminates, some Mg–9 wt% Li-based laminates have been prepared by press-bonding techniques and, subsequently, tested for their superplastic properties [18, 19, 39].

References

1. K.K. Chawla, *Composite Materials*, Springer-Verlag New York Inc., 1987.

2. M. Taya and R.J. Arsenault, *Metal Matrix Composites – Thermomechanical Behavior*, Pergamon Press, New York, 1989.

3. T.W. Clyne and P.J. Withers, *Introduction of Metal Matrix Composites*, Cambridge University Press, 1993.

4. *Fundamental of Metal Matrix Composites*, ed. S. Suresh, A. Mortensen, and A. Needleman, Butterworth-Heiemann, Boston, 1993.

5. J.E. Allison and G.S. Cole, 'Metal-Matrix Composites in Automotive Industry : Opportunities and Challenges,' *JOM*, **45**(1) (1993), pp. 19–24.

6. M.A. Martinez, A. Martin, and J. Llorca, 'Wear of Al-Si Alloys and Al-Si/SiC Composites at Ambient and Elevated

Temperatures,' *Scr. Metall. Mater.*, **28** (1993), pp. 207–212.

7. A. Wang and H.J. Rack, 'Dry Wear in 2124 Al-SiC$_w$/17–4 PH Stainless Steel Systems,' *Wear*, **147** (1991), pp. 355–374.

8. R.U. Vaidya and K.K. Chawla, 'Thermal Expansion of Metal-Matrix Composites,' *Comp. Sci. Technol.*, **50** (1994), pp. 13–22.

9. T.G. Nieh, C.A. Henshall, and J. Wadsworth, 'Superplasticity at High Strain Rate in SiC–2124 Al Composite,' *Scr. Metall.*, **18** (1984), pp. 1405–1408.

10. M.Y. Wu, J. Wadsworth, and O.D. Sherby, 'Superplasticity in a Silicon Carbide Whisker Reinforced Aluminum Alloy,' *Scr. Metall.*, **18** (1984), pp. 773–776.

11. G. Gonzales-Doncel, S.D. Karmarkar, A.P. Divecha, and O.D. Sherby, 'Influence of Anisotropic Distribution of Whiskers on the Superplastic Behavior of Aluminum in a Back-Extruded 6061 Al–20% SiC$_w$ Composite,' *Comp. Sci. Technol.*, **35** (1989), pp. 105–120.

12. M.W. Mahoney and A.K. Ghosh, *Silicon Carbide Reinforced Aluminum Structure*, Air Force Contract, AFWAL-TR–82–3051, Rockwell International, 1982.

13. M.W. Mahoney and A.K. Ghosh, 'Superplasticity in a High Strength Powder Aluminum Alloy with and without SiC Reinforcement,' *Metall. Trans.*, **18A** (1987), pp. 653–661.

14. T. Imai, M. Mabuchi, Y. Tozawa, Y. Murase, and J. Kusui, 'Effect of Dislocation and Recovery on Si$_3$N$_4$ Whisker Reinforced Aluminum P/M Composite,' in *Metal & Ceramic Matrix Composites: Processing, Modeling & Mechanical Behavior*, pp. 235–242, ed. R.B. Bhagat, A.H. Clauer, P. Kumar, and A.M. Ritter, TMS-AIME, Warrendale, PA, 1990.

15. T. Imai, M. Mabuchi, Y. Tozawa, and M. Yamada, 'Superplasticity in β-Silicon Nitride Whisker-Reinforced 2124 Al Composite,' *J. Mater. Sci. Lett.*, **9** (1990), pp. 255–257.

16. M. Mabuchi and T. Imai, 'Superplasticity of Si$_3$N$_4$ Whisker Reinforced 6061 Al Composite at High Strain Rate,' *J. Mater. Sci. Lett*, **9** (1990), pp. 763–765.

17. K. Higashi, T. Okada, T. Mukai, S. Tanimura, T.G. Nieh, and J. Wadsworth, 'Superplastic Behavior in a Mechanically-Alloyed Aluminum Composite Reinforced with SiC Particulates,' *Scr. Metall. Mater.*, **26**(2) (1992), pp. 185–190.

18. G. Gonzalez-Doncel, J. Wolfenstine, P. Metenier, O.R. Ruano, and O.D. Sherby, 'The Use of Foil Metallurgy Processing to Achieve Ultrafine Grained Mg–9Li Laminates and Mg–9Li–5B$_4$C Particulate Composites,' *J. Mater. Sci.*, **25** (1990), pp. 4535–4540.

19. J. Wolfenstine, G. Gonzalez-Doncel, and O.D. Sherby, 'Processing and Elevated Temperature Properties of Mg-Li Laminates,' in *Metal & Ceramic Matrix Composites: Processing, Modeling & Mechanical Behavior*, pp. 263–270, ed. R.B. Bhagat, A.H. Clauer, P. Kumar, and A.M. Ritter, The Minerals, Metals & Materials Society, Warrendale, PA 1990.

20. J. Wolfenstine, G. Gonzalez-Doncel, and O.D. Sherby, 'Tension versus Compression Superplastic Behavior of a Mg–9wt%Li–5wt% B$_4$C Composite,' *Mater. Lett.*, **15** (1992), pp. 305–308.

21. S.-W. Lim, T. Imai, Y. Nishida, and I. Chou, 'High Strain Rate Superplasticity of a TiC Particulate Reinforced Magnesium Alloy Composite by Vortex Method,' *Scr. Metall. Mater.*, **32** (1995), pp. 1713–1718.

22. T.G. Nieh and J. Wadsworth, 'Superplasticity in a Powder Metallurgy Magnesium Composite,' *Scr. Metall. Mater.*, **32** (1995), pp. 1133–1138.

23. M. Mabuchi, K. Kubota, and K. Higashi, 'High Strength and High Strain Rate Superplasticity in a Mg-Mg$_2$Si Composite,' *Scr. Metall. Mater*, **33** (1995), pp. 331–335.

24. M.Y. Wu, J. Wadsworth, and O.D. Sherby, 'Elimination of the Threshold Stress for Creep by Thermal Cycling in Oxide-Dispersion-Strengthened Materials,' *Scr. Metall.*, **21** (1987), pp. 1159–1164.

25. H. Ohsawa and M. Yoshizawa, 'Processing of Zn–22Al/Al$_2$O$_3$ Composite Sintered Sheet and Mechanical Characterization,' in *International Conference on Superplasticity in Advanced Materials (ICSAM–91)*, pp. 355–360, ed. S. Hori, M. Tokizane, and N. Furushiro, The Japan Society for Research on Superplasticity, Osaka, Japan, 1991.

26. J.S. Kim, J. Kaneko, and M. Sugamata, 'Elevated Temperature Deformation of SiC Whisker/Zn–22%Al Alloy Composites,' in *International Conference on Superplasticity in Advanced Materials (ICSAM–91)*, pp.

391–398, ed. S. Hori, M. Tokizane, and N. Furushiro, The Japan Society for Research on Superplasticity, Osaka, Japan, 1991.

27. X. Huang, Q. Liu, C. Yao, and M. Yao, 'Superplasticity in a SiC_w–6061 Al Composite,' *J. Mater. Sci. Lett.*, **10** (1991), pp. 964–966.

28. Z. Wei, B. Zhang, and Y. Wang, 'Microstructure and Superplasticity in a Stir-Cast SiC_p/2024 Aluminum Composite,' *Scr. Metall. Mater.*, **30** (1994), pp. 1367–1372.

29. T. Hikosaka, T. Imai, T.G. Nieh, and J. Wadsworth, 'High Strain Rate Superplasticity of a SiC Particulate Reinforced Aluminum Alloy Composite by a Vortex Method,' *Scr. Metall. Mater.*, **31** (1994), pp. 1181–1186.

30. J. Pilling, 'Superplasticity in Aluminum Base Metal Matrix Composites,' *Scr. Metall.*, **23** (1989), pp. 1375–1380.

31. R.H. Johnson, 'Superplasticity,' *Metall. Rev.*, **15** (1970), pp. 115–134.

32. J.P. Poirier, 'On Transformation Plasticity,' *J. Geophys. Res.*, **87** (1982), pp. 6791–6797.

33. A.A. Bochvar and Z.A. Sviderskaya, 'Superplasticity in Zinc-Aluminum Alloys,' *Izv. Akad. Nauk SSSR, Otdel. Tekh. Nauk*, **9** (1945), pp. 821–827.

34. B.C. Snyder, J. Wadsworth, and O.D. Sherby, 'Superplastic Behavior in Ferrous Laminated Composites,' *Acta Metall*, **32** (1984), pp. 919–923.

35. G. Daehn, D.W. Kum, and O.D. Sherby, 'Superplasticity of a Stainless Steel Clad Ultrahigh Carbon Steel,' *Metall. Trans.*, **17A** (1986), pp. 2295–2298.

36. G. Daehn, Ph.D. Thesis, Dept. of Materials Science and Engineering, Stanford University, (1987).

37. C.K. Syn, D.R. Lesuer, K.L. Cadwell, O.D. Sherby, and K.R. Brown, 'Microstructures and Deformation Properties from Room Temperature to 1400 °C of an Al_2O_3-Ni_3Al Composite,' in *Developments in Ceramic and Metal-Matrix Composites*, pp. 85–96, ed. K. Upadhya, The Minerals, Metals & Materials Society, Warrendale, PA, 1992.

38. C.K. Syn, D.R. Lesuer, J. Wolfenstine, and O.D. Sherby, 'Layer Thickness Effect on Ductile Tensile Fracture of Ultrahigh Carbon Steel-Brass Laminates,' *Metall. Trans.*, **24A** (1993), pp. 1647–1653.

39. E.M. Taleff, O.A. Ruano, J. Wolfenstine, and O.D. Sherby, 'Superplastic Behavior of a Fine-Grained Mg–9Li Material at Low Homologous Temperature,' *J. Mater. Res.*, **7**(8) (1992), pp. 2131–2135.

Chapter 9

High-strain-rate superplasticity

It is often thought that superplasticity is only found at relatively low strain rates, typically about 10^{-4} to 10^{-3} s^{-1}. Several recent studies have indicated, however, that superplasticity can exist at strain rates considerably higher than 10^{-2} s^{-1}. This high-strain-rate superplasticity (HSRS) phenomenon has now been observed in metal-matrix composites [1–4], mechanically alloyed materials [5, 6], and even the more conventionally produced metallic alloys [7–9]. We will discuss the phenomenon in detail in the following.

9.1   Experimental observations

9.1.1   Metal-matrix composites

The phenomenon of HSRS was initially observed in Al-based metal-matrix composites and has continued to be studied mainly in Al-based alloys. Composite reinforcements include SiC and Si_3N_4 whiskers and SiC particles; matrix alloys include 2000, 6000, and 7000 series Al. A list of published HSRS results is presented in Table 9.1. Despite the differences in the type of reinforcement and matrix composition, all of these composites are noted to exhibit approximately similar deformation and microstructural characteristics. In the following, we use a powder-metallurgy 20%SiC whisker-reinforced 2124Al composite (SiC$_w$/2124Al) as an example to reveal the key experimental observations of HSRS. This composite was the first material observed to exhibit HSRS [1].

9.1.1.1 **SiC whisker-reinforced 2124Al composite**

To the present time, reports on HSRS are found in aluminum composites mainly produced by powder-metallurgy methods. High-temperature deformation investigations of the $SiC_w/2124Al$ indicated that the material was not superplastic in as-extruded conditions; over the conventional strain-rate range of 1.7×10^{-3} to

Table 9.1 HSRS aluminum alloys and composites

Material/reference[a]	Test temperature (°C)	Solidus (°C)	Strain rate (s⁻¹)	Stress (MPa)	Elongation (%)
Composites					
β SiC_w/2124Al [1]	525	502	0.3	~10	~300
β $Si_3N_{4(w)}$/2124Al [3]	525	502	0.2	~10	~250
α $Si_3N_{4(w)}$/7064Al [10]	525	~525	0.2	~15	~250
β $Si_3N_{4(w)}$/6061Al [4]	545	582	0.5	~20	~450
β SiC_w/6061Al [11]	550	582	0.2	6.5	300
27 wt% β SiC_w/7075Al [12, 13]	~500	<538	0.2	40	—
AlN/6061Al [14]	600	582	0.5	10	350
TiC/2014Al [14]	545	507	0.2	15	250
Mechanically alloyed Al					
IN9021 [5]	475	495	1	30	~300
IN9021 [15]	550	495	50	22	~1250
IN90211 [6]	475	495	2	30	~500
MA $SiC_{(p)}$/IN9021 [16]	550	495	10	~7	~500
IN9052 [17]	590	~580	10	18	330
IN905XL [18]	550	—	20	5	200
Alloys					
7475Al–0.9 wt% Zr [8]	520	<538	0.3	13	~900
7475Al–0.7 wt% Zr [8]	520	<538	0.05	5	~900
2124Al–0.6 wt% Zr [9, 19]	475	502	0.3	34.5	~500
Al–3.2 wt% Li–1 wt% Mg–0.3 wt% Cu–0.18 wt% Zr [20]	570	562	0.1	6	250

[a] w=whisker; p=particulate.

3.3×10^{-1} s^{-1}, elongation-to-failure values of 30 to 40% were recorded. After proper thermomechanically processing the composite became superplastic; the upper part of Figure 9.1 shows the elongation-to-failure value at 525 °C as a function of the initial strain rate. The strain-rate range examined covered about four orders of magnitude. At the slowest strain rate in this range, 9.3×10^{-5} s^{-1}, a non-superplastic value of elongation-to-failure of 40% was measured. This strain rate is typical of that for optimum superplasticity in many aluminum alloys. As the strain rate increased, the tensile ductility increased for the case of the SiC$_w$/Al composites. At a strain rate of 1.3×10^{-2} s^{-1}, values of about 100% were observed both in the longitudinal and transverse directions. At even higher strain rates, the elongation-to-failure values continued to increase until a maximum value of 300% was observed at a very high strain rate of 3.3×10^{-1} s^{-1} (2000% min^{-1}). Beyond this, a decrease in elongation occurred. It is pointed out that at these high strain rates, no significant difference was observed between samples from the longitudinal and transverse directions of the processed material.

The lower part of Figure 9.1 shows the corresponding flow stress (at a fixed strain of 0.3) as a function of strain rate. As may be observed, the maximum in elongation corresponds to an increase in the strain-rate-sensitivity value (m) from 0.1 for the low-strain-rate region to 0.33 for the high-strain-rate region. This corresponding increase in m with higher elongations-to-failure is consistent with the observations in conventional superplastic metals.

The data in Figure 9.1 are noted to have been measured at a temperature of 525 °C, which is above the solidus temperature of the 2124Al matrix. (The solidus temperature for 2124Al is 502 °C [21].) Therefore, mechanical tests were carried out on the composite at temperatures of 475 and 500 °C (below the solidus temperature) and 550 °C (yet further above the solidus temperature); these results are depicted in Figure 9.2. In a manner similar to that found at 525 °C, the data in Figure 9.2 show that at high strain rates, m increased to about 0.3 at all temperatures, and corresponding high values of elongation-to-failure were also recorded in these strain-rate regimes. It should be emphasized particularly that 550 °C is nearly 50 °C above the solidus temperature of the 2124Al matrix. The data in Figure 9.2, however, indicate that the flow characteristics do not exhibit a sharp discontinuity across the solidus temperature, implying that the presence of a partial liquid phase may not be the main contributing factor to the observed superplasticity. Rather, low-strain-rate sensitivities ($m \sim 0.1$), and therefore, low elongations to failure (<100%), were generally observed at low-strain-rate regions (<10^{-3} s^{-1}) at each temperature. Low-strain-rate sensitivity in the low-strain-rate regions has been suggested to be associated with the existence of a threshold stress [22–26].

The above transition in plastic behavior from low to high strain rates is also manifested by a change in the apparent activation energy for deformation. The apparent activation energies of SiC$_w$/2124Al at the two regions are computed to be 920 and 218 kJ/mol for the low- and high-strain-rate regions, respectively; this is shown in Figure 9.3. Pandey *et al.* [24] argued that, analogous to oxide dispersion strengthened alloys [27], the high apparent activation energy at the low-strain-rate region was a result of the existence of a threshold stress in metal-matrix composites. Although uncertainty remains regarding the origin of threshold stresses, a relatively high activation energy at low strain rates is noted to occur in all the HSRS materials. In fact, in the present case, even the activation energy of 218 kJ/mol measured at the high-strain-rate region, where superplasticity is observed, is notably higher than the value of 140 kJ/mol, the activation energy for self-diffusion of aluminum [28].

A general question regarding the data in Figure 9.2, specifically the data at 475 and 500 °C, is that of adiabatic heating. Although these test temperatures are below the solidus temperature for the 2124Al matrix, the question arises as to whether adiabatic heating during high-strain-rate deformation may have

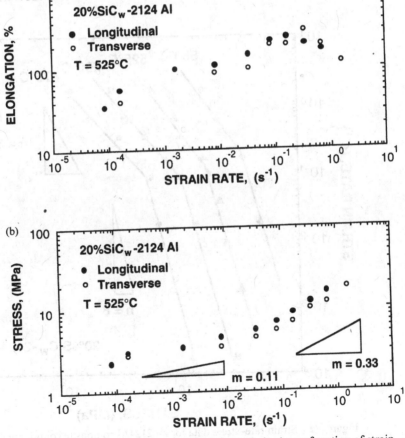

Figure 9.1 Elongation (above) and flow stress (below) as a function of strain rate for an Al 2124 composite containing 20 vol % SiC whisker at 525 °C.

caused partial melting of the alloy matrix. By assuming that all the strain energy (i.e., the area under the stress–strain curve) was converted into heat, which was then considered to result in a uniform temperature increase in the test sample, the temperature rise in the test sample was estimated to be approximately 15 °C. Macroscopic localized heating can result in a higher local temperature rise.

Only limited microstructural characterization has been performed on the superplastic SiC$_w$/2124Al composite. This is primarily because the micro-structures of metal-matrix composites are generally quite complex. A transmission electron micrograph (TEM) taken from the gage lengths of the SiC$_w$/2124Al sample, deformed at 525 °C at a strain rate of 0.16 s^{-1} to an elongation of about 100%, is shown in Figure 9.4. It can be seen in the figure that some clustering of the whiskers is apparent; also, a range of diameters from less than 1 μm to about 2 μm is found for the individual whiskers. Although not obvious in this partic-ular figure, some examinations have indicated the presence of a zone adjacent to the whiskers that is free of dislocations [21, 29].

Another important microstructural feature is that the fracture surface of a

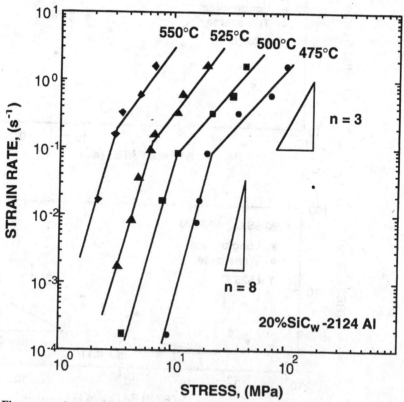

Figure 9.2 Strain rate–stress data for an 2124Al composite containing 20 vol % SiC whiskers at four different temperatures that bracket the solidus temperature for the matrix.

superplastically deformed SiC$_w$/2124Al specimen is quite different from those of specimens that fail in a nonsuperplastic fashion. A direct comparison between two such fracture surfaces is shown in Figure 9.5. The chemical bonding between the SiC whisker and Al matrix is generally recognized to be good [30]. As a result, the fracture surface of a conventional SiC$_w$/Al sample normally exhibits localized ductile fracture of the Al matrix with only limited fiber pull out. This is the major characteristic noted on the fracture surface of a nonsuperplastic specimen, as shown in Figure 9.5(a). The fracture surface of a superplastically deformed specimen, however, tested at the same temperature (Figure 9.5(b)) often exhibits extensive fiber pull out, suggesting substantial debonding during deformation. This fracture surface appearance may be interpreted as a result of extensive interfacial sliding between the whisker and the matrix (but may also be interpreted as a result of the high stress–high strain rate deformation condition).

One of the most peculiar observations is the fact that HSRS is not universally observed in SiC$_w$/Al composites. For example, elongations-to-failure for a matrix of 6061Al with also 20 vol % SiC$_w$ reinforcement were never observed to

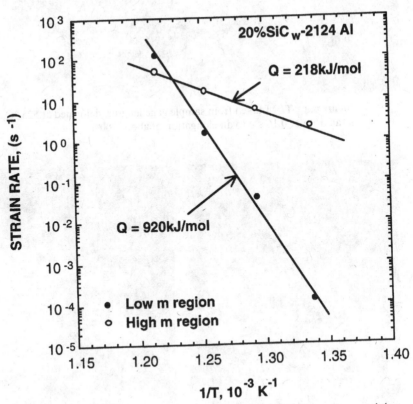

Figure 9.3 Apparent activation energies for the 2124Al composite containing 20 vol % SiC whisker at four different temperatures that bracket the solidus temperature for the matrix.

exceed 100%, even though the material had been extensively thermomechanically processed and tested under similar conditions to the SiC$_w$/2124Al composite [1]. On the other hand, Huang *et al.* [11] demonstrated that a SiC$_w$/6061Al prepared by a squeeze-cast technique can be superplastic (300%) at 550 °C with a strain rate of 1.7×10^{-1} s^{-1}; m is about 0.32. Apparently, both the matrix ele-

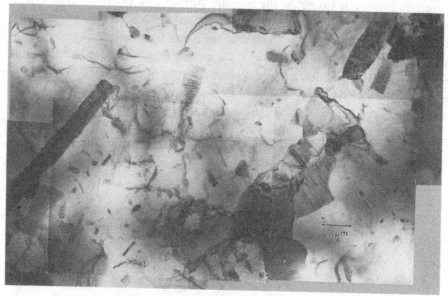

Figure 9.4 TEM taken from sample gage lengths, deformed at 525 °C at a strain rate of 0.16 s^{-1} to an elongation of about 100%.

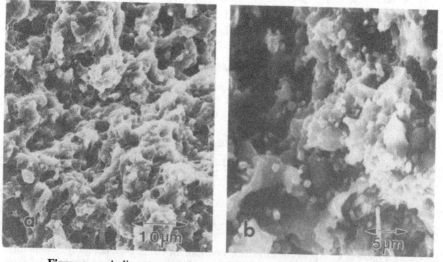

Figure 9.5 A direct comparison of the fracture surfaces between (a) nonsuperplastic and (b) superplastic samples.

ments and thermomechanical processing routes can play an important role in determining whether or not superplasticity occurs in these composites.

9.1.1.2 Si_3N_4 whisker-reinforced Al composites

In addition to SiC whisker-reinforced aluminum composites, HSRS has been further demonstrated in a number of Si_3N_4 (both in α and β forms) whisker-reinforced composites [2, 4, 10, 31–34]. These composites were also mainly prepared using powder-metallurgy methods. It is interesting to note that some of these composites are superplastic but some are not. Table 9.2 lists the $Si_3N_{4(w)}$-reinforced composite combinations that are superplastic and nonsuperplastic. Results from SiC_w-reinforced aluminum are included for comparison. For discussion purposes, the $Si_3N_{4(w)}$-reinforced composites listed in Table 9.2 are hereafter abbreviated as Si_3N_4/matrix alloy, e.g., 20 vol % β Si_3N_4/2124Al becomes β Si_3N_4/2124.

From Table 9.2, it is apparent that a simple criterion for HSRS based on the individual types of whisker or alloy matrix is not possible. This viewpoint is supported by the observation that β Si_3N_4/6061 is superplastic but β SiC/6061 is not, which illustrates the importance of the reinforced whisker selection. However, the α Si_3N_4/7064 is superplastic, but α Si_3N_4/2124 fails to exhibit superplasticity, illustrating the importance of the alloy matrix selection.

For those composites that do exhibit HSRS, superplasticity is only observed at strain rates of about 0.1 to 0.2 s^{-1}. The elongation of the β Si_3N_4/2124 composite, plotted as a function of strain rate, is given in Figure 9.6. The strong resemblance between Figure 9.6 and the upper part of Figure 9.1, which was obtained from a β SiC/2124 composite, is apparent. (The morphology of the β Si_3N_4 whisker is similar to that of β SiC_w.)

To illustrate further the similarity of deformation behavior, a comparison of the strain rate as a function of flow stress for β Si_3N_4/2124, β SiC/2124, and β SiC/6061 is depicted in Figure 9.6. It is evident that the data from the two 2124Al metal-matrix composites not only agree qualitatively (namely, both exhibit a

Table 9.2 Summary of HSRS whisker-reinforced aluminum composites

Matrix	Whisker	Superplastic?
2124Al	β Si_3N_4	Yes
6061Al	β Si_3N_4	Yes
7064Al	α Si_3N_4	Yes
2124Al	α Si_3N_4	No
2124Al	β SiC	Yes
6061Al	β SiC	No

transition in flow behavior at a strain rate of about 10^{-1} s^{-1}), but also the absolute values of stress and strain rate are consistent. The similarity of the two sets of data suggests that HSRS in these two composites is dominated by the behavior of the alloy matrix 2124Al and the reinforcement–matrix interfaces. Because the unreinforced 2124Al, 7064Al, and 6061Al matrix alloys were not superplastic at high strain rates, the above results indicate that the alloy matrix itself is not solely responsible for determining whether or not a composite is superplastic at high strain rates. The reinforcement–matrix interfaces play perhaps a more important role.

As is the case for SiC/Al composites, only a limited amount of microstructural information is available for the Si_3N_4/Al composite systems. High-strain-rate deformation of the Si_3N_4/Al composites results in a tangled dislocation network, whereas low-strain-rate deformation results in a low dislocation density [29]. The mechanistic implications of this observation are not yet obvious. Notably, the fracture surface of superplastic β Si_3N_4/2124Al is remarkably similar to the fracture surface of the superplastic β SiC/2124Al, suggesting that a similar deformation mode operates in both materials.

9.1.2 Mechanically alloyed alloys

Several commercial mechanically alloyed (MA) materials have been observed to exhibit HSRS, including the Al-based alloys IN9021 [5], IN90211 (a modification of IN9021) [6], IN9052 [15], IN905XL [18], and MA SiC_p/IN9021 [16] and MA SiC_p/7075Al [13] composites. In addition, two Ni-based alloys MA 754 and MA 6000 [35] have been found to be superplastic. The MA materials usually contain oxide and carbide dispersions approximately 30 nm in diameter that have an interparticle spacing of about 60 nm. The grain size, chemical compositions, and volume fraction of dispersoids in IN9021, IN9052, and IN905XL are summarized in Table 9.3. Figure 9.7 is a TEM taken from mechanically alloyed IN9021, showing the fineness and complexity of the microstructure.

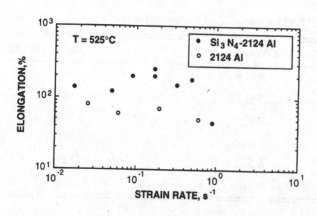

Figure 9.6 Elongations of 2124Al and β Si_3N_4/2124Al composite, plotted as a function of strain rate.

9.1.2.1 Aluminum-based alloys

IN9021 was the first MA Al-based alloy demonstrated to be superplastic at an extremely high strain rate [5]. The elongation-to-failure for IN–9021, as a function of the initial strain rate is shown in Figure 9.8 for the temperature range of 425 to 550 °C. The alloy shows a ductility typical of most pure metals and alloys (30–40%) at the strain rates at which most aluminum alloys exhibit superplasticity, i.e., 10^{-4} to 10^{-3} s^{-1}. Nonetheless, similar to metal-matrix composites, the elongation-to-failure values for IN9021 increase continuously with strain rate. At extremely fast strain rates, 50 s^{-1}, a maximum value of elongation of 1250% at 550 °C was recorded. This strain rate is noted to be much higher than that for metal-matrix composites ($\sim 10^{-1}$ to 10^{0} s^{-1}), apparently resulting from the finer grain size in the MA materials.

The plastic behavior of IN9021 is plotted in Figure 9.9. The general characteristics of the curves in both Figures 9.8 and 9.9 are noted to be similar to those for the SiC$_w$/2124Al composite shown in Figure 9.1. At high strain rates, which correspond to high elongation-to-failure values, a relatively high strain-rate-sensitivity value ($m=0.3$–0.5) was observed. It is again noted that the optimum superplastic temperature of 550 °C is above the solidus temperature of IN9021, which is 495 °C.

The observation of HSRS on IN9021 has been extended to include IN90211 [6]. In the case of IN90211, it was argued that the deformation mechanism is dislocation glide associated with a temperature-dependent threshold stress [36]. The origins of the drag force for dislocation glide and threshold stress are, however, unclear. Based on the observation of scratch lines on the surface of deformed samples, cooperative grain-boundary sliding has been recently proposed as the possible deformation mechanism [37]. While these models can qualitatively explain some aspects of the experiments, neither model offers an explanation for the grain-size dependence of the strain rate. Specifically, why does superplasticity in MA alloys occur at such high strain rates?

In addition to IN9021 and IN90211 alloys, HSRS has been observed in IN9052 [15], IN905XL [18], and a MA SiC$_p$/IN9021 [16] composite. In the case

Table 9.3 Grain size, chemical compositions, and volume fraction of dispersoids in various MA aluminum alloys

Composition (wt%)	Grain size (μm)	Al$_4$C$_3$	Al$_2$O$_3$
IN9021 (Al–4.0%Cu–1.5%Mg–1.1%C–0.8%O)	0.5	4.1	1.2
IN9052 (Al–4.0%Mg–1.1%C–0.8%O)	0.5	4.1	1.2
IN905XL (Al–4.0%Mg–1.5%Li–1.2%C–0.4%O)	0.4	4.5	1.2

of IN9052, the alloy is fine-grained but exhibited a relatively poor ductility; over the strain-rate range of 8.3×10^{-4} to 1.3 s^{-1} and the temperature range of 400 to 500 °C, the elongation-to-failure values for this IN9052 alloy were less than 70% [5]. By raising the test temperatures (above the solidus) and strain rates, Higashi *et al.* [17] have illustrated that the alloy exhibited an elongation value over 300%. In a previous study by Kim and Bidwell [38], values of extended ductility up to 130% were observed in an IN9051 alloy (Al–4.0%Mg–0.8%C–0.6%O) but at temperatures of about 300 °C and a low strain rate of 5×10^{-2} s^{-1}. For the other two mechanically alloyed materials, IN905XL and MA SiC$_p$/IN9021, the general characteristics of HSRS are fully manifested despite slight differences in the optimum elongation and strain-rate values. A summary of the superplastic properties of mechanically alloyed aluminum is listed in Table 9.4.

9.1.2.2 Nickel-based alloys

In the case of the nickel-based alloys MA 754 and MA 6000, which contain very fine grain sizes of 0.67 and 0.26 μm, respectively, HSRS was observed at strain rates of approximately 0.1 s^{-1} [35]. The maximum elongation values of 200 and 300% with corresponding *m* values of about 0.3 and 0.5 were recorded at 1000 °C ($\sim$0.8 T_m) for MA 754 and MA 6000, respectively. These data are consistent with the general trend – the finer the grain size, the greater the *m* value, and thus

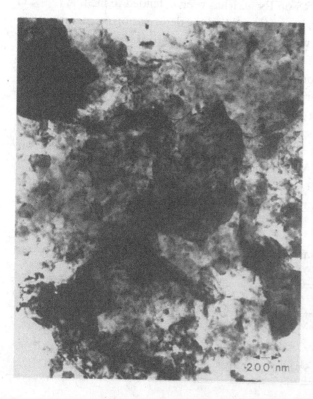

Figure 9.7
Microstructure of MA IN9021, showing the fineness of the microstructure.

elongation. Gregory *et al.* [35] have argued that a combination of slip with Coble creep, in which the Coble creep exhibits a threshold stress, can explain the above observation. Although the origin of the threshold stress was not given, the model predictions and experimental data for MA 6000 appear to be in good agreement; this is shown in Figure 9.10.

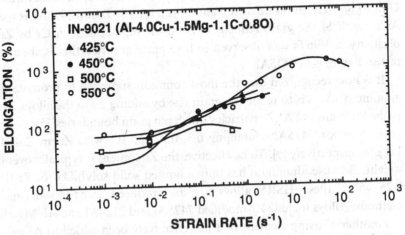

Figure 9.8 Elongation-to-failure values for IN9021 alloy, as a function of strain rate, for the temperature range from 425 to 550 °C.

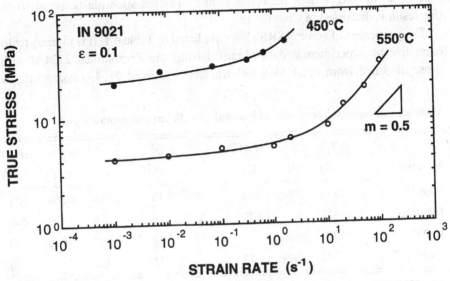

Figure 9.9 Stress–strain rate relationship in mechanically alloyed IN9021.

9.1.3   Metal alloys

9.1.3.1   Aluminum

In 1978, a high-strength Al alloy designated as V96Ts (composition by weight: Al–7.62%Zn–2.75%Mg–2.3%Cu–0.15%Zr), which is essentially a 7000 series Al alloy with Zr rather than Cr as a grain refiner, was developed in the former Soviet Union [39]. (The grain sizes are 5 and 12 μm for V96Ts and 7475Al, respectively.) As a result of the grain refinement from the replacement of Cr by Zr, superplasticity in V96Ts was observed to take place at a rate that is about 10 times faster than that in 7475Al.

It is now recognized that the most common approach to promote HSRS in aluminum alloys is to refine the grain size by adding Zr to the alloys. This leads to the formation of Al_3Zr particles which pin grain boundaries. For example, the grain sizes for 7475Al containing 0.3, 0.7, and 0.9 wt% Zr are 2.4, 1.9, and 1.3 μm, respectively [7]. To be effective, the Zr content is typically over 0.5% by weight. Because aluminum has only a limited solid solubility for Zr (less than 0.25 wt%), these HSRS alloys are all produced using powder-metallurgy methods. Alloys include Zr-modified 7475Al and 2124Al and Al–Mg alloys.

Another alloying element that has effectively been added to Al to promote HSRS is scandium [40]. In contrast to Zr, Sc has some solubility ($\sim$0.5% by weight) and forms coherent Al_3Sc precipitates that stabilize the substructure of the alloy. For example, several Al–Mg–0.5%Sc alloys have been shown to exhibit superplastic elongation in excess of 1000% at a strain rate of 0.01 s^{-1}. Deformation in Al–Mg–Sc and Al–Mg–Zr alloys is believed to follow Class I type behavior, rather than the classical fine-structure superplasticity ($m=0.5$); this result is discussed in Chapter 12.

The properties of some HSRS alloys are listed in Table 9.1. It is interesting to compare the experimental data obtained from the Zr-modified 2124Al with those obtained from the HSRS SiC whisker-reinforced 2124Al composite. A

Table 9.4   Superplastic properties of mechanically alloyed aluminum alloys

Material	T_i (K)	T_s (K)	Test T (K)	ε (s^{-1})	σ^* (MPa)	m	e_f (%)	d (nm)
IN9052	837	866	863	10	15	0.6	330	500
IN905XL	818	851	848	20	12	0.6	190	400
IN9021	754	860	823	50	18	0.5	1250	500
IN9021/SiC/15p	751	849	823	5	15	0.5	610	500

*True stress at $\epsilon=0.1$.

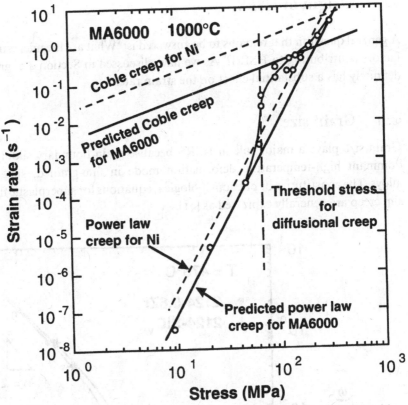

Figure 9.10 Comparison of the flow stress data for MA 6000 at 1000 °C with the predicted dislocation and Coble creep behavior of Ni (from Ref. [35]).

comparison of the plastic deformation behavior between these two materials is shown in Figure 9.11. In the low-strain-rate regime, the composite is noted to be more resistant to deformation than the matrix alloy. Also, in this regime, the n value of 5 for the alloy is lower than that for the composite, which is about 8. A stress exponent of 5, often observed during creep of metal alloys, is generally attributed to the dislocation climb mechanism [41]. The high stress exponent for the composite, however, may be associated with the existence of a threshold stress [24, 26].

Although the composite is stronger than the alloy in the low-strain-rate regime, both materials exhibit a similar strength in the high-strain-rate regime. Also, in this latter regime, the stress exponents of the two materials are noted to decrease and both approach a value of 2 (i.e., $m=0.5$). This suggests that a similar deformation mechanism, specifically grain-boundary sliding, dominates in both materials.

9.2   Origin of HSRS

A general question that remains to be answered is: 'What are the microstructural factors contributing to HSRS?' As we have discussed in Section 4.2, grain size definitely has a substantial effect on the strain rates.

9.2.1   Grain size

Grain size plays a major role in HSRS because grain-boundary sliding is the dominant high-temperature deformation mode in fine-grained, superplastic alloys. The constitutive or phenomenological equations for superplastic flow and slip creep are generally expressed as [41],

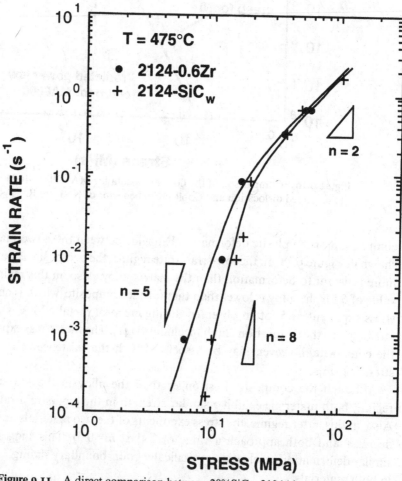

Figure 9.11   A direct comparison between 20%SiC$_w$–2124Al composite and 0.6%Zr–2124Al alloy. The transition in flow behavior occurs at a strain rate of about 10^{-1} s^{-1} for both materials.

$$\dot{\varepsilon} = Ad^{-p}D\left(\frac{\sigma}{E}\right)^n, \tag{9.1}$$

where $\dot{\varepsilon}$ is the strain rate, A is a material constant, d is the grain size, p is the grain size dependence (which is typically 2 to 3), D is the diffusivity, σ is the stress, E is the modulus, and n is the stress exponent. According to Eq. (9.1), refinement of the grain size by a factor of 2 would be expected to increase the optimum strain rate for superplastic flow by a factor of 4 to 8 depending on the precise grain-size relationship as described above.

An overview of the superplastic behavior of aluminum alloys to demonstrate the grain-size effect is shown in Figure 9.12. In this figure, the superplastic elongation-to-failure is shown as a function of strain rate for 7475Al; Al–Li 2090 alloy; commercial SUPRAL alloys; SiC_w/Al and $Si_3N_{4(w)}$/Al composites; Zr-modified

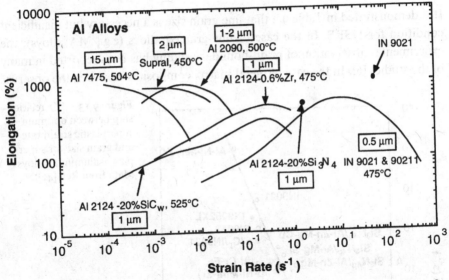

Figure 9.12 Overview of superplastic elongation as a function of strain rate for a wide range of Al-based alloys.

Al 2124; and MA IN9021. The grain-size ranges of these alloy groups are 10 to 20 μm for the 7475Al alloys, 2 to 3 μm for the 2090 Al and SUPRAL alloys, 1 μm or less for the SiC$_w$/Al composites and Zr-modified Al 2124, and 0.5 μm for IN9021. Figure 9.12 clearly illustrates a general trend; namely, there is an increased strain rate for optimal superplastic flow with a decrease in grain size.

To further illustrate the effect of grain size on the optimum superplastic strain rate, Figure 9.13 shows the quantitative relation between the logarithm of the optimum strain rate for superplastic flow and the grain size (plotted as the logarithm of reciprocal grain size) [42]. Note that an optimistic prediction is made for two-phase, aluminum-based materials with nanometer-size grains, where superplastic flow at 10^3 to 10^4 s^{-1} can be expected.

9.2.2 Interfaces

It is demonstrated in Table 9.2 that fine grain size is a necessary but insufficient condition for HSRS. In the case of fine-grained alloys (e.g., MA alloys), the experimental observation of grain-boundary sliding has been reported in many of the studies [6]. In the case of metal-matrix composites, however, the observa-

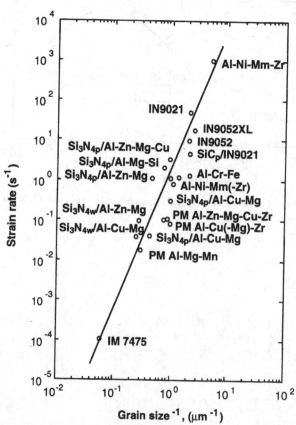

Figure 9.13 The relationship between optimum superplastic strain rate and grain size for superplastic aluminum alloys (data from Ref. [42]).

tion that fracture surfaces of the HSRS metal-matrix composites usually exhibit extensive fiber pull out indicates that reinforcement–matrix interface sliding may be just as important as grain-boundary sliding during superplastic deformation. In both cases, the nature of the boundary, whether it is an interface or a grain boundary, is apparently critical in determining the sliding properties. The question remains as to why these boundaries are highly mobile.

It must be emphasized in Table 9.1 that despite some subtle differences in the experimental procedures and results, all of the tests were performed at temperatures near to or above the matrix solidus temperatures of the materials. In other words, materials were usually tested in a semi-solid state. (Some materials may have been tested at a nominal temperature, which is slightly lower than the solidus. Because of the extremely fast strain rates, however, the effective temperature is expected to be significantly higher than the nominal test temperature as a result of adiabatic heating.)

Mechanistically, the deformation properties (or viscous properties) of a material can be altered dramatically with the presence of even a small amount of liquid phase. For example, Pharr and Ashby [43] demonstrated that the deformation characteristics of crystalline solids containing liquid phases, in which the liquid phases act as lubricants to promote grain and particle sliding, can be drastically changed. Such a result is expected to prevail, particularly in fine-grained materials, because grain rearrangement, which involves sliding and rotation, is easier between fine grains compared to coarse grains. The situation is similar to that occurring during the rheocasting (or semi-solid metal forming) of a metal [44, 45]. Both cases can be described in rheological terms.

From a rheological viewpoint then the shear viscosity η of a semi-solid is defined by

$$\eta = \frac{\tau}{\dot{\gamma}}, \tag{9.2}$$

where τ is the shear stress and $\dot{\gamma}$ is the shear strain rate.

A fluid containing a suspension of particles normally behaves like a non-Newtonian fluid. For a non-Newtonian fluid, η is not a constant but instead is itself a function of shear rate (or stress), i.e.,

$$\eta = fn(\dot{\gamma}). \tag{9.3}$$

(Strictly speaking, the rheological behavior of a non-Newtonian fluid is also dependent upon 'structural parameters,' e.g., the volume fraction, shapes, size distribution, and, sometimes, charges on the particles.) Equation (9.3) can usually be expressed by a power law

$$\eta = K \cdot \dot{\gamma}^{-p} \tag{9.4}$$

where K and p (the power law index) are both material constants. A fluid is classified as shear thinning (pseudoplastic) when $p>0$ and as shear thickening when $p<0$. (For a Newtonian fluid, $p=0$.) Combining Equations (9.2) and (9.4) yields

$$\tau=K\cdot\dot{\gamma}^{m}, \tag{9.5}$$

where $m=1-p$ is the strain-rate sensitivity of a material. This type of equation is commonly used to describe the high-temperature deformation of structural materials.

Experimentally, the steady-state viscosity of an Al–6.5 wt% Si alloy at a temperature at which the solid volume fraction was 0.4 has been measured [46]. In addition, Moon measured the viscosities of several SiC particulate-reinforced Al–6.5 wt% Si metal-matrix composites (with 10, 20, and 30 vol % SiC) as a function of shear rate at 700 °C. The results obtained from these composites are represented in Figure 9.14. A shear thinning effect occurred in both the alloy and the composites. Also, despite a slight difference in their apparent viscosity values, the p values for the alloy and the 30 vol % SiC-containing composite were computed to be 0.88 and 0.61, respectively. (The p values for the 10 and 20 vol % SiC-containing composites are 0.40 and 0.58, respectively.) These p values, in turn, give rise to m values of 0.6, 0.42, and 0.39 for the 10, 20, and 30 vol % SiC-containing composites, respectively, whereas the m value for the alloy itself is only 0.12. It is especially worth noting that the above results were all obtained at very

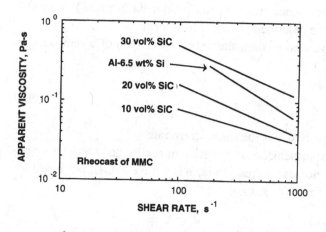

Figure 9.14 Viscosity of Al–6.5wt%Si alloy and its composites containing 10, 20, and 30 vol % SiC particulates (from Ref. [46]).

high shear strain rates $\dot{\gamma} \sim 200$–1000 s^{-1}). These shear strain rates are similar to those at which HSRS has been observed as listed in Table 9.1. The above m values for the composites are noted to be very close to those (~ 0.3–0.5) obtained from HSRS studies.

Thanh and Suery [47] studied the compressive properties of an Al 356 composite in a solid–liquid state. Their results are summarized in Figure 9.15. It is readily observed that the composite has a higher strain-rate-sensitivity value than its unreinforced matrix counterpart. In fact, the strain-rate-sensitivity value is close to 0.33, consistent with the above observation of Moon [46].

As mentioned previously, large tensile elongations occur in metals exhibiting HSRS values (typically >0.3); the high elongations are attributed to the material's resistance to local necking upon deformation because of the HSRS [48]. Assuming such a criterion also holds in the semi-solid state, then a composite in such a state would be expected to show large elongations under appropriate test conditions. Furthermore, in terms of rheological flow, the exact forms of the solid phase in the solid–liquid mixture are not of primary importance. If the presence and distribution of the solid phase yields an appropriate fluidity (or viscosity) in a solid–liquid mixture, i.e., it shows a p value that is approximately less than 0.7, then such a mixture would also be expected to exhibit a large elongation, provided a fracture mechanism such as gross decohesion does not intervene. This may be the reason that a very fine grain size is required for HSRS.

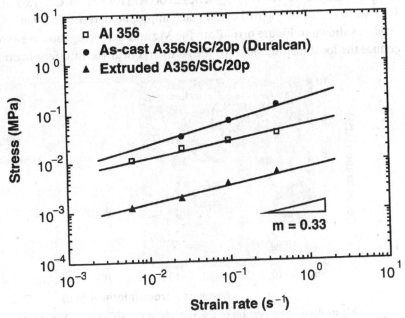

Figure 9.15 Comparison of the compressive properties of Al 356 alloy and its composites at 585 °C (solid–liquid) (from Ref. [47]).

For a given volume fraction of liquid phase, and for the case in which this volume fraction is small, a fine grain size can produce a higher capillarity force between crystalline grains, as the result of a reduced amount of liquid per grain-boundary area. This offers a possible explanation for the observations of HSRS in the fine-grained materials listed in Table 9.1.

The data in Figure 9.2, particularly the data for 475 °C, are below the solidus of the 2124Al matrix (505 °C). The deformation properties of 475 and 525 °C are, therefore, expected to be somewhat different. However, the data in Figure 9.2 clearly indicate that the flow characteristics do not exhibit any marked discontinuity across the solidus temperature. The discrepancy may be attributable to two possible factors: (1) adiabatic heating during high-rate deformation and (2) solute redistribution. As mentioned earlier, the temperature rise during the high-strain-rate testing is about 15 °C (if no localized heating takes place). Therefore, although the apparent test temperature is 475 °C, the actual temperature is certainly higher.

Another factor that can also contribute to the presence of a liquid phase is related to solute segregation at a reinforcement-matrix interface. Numerous reports have shown that the presence of ceramic reinforcements can influence solute distributions in an Al alloy [49]. For example, a special FEM/STEM (field emission microscopy/scanning transmission electron microscopy) technique has been used to study solute segregation to SiC/Al alloys interfaces [50] and showed that the Mg and Cu concentrations at the SiC/matrix interfaces in an underaged 15 vol % SiC$_p$–series 2000 Al (1.45 at. % Cu, 1.67 at. % Mg, 0.12 at. % Zr, 0.1 at. % Mn) could reach approximately 4.5 and 9 at. %, respectively, as shown in Figure 9.16. Both the Mg and Cu segregation are expected to reduce the local melting point of the Al matrix near the interface region.

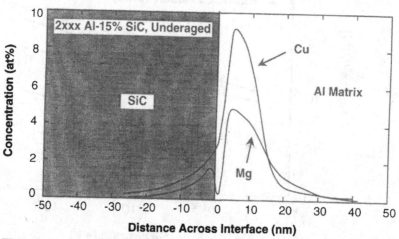

Figure 9.16 Segregation of Cu and Mg at the SiC–series 2000 Al interface. The melting point of the interfacial region is expected to be reduced as a result of such segregation (data from Ref. [50]).

Therefore, although the test temperature of 475 °C in Figure 9.2 (for $SiC_w/2124Al$) appears to be within the solid range of 2124Al because of the Mg segregation, nonetheless it may be sufficiently high to result in local melting at the SiC/Al interface. For example, a Mg concentration of over 6.5 at. % at the interface can by itself result in localized melting at 475 °C, even in the absence of adiabatic heating from high-strain-rate deformation. (The effects of Mg and Cu on the melting point of Al are estimated to be approximately –6 °C/at. % Mg and –6.5 °C/at. %Cu [51]). Therefore, the HSRS phenomenon may occur in the presence of some liquid, even though the temperature appears to be below the solidus in a SiC/Al or Si_3N_4/Al composite.

Solute segregation and local melting phenomena in various Si_3N_4/Al alloy composites was recently illustrated experimentally by Koike *et al.* [52–55]. For example, in addition to using differential scanning calorimetry (DSC) to characterize the incipient melting point, an *in situ* TEM technique was used to examine the microstructure (Figure 9.17) and chemical composition (Table 9.5) of both the Si_3N_4/Al-Mg (5052 Al) interface and Al alloy matrix grain boundaries at the optimum temperature for superplasticity. Local melting, as a result of solute segregation, was clearly demonstrated. Also, solute distribution at interface boundaries was apparently different than that at grain boundaries.

To illustrate the temperature effect, Higashi *et al.* [56] recently performed experiments with several MA materials, including IN9021, IN9052, and IN905XL. The results, as shown in Figure 9.18, clearly show that the m values for these alloys are dependent upon the testing temperature. It is readily observed in the figure that all of the m–T curves exhibit a sigmoidal shape. In the low-temperature region the m value is generally a constant and is relatively low (0.25–0.33; i.e. $3<n<4$). The m value increases rapidly in the intermediate temperature range and reaches another plateau at high temperatures. In this high-temperature region the m value is normally over 0.5 ($n<2$), a typical value for a highly superplastic material. For example, the m value for IN9052 is a constant and is less than 0.3 at temperatures below 500 °C. It increases quickly above 540 °C, however, and approaches another constant value of about 0.6 at $T>600$ °C. In the case of IN9021, m even reaches a value of as high as 0.8 in the high-temperature region. For the purposes of illustration, all curves in Figure 9.18 are marked by an arrow indicating the incipient melting points, T_i, for the material. It is evident that values of m increase drastically at $T>T_i$. A change in m from low- to high-temperature regions indicates a corresponding change in deformation mechanisms.

Higashi *et al.* [57] further demonstrated that testing temperature also strongly influences tensile elongation. For example, the total elongations-to-failure as a function of deformation temperature for the above superplastic MA materials are plotted in Figure 9.19. The elongation values are noted to be measured at

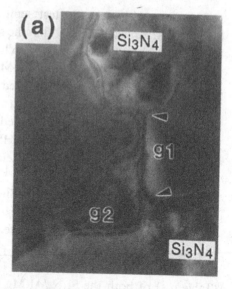

Figure 9.17 (a) TEM bright-field and (b) dark-field images, and (c) selected area diffraction pattern from a 20 vol% Si_3N_4/5052 Al composite at 548 °C (from Ref. [55]).

their individual optimal strain rates as listed in Table 9.4. Although there are different maximum values of elongation-to-failure for the different materials, it is readily seen that for a given material there is a maximum value of elongation-to-failure and that this maximum occurs at a characteristic temperature. Again, for the purposes of illustration all curves in Figure 9.19 are marked by two arrows. The left-hand-side arrow indicates T_i, and the right-hand-side arrow indicates the solidus temperature, T_s. Generally the elongation-to-failure values are noted to increase at temperatures greater than T_i. Once the elongation-to-failure value reaches a maximum, it decreases very rapidly with further increase in temperature. The maximum value of elongation-to-failure is noted to occur always at a temperature that is between T_i and T_s. This general trend also prevails in several HSRS Al composites [57], as shown in Figure 9.20.

At $T \geq T_i$, but at temperatures near T_i, only a small amount of liquid phase is present. The liquid phase would be expected to segregate to grain boundaries, and particularly at grain triple junctions. As shown in Figure 9.21, when the liquid phase is very thin, atoms in the solid state across two adjacent grains can still experience a traction force, T_{gb}, from each other. In other words, the grain boundary can sustain an applied tensile force. Also, shear stresses can be transferred across the boundary. When $T \gg T_i$, and T is close to T_s, macroscopic melting begins to occur, the liquid phase is thick, and atoms across two neighboring grains can no longer experience traction from each other. In other words, the grain boundary can no longer support an applied tensile force and shear stresses cannot be transferred across the boundary. Also, grain-boundary cavitation becomes extensive. Ductility in this latter case is therefore limited.

There apparently exists a critical amount of liquid phase for the optimization of grain/interface boundary sliding during superplastic deformation. The

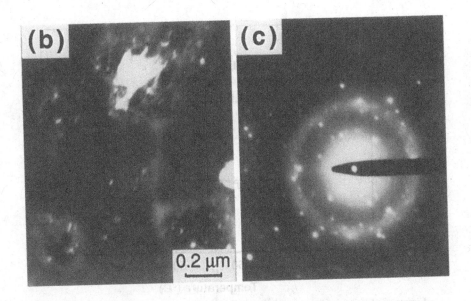

0.2 μm

Table 9.5 Concentration of major solute elements at a grain boundary, an interface, and the matrix in the Si₃N₄/5052Al composite

	Mg	Si	Cr	O	N
Grain boundary	—[a]	2.5	0.13	1.0	3.9
Interface	9.3	—[b]	0.10	3.2	—[b]
Grain interior	—[a]	—[a]	0.18	—[a]	—[a]

[a]Too low.
[b]Strong peak.

optimum amount of liquid phase may depend upon the precise material composition and the precise nature of a grain boundary or interface, such as local chemistry (which determines the chemical interactions between atoms in the liquid phase and atoms in its neighboring grains) and misorientation. The existence of an equilibrium thickness of intergranular liquid phase in ceramics has been discussed [58]. This area of detailed study in metal alloys has not been addressed.

The mechanistic description of HSRS is still unavailable, despite some limited efforts [59, 60]. It is unclear whether a model based on conventional superplasticity modified by the presence of a liquid phase can describe the phenomenon or it is necessary to invoke a completely new model.

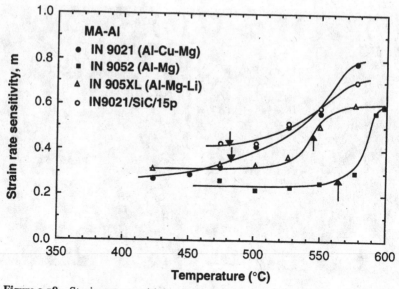

Figure 9.18   Strain-rate-sensitivity – temperature behavior of IN9052, IN9052, and IN905XL, showing the increase of m values at temperatures corresponding to the incipient melting point for each alloy (data from Ref. [57]).

9.3   Cavitation in HSRS materials

Only limited studies have been carried out on the fracture and cavitation of HSRS alloys [61–63]. For example, the cavitation behavior of the SiC_p/IN9021 composite and IN9021 alloy has been investigated, and the results are summarized in Figure 9.22, plotted as the amount of cavitation at different strain rates as a function of true strain. Data in Figure 9.22 can be best described by an exponential relationship:

$$C_v = C_{vo} exp(\eta \varepsilon),\tag{9.6}$$

where C_v is the volume fraction of cavities, ε is the plastic strain, C_{vo} is the volume of cavities at zero strain. The cavity growth-rate parameter, η, usually ranges from 2 to 4 for many superplastic aluminum alloys and is dependent on material, strain rate, temperature, and grain size. The values of η and C_{vo} and some of the mechanical and microstructural properties for SiC_p/IN9021 and IN9021 are summarized in Table 9.6. Also included are data from SiC_p/PM–64 composites and PM–64 alloy, both of which exhibit superplasticity, but at low strain rates [64]. The measured values of the initial cavities, C_{vo}, are higher for the composites than those for the unreinforced materials. Also, cavitation is more sensitive to plastic strain, indicated by a higher η value, for the HSRS SiC_p/IN9021 than for IN9021. These results are consistent with the fact that the composites are less superplastic than the alloys.

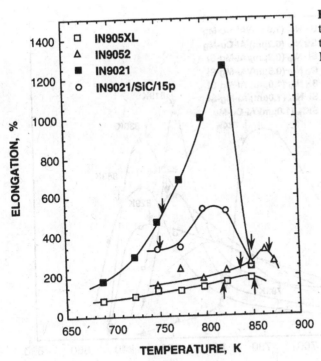

Figure 9.19 Elongation-to-failure as a function of testing temperature for MA materials.

Equation (9.6) suggests that cavity growth is plasticity controlled. In this case, as the materials are deformed plastically, voids also increase in volume as a result of increasing strain wherein [65, 66]

$$\frac{dr}{d\varepsilon}=\frac{\eta}{3}\left(r-\frac{3\tau}{2\sigma}\right)$$
(9.7)

where r is the cavity radius; ε the strain; σ the applied stress; τ the surface energy; and η the cavity growth rate parameter that is dependent upon both the applied stress state and the geometry of deformation.

The effect of tesing temperature on cavitation behavior in IN9021, i.e. the variation of the volume fraction of cavities with superplastic strain, was quite significant. Shown in Figure 9.23 is C_v as a function of superplastic deformation strain at 450 and 550 °C, and at a fixed true stress of 30 MPa. It is apparent that both the cavity nucleation and growth rates at 550 °C are significantly lower than those at 450 °C. This is consistent with the fact that IN9021 exhibited a maximum value of elongation-to-failure of 1250% at 550 °C but only 500% at 450 °C. It is, again, pointed out that 450 °C is lower (but 550 °C is higher) than the incipient melting point (481 °C) of IN9021. Therefore, the presence of a liquid phase, in particular at grain triple junctions, can certainly lower the local stresses and, thus, reduce the rate of cavity nucleation generated during grain-boundary sliding.

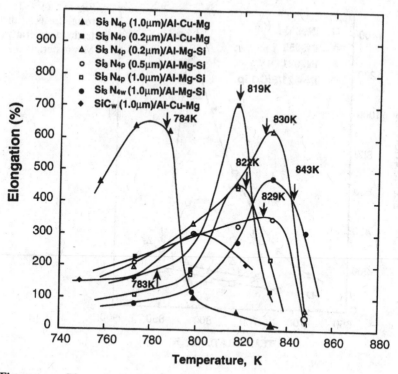

Figure 9.20 Elongation to failure as a function of testing temperature for Al-matrix composites. Incipient melting points for each alloy are indicated by arrows.

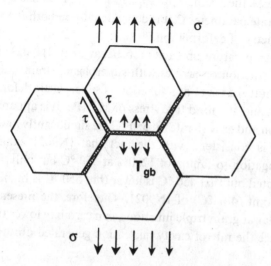

Figure 9.21 Schematic representation of grain structure in the presence of grain-boundary liquid phases.

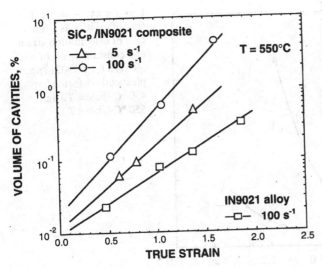

Figure 9.22 Variation of volume of cavities with superplastic true strain at the optimum superplastic temperature of 550 °C for the SiC$_p$/IN9021 composite and IN9021 alloy deformed at high strain rates.

Table 9.6 Values in η and Cvo under optimum superplastic conditions for superplastic aluminum composites and alloys

Materials	T (°C)	Strain rate (s⁻¹)	Grain size (μm)	η	C_{vo} (%)
15 vol % SiC$_p$/IN9021 [61]	550	5	0.5	2.7	0.03
IN9021 alloy [61]	550	100	0.5	1.9	0.01
10 vol % SiC$_p$/PM–64 [64]	500	2×10⁻⁴	8	3.2	0.07
PM–64 [64]	500	2×10⁻⁴	6	3.3	0.01

9.4   Perspective of HSRS and deformation map

Deformation mechanism maps are useful tools to assess the mechanisms that will control plastic flow depending on temperature, strain rate, and grain size. An example of such a map is given in Figure 9.24. The four plots show the strain rate as a function of homologous temperature for four different grain sizes (100, 10, 1, and 0.1 μm). The maps are based on constitutive equations that describe plastic flow controlled by diffusion-controlled dislocation creep (power law creep or Harper–Dorn creep), by grain-boundary sliding, and by diffusional creep (Nabarro–Herring creep). The constitutive equations used to develop these

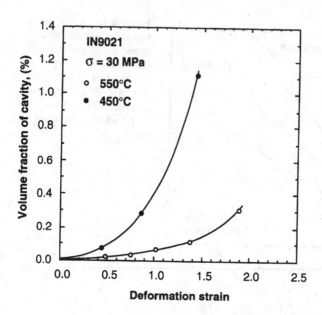

Figure 9.23 Volume fraction of cavities as a function of deformation strain at a constant stress of 30 MPa for IN9021 superplastically deformed at 450 °C (below T_i) and 550 °C (above T_i).

maps are given in Table 4.2. The shaded area in each of the four diagrams in Figure 9.24 refers to plastic flow controlled by grain-boundary sliding and, therefore where fine-grained superplasticity can be expected. A dotted horizontal line is shown at 10% per second, which can be considered as the beginning of HSRS. As can be seen for a grain size of 100 μm, superplastic flow can never be achieved under any conditions. At 10 μm, superplastic flow at 10^{-1} s^{-1} can only be achieved at a very high homologous temperature ($0.85T_m$), where grain growth and oxidation, however, can eliminate the potential of attaining superplastic behavior. At 1 μm, superplasticity at 10^{-1} s^{-1} can be expected at temperatures as low as $0.6T_m$. At 0.1 μm, the exciting prospect of achieving superplasticity at high strain rates at low homologous temperatures ($0.4T_m$) is predicted.

An application of deformation mechanism maps to assess the new field of HSRS in aluminum-based alloys is shown in Figure 9.25. This figure illustrates deformation mechanism maps at a prescribed homologous temperature (0.8–$0.9T_m$), with the grain size plotted as a function of modulus-compensated stress, in the upper part of Figure 9.25, and as a function of strain rate, in the lower part of Figure 9.25. The three different types of aluminum alloys are seen to fall in the predicted range where grain-boundary sliding is expected, and therefore where superplastic flow will be observed. As can be seen in the figure, fine-grained, ingot-processed aluminum alloys exhibit superplasticity only in the strain-rate range of about 10^{-4} s^{-1}, whereas the finer-grained, powder-processed aluminum alloys exhibit superplasticity at higher strain rates (10^{-2} to 1 s^{-1}). The recently developed MA ultrafine-grained aluminum alloys exhibit superplasticity at strain rates as high as 10^2 s^{-1}.

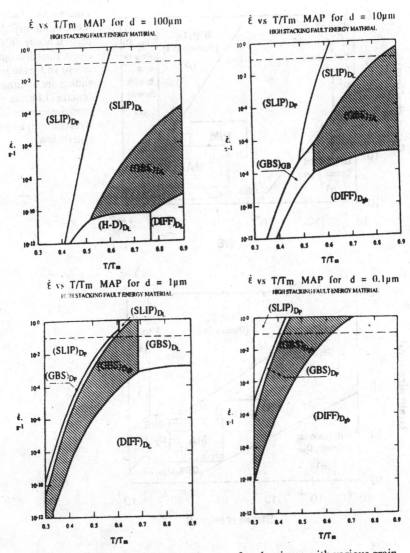

Figure 9.24 Deformation mechanism maps for aluminum with various grain sizes constructed at 800 K (= 0.85 T_m).

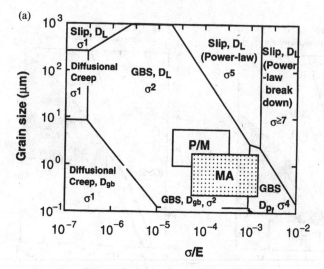

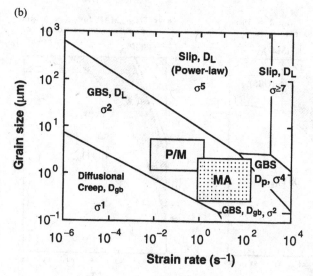

Figure 9.25 Deformation mechanism map showing regions where diffusional creep, slip (dislocation creep) and grain boundary sliding are obtained as a function of grain size and modulus-compensated stress (above) and strain rate (below).

References

1. T.G. Nieh, C.A. Henshall, and J. Wadsworth, 'Superplasticity at High Strain Rate in SiC–2124 Al Composite,' *Scr. Metall.*, **18** (1984), pp. 1405–1408.

2. T. Imai, M. Mabuchi, Y. Tozawa, Y. Murase, and J. Kusui, 'Effect of Dislocation and Recovery on Si₃N₄ Whisker Reinforced Aluminum P/M Composite,' in *Metal & Ceramic Matrix Composites: Processing, Modeling & Mechanical Behavior*, pp. 235–242, ed. R.B. Bhagat, A.H. Clauer, P. Kumar, and A.M. Ritter, TMS-AIME, Warrendale, PA, 1990.

3. T. Imai, M. Mabuchi, Y. Tozawa, and M. Yamada, 'Superplasticity in β-Silicon Nitride Whisker-Reinforced 2124 Al Composite,' *J. Mater. Sci. Lett.*, **9** (1990), pp. 255–257.

4. M. Mabuchi and T. Imai, 'Superplasticity of Si₃N₄ Whisker Reinforced 6061 Al Composite at High Strain Rate,' *J. Mater. Sci. Lett*, **9** (1990), pp. 763–765.

5. T.G. Nieh, P.S. Gilman, and J. Wadsworth, 'Extended Ductility at High Strain Rates in a Mechanically Alloyed Aluminum Alloy,' *Scripta Metall.*, **19** (1985), pp. 1375–1378.

6. T.R. Bieler, T.G. Nieh, J. Wadsworth, and A.K. Muhkerjee, 'Superplastic-Like Behavior at High Strain Rates in a Mechanically Alloyed Aluminum,' *Scr. Metall.*, **22** (1988), pp. 81–86.

7. N. Furushiro and S. Hori, 'Significance of High Rate Superplasticity in Metallic Materials,' in *Superplasticity in Metals, Ceramics, and Intermetallics, MRS Proceeding No. 196*, pp. 385–390, ed. M.J. Mayo, J. Wadsworth, and M. Kobayashi, Materials Research Society, Pittsburgh, PA, 1990.

8. N. Furushiro, S. Hori, and Y. Miyake, 'High Strain Rate Superplasticity and its Deformation Mechanism in Aluminum Alloys,' in *International Conference on Superplasticity in Advanced Materials (ICSAM-91)*, pp. 557–562, ed. S. Hori, M. Tokizane, and N. Furushiro, The Japan Society for Research on Superplasticity, Osaka, Japan, 1991.

9. T.G. Nieh and J. Wadsworth, 'Effects of Zr on the High Strain Rate Superplasticity of 2124 Al,' *Scr. Metall. Mater.*, **28** (1993), pp. 1119–1124.

10. T. Imai, M. Mabuchi, and Y. Tozawa, 'Superplasticity in Si₃N₄ Whisker-Reinforced 7064 Al Alloy Composite,' in *International Conference on Superplasticity in Advanced Materials (ICSAM-91)*, pp. 373–378, ed. S. Hori, M. Tokizane, and N. Furushiro, The Japan Society for Research on Superplasticity, Osaka, Japan, 1991.

11. X. Huang, Q. Liu, C. Yao, and M. Yao, 'Superplasticity in a SiCw–6061 Al Composite,' *J. Mater. Sci. Lett.*, **10** (1991), pp. 964–966.

12. A. Sakamoto, C. Fujiwara, and T. Tsuzuku, 'Development of Fabrication Process for High Strength Metal Matrix Composites,' in *Proceedings of the 33rd Japan Congress on Materials Research*, pp. 73–79, The Society of Materials Science, Japan, 1990.

13. K. Matsuki, H. Matsumoto, M. Tokizawa, N. Takatsuji, M. Isogai, and Y. Murakami, 'Effect of SiC Particulate Content on Stress-Strain Rate Relationship during Hot Compression of SiCp/7075Al Composites,' *J. JIM*, **8** (1993), pp. 876–883.

14. T. Imai, G. L'Esperance, and B.D. Hong, 'High Strain Rate Superplasticity of AlN Particulate Reinforced Aluminum Alloy Composites,' *Scr. Metall. Mater.*, **31** (1994), pp. 321–326.

15. K. Higashi, T. Okada, T. Muka, and S. Tanimura, 'Positive Exponent Strain-Rate Superplasticity in Mechanically Alloyed Aluminum IN 9021,' *Scr. Metall. Mater.*, **25** (1991), pp. 2503–2506.

16. K. Higashi, T. Okada, T. Mukai, S. Tanimura, T.G. Nieh, and J. Wadsworth, 'Superplastic Behavior in a Mechanically-Alloyed Aluminum Composite Reinforced with SiC Particulates,' *Scr. Metall. Mater.*, **26**(2) (1992), pp. 185–190.

17. K. Higashi, T. Okada, T. Mukai, and S. Tanimura, 'Very High Strain Rate Superplasticity in a Mechanically Alloyed IN 9052 Aluminum Alloy,' *Mater. Sci. Eng.*, **159A** (1992), pp. L1-L4.

18. K. Higashi, T. Okada, T. Mukai, and S. Tanimura, 'Superplastic Behavior at High Strain Rates of a Mechanically-Alloyed Al-Mg-Li Alloy,' *Scr. Metall. Mater.*, **26** (1992), pp. 761–766.

19. T.G. Nieh and J. Wadsworth, 'Superplasticity in Aerospace Aluminum Alloys,' in *Superplasticity in Aerospace-Aluminum*, pp. 194–214, ed. R. Pearce and L. Kelly, Ashford Press, Curdridge, Southampton, Hampshire, 1985.

20. Z. Cui, W. Zhong, and Q. Wei, 'Superplastic Behavior at High Strain Rate of Rapidly Solidified Powder Metallurgy Al-Li Alloy,' *Scr. Metall. Mater.*, **30** (1994), pp. 123–128.

21. J. Wadsworth, C.A. Henshall, T.G. Nieh, A.R. Pelton, and P.S. Gilman, 'Extended Ductility in Whisker Reinforced Aluminum Alloys and Mechanically Alloyed Aluminum Alloys,' in *High Strength Powder Metallurgy Aluminum Alloys II*, pp. 137–154, ed. G.J. Hildeman and M.J. Koczak, TMS-AIME, Warrendale, PA, 1986.

22. R.S. Mishra and A.B. Pandey, 'Some Observations on the High-Temperature Creep Behavior of 6061 Al-SiC Composites,' *Metall. Trans.*, **21A** (1990), pp. 2089–2090.

23. F.H. Mohamed, J.-T. Park, and E.J. Lavernia, 'Creep Behavior of Discontinuous SiC-Al Composites,' *Mater. Sci. Eng.*, **A150** (1992), pp. 21–35.

24. A.B. Pandey, R.S. Mishra, and Y.R. Mahajan, 'Steady State Creep Behavior of Silicon Carbide Particulate Reinforced Aluminum Composites,' *Acta Metall. Mater.*, **40** (1992), pp. 2045–2052.

25. J. Cadek and V. Sustek, 'Comment on "Steady State Creep Behavior of Silicon Carbide Particulate Reinforced Aluminum Composites",' *Scr. Metall. Mater.*, **30** (1994), pp. 277–282.

26. G. Gonzales-Doncel and O.D. Sherby, 'High Temperature Creep Behavior of Metal Matrix Aluminum-SiC Composites,' *Acta Metall. Mater.*, **41** (1993), pp. 2797–2805.

27. R.W. Lund and W.D. Nix, 'High Temperature Creep of Ni-20Cr-2ThO$_2$ Single Crystal,' *Acta Metall.*, **24** (1976), pp. 469–481.

28. A. Seeger, D. Wolf, and H. Mehrer, 'Analysis of Tracer and Nuclear Magnetic Resonance Measurement of Self-Diffusion in Aluminum,' *Phys. Status Solidi*, **48** (1971), pp. 481–186.

29. G. L'Esperance and T. Imai, 'Superplastic Behavior and Microstructural Characterization of a β-Si$_3$N$_4$ Whisker Reinforced 2124 Aluminum Composite,' in *International Conference on Superplasticity in Advanced Materials (ICSAM-91)*, pp. 379–384, ed. S. Hori, M. Tokizane, and N.

Furushiro, The Japan Society for Research on Superplasticity, Osaka, Japan, 1991.

30. T.G. Nieh, J. Wadsworth, and D.J. Chellman, 'In Situ Auger Fractographic Study of a SiC-Reinforced Al Alloy,' *Scripta Metall.*, **19** (1985), pp. 181–184.

31. M. Mabuchi, T. Imai, K. Kubo, K. Higashi, Y. Okada, and S. Tanimura, 'Superplasticity in As-extruded Si$_3$N$_4$ Whisker/6061 Aluminum Composites Processed at a Ratio of 100:1,' *Mater. Lett.*, **10** (1990), pp. 339–342.

32. M. Mabuchi, K. Higashi, Y. Okada, S. Tanimura, T. Imai, and K. Kubo, 'Superplastic Behavior at High Strain Rates in a Particulate Si$_3$N$_4$ whisker/6061 Aluminum Composite,' *Scr. Metall. Mater.*, **25** (1991), pp. 2003–2006.

33. M. Mabuchi, T. Imai, and K. Kubo, 'Processing of 6061 Aluminum Matrix Composites Reinforced with Si$_3$N$_4$ Whisker and Their Superplastic Behavior,' in *International Conference on Superplasticity in Advanced Materials (ICSAM-91)*, pp. 367–372, ed. S. Hori, M. Tokizane, and N. Furushiro, The Japan Society for Research on Superplasticity, Osaka, Japan, 1991.

34. M. Mabuchi, T. Imai, and K. Higashi, 'Production of Superplastic Aluminum Composites Reinforced with Si$_3$N$_4$ by Powder Metallurgy,' *J. Mater. Sci.*, **28** (1993), pp. 6582–6586.

35. J.K. Gregory, J.C. Gibeling, and W.D. Nix, 'High Temperature Deformation of Ultra-Fine-Grained Oxide Dispersion Strengthened Alloys,' *Metall. Trans.*, **16A** (1985), pp. 777–787.

36. T.R. Bieler and A.K. Mukherjee, 'The High Strain Rate Superplastic Deformation Mechanisms of Mechanically Alloyed Aluminum IN90211,' *Mater. Sci Eng.*, **A128** (1990), pp. 171–182.

37. M.G. Zelin, T.R. Bieler, and A.K. Muhkerjee, 'Cooperative Grain-Boundary Sliding in Mechanically Alloyed IN 90211 Alloy during High Strain Rate Superplasticity,' *Metall. Trans.*, **24A** (1993), pp. 1208–1212.

38. Y.W. Kim and L.R. Bidwell, 'Tensile Properties of a Mechanically-Alloyed Al–4.0Mg Powder Alloy,' *Scr. Metall.*, **16** (1982), pp. 799–802.

39. M.K. Rabinovich, O.A. Kaibyshev, and V.G. Tufinov, 'Superplasticity of the Aluminum Alloy V96Ts,' *Metalloved Term Obrab Met.*, **3** (1978), pp. 55–59.

40. R.R. Sawtell and C.L. Jensen, 'Mechanical Properties and Microstructures of Al-Mg-Sc Alloys,' *Metall. Trans.*, **21A** (1990), pp. 421–430.

41. O.D. Sherby and J. Wadsworth, 'Development and Characterization of Fine Grain Superplastic Material,' in *Deformation, Processing and Structure*, pp. 355–389, ed. G. Krauss, ASM, Metal Park, Ohio, 1984.

42. M. Mabuchi, J. Koike, H. Iwasaki, K. Higashi, and T.G. Langdon, 'Processing and Development of Superplastic Metals Matrix Composites,' in *Materials Science Forum Vols. 170–172*, pp. 503–512, ed. T.G. Langdon, Trans Tech Publications, Switzerland, 1994.

43. G.M. Pharr and M.F. Ashby, 'On Creep Enhanced by a Liquid Phase,' *Acta Metall.*, **31** (1983), pp. 129–138.

44. M.C. Fleming, 'Behavior of Metal Alloys in the Semisolid State,' *Metall. Trans.*, **22A** (1991), pp. 957–981.

45. C.J. Quaak and W.H. Kool, 'Properties of Semisolid Aluminum Matrix Composites,' *Mater. Sci. Eng.*, **A188** (1994), pp. 277–282.

46. H.-K. Moon, Ph.D. Thesis, Massachusetts Institute of Technology, MA, (1990).

47. L.N. Thanh and M. Suery, 'Compressive Behavior of Partially Remelted A356 Alloys Reinfgorced with SiC Particles,' *Mater. Sci. Technol.*, **10** (1994), pp. 894–901.

48. D.A. Woodford, 'Strain Rate Sensitivity as a Measure of Ductility,' *Trans. ASM*, **62** (1969), pp. 291–299.

49. S.R. Nutt and R.W. Carpenter, 'Non-Equilibrium Phase Distribution in an Al-SiC Composite,' *Mater. Sci. Eng.*, **75** (1985), pp. 169–176.

50. M. Strangwood, C.A. Hippsley, and J.J. Lewandowski, 'Segregation to SiC/Al Interfaces in Al Based Metal Matrix Composites,' *Scr. Metall. Mater.*, **24** (1990), pp. 1483–1487.

51. *Binary Alloy Phase Diagrams, Vol. 1*, ed. T. Massalski, J.L. Murray, L.H. Bennett, and H. Baker, American Society for Metals, Metals Park, Ohio, 1986.

52. M. Mabuchi and K. Higashi, 'Superplastic Deformation Mechanism Accommodated by Liquid Phase in Metal Matrix Composites,' *Phil. Mag. Lett.*, **40**(1) (1994), pp. 1–6.

53. Y. Nakatani, T. Ohnishi, and K. Higashi, 'Superplastic Behavior of Commercial Aluminum Bronze,' *Japan Inst. of Metals*, **48** (1984), pp. 113–114.

54. J. Koike, M. Mabuchi, and K. Higashi, 'In-Situ Observation of Partial Melting in Superplastic Aluminum Alloy Composites at High Temperatures,' *Acta Metall Mater.*, **43**(1) (1994), pp. 199–206.

55. J. Koike, M. Mabuchi, and K. Higashi, 'Partial Melting and Segregation Behavior in a Superplastic Si_3N_4/Al-Mg Alloy Composite,' *J. Mater. Res.*, **10**(1) (1995), pp. 133–138.

56. K. Higashi, T.G. Nieh, and J. Wadsworth, 'Effect of Temperature on the Mechanical Properties of Mechanically–Alloyed Materials at High Strain Rates,' *Acta Metall. Maters.*, **32** (1995), pp. 3275–3282.

57. K. Higashi, T.G. Nieh, M. Mabuchi, and J. Wadsworth, 'Effect of Liquid Phases on the Tensile Elongation of Superplastic Aluminum Alloys and Composites,' *Scr. Metall. Mater.*, **32** (1995), pp. 1079–1084.

58. D.R. Clarke, 'On the Equilibrium Thickness of Intergranular Glass Phases in Ceramic Materials,' *J. Am. Ceram. Soc.*, **70**(1) (1987), pp. 15–22.

59. T.G. Nieh, J. Wadsworth, and T. Imai, 'A Rheological View of High-Strain-Rate Superplasticity in Metallic Alloys and Composites,' *Scr. Metall. Mater.*, **26**(5) (1992), pp. 703–708.

60. B.Q. Han, K.C. Chan, T.M. Yue, and W.S. Lau, 'A Theoretical Model for High-Strain-Rate Superplastic Behavior of Particulate Reinforced Metal Matrix Composites,' *Scr. Metall. Mater.*, **33** (1995), pp. 925–930.

61. K. Higashi, T. Okada, T. Mukai, S. Tanimura, T.G. Nieh, and J. Wadsworth, 'An Investigation of Cavitation in a Mechanically Alloyed 15 vol% SiC_p/IN9021 Aluminum Composite,' in *Transaction of the Materials Research Society of Japan Vol 16B – Composites, Grain Boundaries and Nanophase Materials*, pp. 989–994, ed. M. Sakai, M. Kobayashi, T. Suga, R. Watanabe, Y. Ishida, and K. Niihara, Elsevier Science, Netherland, 1994.

62. K. Higashi, T.G. Nieh, and J. Wadsworth, 'A Comparative Study of Superplasticity and Cavitation in Mechanically-Alloyed IN9021 and a SiC_p/IN9021 Composite,' *Mater. Sci. Eng.*, **188A** (1994), pp. 167–173.

63. H. Iwasaki, M. Taceuchi, T. Mori, M. Mabuchi, and K. Higashi, 'A Comparative

Study of Cavitation in Characteristics in Si$_3$N$_4$/Al-Mg-Si Composite and 7475 Aluminum Alloy,' *Scr. Metall. Mater.*, **31** (1994), pp. 255–260.

64. M.W. Mahoney and A.K. Ghosh, 'Superplasticity in a High Strength Powder Aluminum Alloy with and without SiC Reinforcement,' *Metall. Trans.*, **18A** (1987), pp. 653–661.

65. M.J. Stowell, D.W. Livesey, and N. Ridley, 'Cavity Coalescence in Superplastic Deformation,' *Acta Metall.*, **32** (1984), pp. 35–42.

66. M.J. Stowell, 'Failure of Superplastic Alloys,' *Metal Sci.*, **17** (1983), pp. 1–11.

Chapter 10

Ductility and fracture in superplastic materials

We have discussed fine-structure superplasticity in various materials, including metals and ceramics, in Chapters 5–9. In these chapters, our discussions focused primarily on experimental observations, deformation mechanisms, and microstructural characteristics. Tensile ductility, which is determined by cavitation and fracture, was not emphasized. In this chapter, we will focus on the latter issue. A fracture mechanics model will be examined with available tensile elongation data for superplastic ceramics and superplastic intermetallics. The analysis permits a broad understanding of tensile elongation behavior of superplastic materials.

10.1 Tensile ductility in superplastic metals

It is well accepted that two competing processes govern the failure of superplastic materials at high temperature. One is related to macroscopic necking, and the other is related to microscopic cavitation and cracking. Macroscopic necking is governed by the strain-rate-sensitivity exponent, m, in the simplified constitutive equation $\sigma = k\dot{\varepsilon}^m$, where σ is the true flow stress, $\dot{\varepsilon}$ is the true strain rate, and k is a material constant. A high m value usually indicates a diffuse neck development and, thus, a delay of the onset of tensile failure, which leads to high tensile elongations [1]. The fracture profile of many superplastic metals with $m \geq 0.4$, however, reveals that there is no sharp pinpoint necking. This is because final fracture is caused by the evolution of cavities at grain boundaries, and in this sense, cavities lead to premature failure of test samples. This cavitation damage therefore results in tensile elongations that are lower than might be expected from a given value of the strain-rate-sensitivity exponent.

It has been well recognized that the strain-rate-sensitivity exponent is directly related to the tensile ductility of superplastic metals. When m is low, the increase in stress at the neck will lead to a large increase in strain rate in that region. Thus, the neck will grow sharply, leading to sudden failure and a low elongation to fracture. Conversely, when m is large, the strain rate increases slowly due to the increased stress in the neck region. As a result, the neck forms gradually. Many reviews on superplasticity in metallic alloys reveal that the strain-rate-sensitivity exponent is directly related to the tensile ductility of metallic alloys [2]. This is demonstrated in Figure 10.1 in which m is correlated with the tensile fracture engineering strain (%) for a variety of metallic materials. The tensile ductility increases with an increase in m. Some scatter of the tensile elongation data in this logarithm plot may be attributed to different specimen size and geometry as well as to different tolerances to cavitation, which, in turn, depend on the microstructure of a specific material and testing conditions [3]. It is quite evident that the m value plays an important role in determining the tensile ductility of most metallic alloys. In fact, Figure 10.1 is a useful tool in predicting the tensile elongation of superplastic metals. Moreover, a high m value is found to be a necessary and usually sufficient requirement to achieve superplastic behavior in metallic alloys.

Various methods [4–9] of applied mechanics (a macroscopic approach) have been developed to predict elongation to fracture for materials that fail by necking. Five of these methods are outlined below.

- Rossard [5] has shown that the strain at the start of necking is given by $\varepsilon_{neck}=N/(1-2m)$, where N is the strain-hardening exponent in $\sigma=K''\varepsilon^N$.

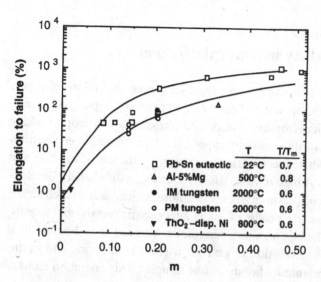

Figure 10.1 Strain-rate-sensitivity vs elongation to failure for various metals (data from Ref. [2]).

	T	T/T$_m$
□ Pb-Sn eutectic	22°C	0.7
▲ Al-5%Mg	500°C	0.8
● IM tungsten	2000°C	0.6
○ PM tungsten	2000°C	0.6
▼ ThO$_2$–disp. Ni	800°C	0.6

His theory would predict infinite plasticity at an m value of 0.5 provided that N has a finite positive value.

■ Morrison [6] demonstrated that the dimensions of the specimen will influence the total elongation observed and showed that

$$\% \text{ elongation to fracture} = bm^2\left(\frac{d_o}{L_o}\right) \times 100 \tag{10.1}$$

where b is a material constant, and d_o and L_o are the initial diameter and initial gage length of the specimen, respectively.

■ Avery and Stuart [7] took into account the possible tapered shape of the specimen. Their equation is given as

$$\% \text{ elongation} = \left(\frac{1 - \beta^{1-m}}{1 - \alpha^{1-m}}\right) \times 100 \tag{10.2}$$

where α is the ratio of the minimum to maximum area at the start of the test and β is the same ratio at some arbitrary stage of the test at which the elongation is measured.

■ Burke and Nix [8] analyzed the influence of strain-rate sensitivity on neck growth using a finite-element method. They predicted elongation to failure as a function of strain-rate sensitivity, as follows:

$$\% \text{ elongation} = [\exp\left(\frac{2m}{1-m}\right) - 1] \times 100 \tag{10.3}$$

■ Ghosh and Ayres [9], based on the assumption that a neck forms (i.e., inhomogeneous deformation occurs) in a region where the initial defect concentration is highest, developed a mathematical model to predict the tensile ductility of metallic alloys. In the model, the fracture strain can be expressed as a strong function of the strain-rate-sensitivity exponent:

$$\varepsilon_f = \pm m \times \ln[1 \pm (f_o)^{1/m}], \tag{10.4}$$

where f_o is the ratio of the initial effective load-bearing cross-sectional area of the neck region to that of an unnecked region.

Equations (10.1)–(10.4) can and have been used to describe reasonably well the tensile ductility of superplastic metals [10]; an example is given in Figure 10.2. However, it is pointed out that they are not suitable to describe superplastic ceramics, as illustrated in Figure 10.3 [12]. There is no apparent correlation between the m value and tensile elongation of ceramics. For example, Wakai and Nagano [13] have shown that the m value increases, while elongation decreases, as the initial grain size in a 2 mol % Y_2O_3-doped zirconia (2YTZP) increases; the m values are 0.5 and 0.8 for materials with grain sizes of 0.55 and 2.63 μm, respectively. The results in Figure 10.3 clearly indicate that the tensile ductility of super-plastic ceramics is not determined by necking stability, as it is in metals. We thus see that, from an applied mechanics approach, a high strain-rate-sensitivity exponent, m, leads to high ductility. A high value of m is a necessary condition for superplasticity. It is, however, not a sufficient condition. Material-embrittling characteristics, such as grain-boundary separation (e.g., in ceramics), cavitation at interphase boundaries and other premature failure modes, can lead to early failure even when the material is highly strain-rate sensitive.

10.2   Tensile ductility in superplastic ceramics

Many fine-grained ceramic materials have been investigated for their tensile ductility [14–36]. These ceramics were tested in tension at temperatures and strain rates where high m values were recorded. Experimental results obtained from these tests indicated that tensile elongations in ceramic materials do not correlate with m values in the same systematic way that they do in metallic alloys, as illustrated in Figure 10.2. Also indicated in it is that, although ceramics gener-

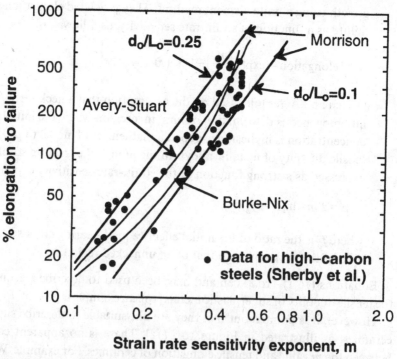

Figure 10.2   Predicted elongation to failure as a function of the strain-rate-sensitivity exponent, m, based on various applied mechanics models. The curves predicted by the Morrison relation were calculated using $b=100$. The elongation and m-value data are from Ref. [11].

ally exhibit high values of m, the tensile ductility of ceramics is usually limited compared with that of metals. Specifically, the tensile elongation of ceramics is considerably less than that of metallic alloys having a similar m value. The fact that tensile elongations in ceramics do not correlate with m values suggests that necking stability does not govern the tensile ductility of ceramics.

It must be especially pointed out, however, that the m value is still an important parameter in determining high-temperature tensile ductility in ceramics. This is because a specific strain-rate-sensitivity exponent is associated with a specific deformation mechanism. In fact, high tensile ductility in ceramics has only been observed in those exhibiting high m values. A high value of m can be beneficial if it is associated with a deformation mechanism that suppresses cavitation damage. For instance, limited slip in polycrystalline ceramics may lead to a high flow stress (beyond the critical stress for cavity nucleation), which gives rise to extensive inter-granular cavitation during plastic flow and results in a limited tensile ductility. If grain-boundary sliding occurs, a low flow stress is expected, and extensive cavitation (cracking) may be avoided. It can be concluded, therefore, that a high m value is a necessary but insufficient condition for achieving high tensile ductility in ceramics. Some other factors must determine the tensile ductility of ceramics.

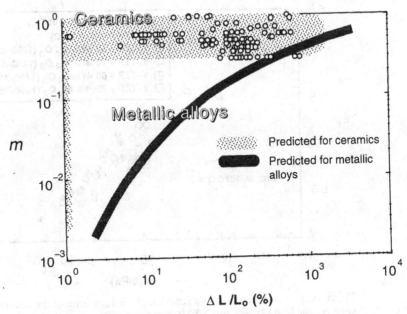

Figure 10.3 Strain-rate-sensitivity vs elongation to failure for superplastic metals and ceramics (data from Ref. [12]).

10.2.1 Tensile elongation as a function of flow stress

Kim *et al.* [12] and Chen and Xue [37] first pointed out that flow stress plays a dominant role in determining the tensile elongation of superplastic ceramics. For example, Kim *et al.* [12] found that the tensile elongation of an iron carbide-based material can be correlated well with flow stress in regions where high m values are observed. Kim *et al.* [38] subsequently showed that a similar correlation could also be found for YTZP and YTZP-based composites. Figure 10.4 shows these data, which illustrate that the true fracture strain is a linearly decreasing function of the logarithm of the flow stress.

Figure 10.5 shows additional data from a number of other ceramics, including hydroxyapatite [16], PbTiO$_3$ [17], Mn–YTZP [18], SiC/Si$_3$N$_4$ composite [19], MgO–Al$_2$O$_3$ [32], pure Al$_2$O$_3$ [32], β-SiAlON [34], Al$_2$O$_3$–Y$_2$O$_3$–SiC/Si$_3$N$_4$ composite [34], 12Ce–CZP [21], 2YTZP [21], and β-spodumene glass [36], to demonstrate the generality of the fracture strain–flow stress relation. It is evident from Figure 10.5 that the fracture strains of fine-grained superplastic ceramics can be generally correlated with the logarithm of flow stress in a manner similar to that shown in Figure 10.4. (Some data anomalies may exist in Figure 10.5, e.g., a con-

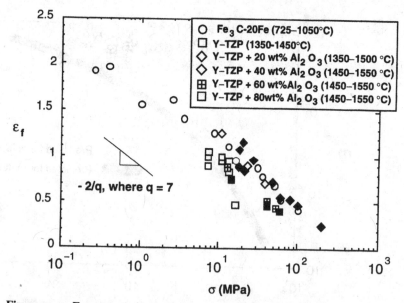

Figure 10.4 Fracture strain as a function of the logarithm of the flow stress for superplastic iron carbide and YTZP-based ceramics.

siderably lower slope than is expected for hydroxyapatite [16], pure Al_2O_3 [32], and 12Ce–CZP [21]. Kim [39] has addressed these anomalies in detail.) The correlation between fracture strain and flow stress is directly related to grain-boundary cavitation during superplastic deformation, particularly during cavity interlinkage in a direction perpendicular to the tensile axis [31]. When the flow stress is lower than the grain-boundary strength of a ceramic, intergranular failures do not occur and the ceramic deforms plastically. As the flow stress is increased, so too is the likelihood that the cohesive strength of grain boundaries will be reached. Once this level of stress is attained, intergranular cavitation and cracking take place, and the elongation to failure is decreased. Consequently, a strong ceramic (i.e., with a high flow stress) is expected to have a low tensile ductility.

To illustrate the reverse effect of strengthening on ductility, Dougherty *et al.* [40] performed experiments with Al_2O_3 particle and SiC whisker- (SiC$_w$-) reinforced fine-grained YTZP. Experimental results showed that the addition of SiC whiskers to YTZP causes a significant strengthening effect, but not for the Al_2O_3 particle-reinforced composite. As shown in Figure 10.6, at 1550 °C and a strain rate of $10^{-3}\,s^{-1}$, the flow stresses of both YTZP and a 20 wt% (28 vol %) Al_2O_3 particle-reinforced YTZP are both less than 30 MPa, whereas the flow stress of

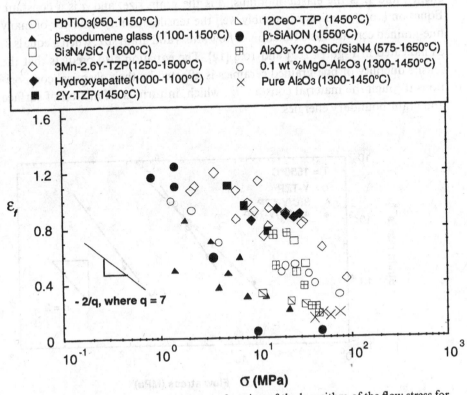

Figure 10.5 Fracture strain as a function of the logarithm of the flow stress for various superplastic ceramics (data from Ref. [39]).

20 vol % SiC whisker-reinforced YTZP is almost 200 MPa. All three materials have a fine-grained matrix (~0.5 μm). Also, the dominant deformation mechanism in the SiC$_w$/YTZP composite was identified to be grain-boundary sliding, similar to that in the monolithic YTZP. The strengthening effect did not result because SiC has a greater strength *per se*, but rather from the inhibition of grain-boundary sliding caused by the presence of SiC whiskers. Because of the strengthening effect, the maximum elongation value of the SiC$_w$/YTZP composite was reduced considerably (~50%). In contrast, both the YTZP and Al$_2$O$_3$ particle-reinforced YTZP composite behave superplastically.

Using the above flow stress–elongation correlation, Kim *et al.* [12] have developed a fracture mechanics model, which predicts that the fracture strain of a ceramic with a high m value is a linear function of the logarithm of flow stress. The developed fracture equation is expressed as:

$$\varepsilon_f = \ln(k' d^{-1/q} \sigma^{-2/q}) \big|_{m \text{ is high}} \tag{10.5}$$

where ε_f is the true tensile fracture strain; k' is a material constant equal to $[(2\gamma_s - \gamma_{gb})E/\pi]^{-1/q}$, where γ_s is the surface energy, γ_{gb} is the grain-boundary energy, and E is the elastic modulus; d is the grain size; and q is a constant. Equation (10.5) predicts reasonably well the tensile ductility behavior of many fine-grained ceramics as a function of flow stress and grain size, when q equals 7, as indicated in Figures 10.3 and 10.4 [12]. The equation also predicts that the tensile ductility of fine-grained ceramics is affected by their chemical compositions through the material constant k', which, in turn, is a function of surface and grain-boundary energies.

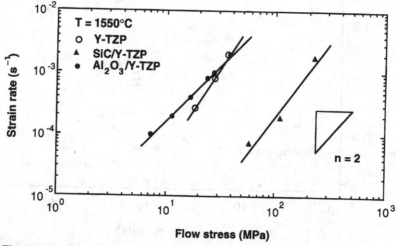

Figure 10.6 Direct comparison of the strengths of YTZP and the composites SiC/YTZP and Al$_2$O$_3$/YTZP at 1550 °C.

10.2.2 Tensile elongation as a function of grain size

It is well-known that the tensile elongation of a superplastic material is a strong function of grain size (e.g., see Figures 6.4 and 7.6). The effect of grain size on tensile ductility can be predicted from the fracture strain equation, Equation (10.5). For the case of constant stress (creep test), the dependence of tensile ductility on grain size can be derived from Equation (10.5) as:

$$\varepsilon_f = \ln(k'' d^{-1/q}) \tag{10.6}$$

where k'' is a constant for a given stress. For the case of a constant value of the Zener–Hollomon parameter ($\dot{\varepsilon} \exp(Q/RT)$), i.e., at a constant strain rate (tensile test) and at a given temperature, the dependence of tensile ductility on grain size is given by [12]:

$$\varepsilon_f = \ln(k^* d^{-1/q(1+2p/n)}) \tag{10.7}$$

where k^* is a constant for a given value of the Zener–Hollomon parameter, and p is the power dependence of grain size in Equation (6.8). Thus, ε_f is expected to be a linear function of the natural logarithm of the grain size with a slope equal to $-1/q$ and $-1/q(1+2p/n)$, for constant stress and for a constant Zener–Hollomon parameter, respectively. For the typical case where $p=3$, $q=7$, and $n=2$, the slope ($d\varepsilon_f/d\ln L$) is -0.14 and -0.57, respectively. The value of $-1/q(1+2p/n)$ is generally much lower than that of $-1/q$. This lower value of the former parameter results from the presence of a component, p/n, which depicts the flow stress dependence on grain size at a given strain rate and temperature. As indicated by Equation (10.6), an increase of tensile ductility with a decrease in grain size, often observed under constant strain-rate conditions, is mainly a result of the decrease of flow stress caused by the decrease in grain size. Also, according to Equations (10.6) and (10.7), tensile ductility has a stronger grain-size dependence in constant strain rate tests than it does in constant stress tests.

Experimental evidence for the validity for each condition of testing is shown by an ε_f-log grain size plot for several ceramics in Figure 10.7. Straight-line relations are apparent, with the predicted low slope (-0.33) for alumina (constant stress) and with the predicted high slope (-1.31) for iron carbide as well as for Si_3N_4/SiC composite [20], 2YTZP [21], (2–6)YTZP [21], (2.3–22.2) wt% Al_2O_3/YTZP composite [41], 20 wt% Al_2O_3/3YTZP composite [26], and 3YTZP [22] (constant strain rate). The tensile elongation data for alumina, even though doped with various impurities (open circle symbol), remain fairly well correlated with grain size. In addition, the tensile elongation data for zirconia containing different amounts of Y_2O_3 (open diamond symbol) are also well correlated with grain size. These good correlations of tensile elongation with grain size for given ceramic materials doped with different additives suggest that the often observed scatter of tensile ductility data with impurities is probably a result of grain-size variation. It is evident in Equation (10.5) that grain size simply overwhelms the other factors, such as grain-boundary and surface energies, crystallographic structure, and diffusion rate.

10.2.3 Cavitation in superplastic ceramics

Cavitation during superplastic deformation is an important practical concern, not only because it affects the optimum elongation but also because of the deterioration in the subsequent room-temperature properties. Experimental

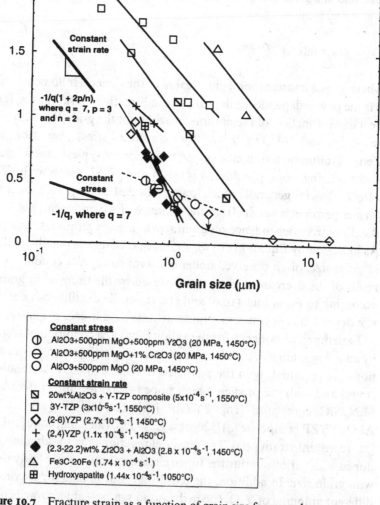

Figure 10.7 Fracture strain as a function of grain size for several superplastic ceramics.

results in metallic alloys indicate that even small levels of cavitation may sometimes lead to a dramatic degradation in mechanical properties [42, 43]. Cavitation during the superplastic deformation of metallic alloys has been extensively studied. Technically, cavitation can be reduced by imposing a hydrostatic back pressure or electric field [44, 45] during superplastic deformation. The suppression of cavitation by the concurrent application of a hydrostatic back pressure has also been investigated [46–48]. Interested readers are referred to Reference [49]. In this section, we will discuss mainly cavitation in ceramics.

Only limited quantitative studies have been conducted on the cavitation of superplastic ceramics, in particular cavitation as a function of experimental parameters such as strain, strain rate, and temperature [31, 50, 51]. Nevertheless, these existing data provide some insights into the fracture mechanisms in superplastic ceramics. In the following, we use YTZP as an example.

For YTZP, cavities nucleate predominantly at triple-point junctions, and they appear to have a quasi-equilibrium, spherical-cap shape [31]. Generally, superplastic ceramics are able to cavitate extensively prior to fracture. Figure 10.8 is a cross-sectional micrograph from a YTZP specimen pulled to an elongation to failure of ~700% at 1550 °C and a strain rate of 8.3×10^{-5} s^{-1} (horizontal tensile axis). This micrograph was obtained in a region about halfway to the fracture tip from the end of the specimen gage length, corresponding to a true local strain of ~2.2, and it shows the formation of cavities in stringers parallel to the tensile axis. Also, there is little cavity interlinkage perpendicular to the tensile axis, although there is evidence for interlinkage parallel to the tensile axis.

The formation of cavity stringers has been reported to occur in many superplastic alloys [42, 52–59]. It has been suggested that the observation of stringers in a microduplex superplastic alloy without particles arise from the relaxation of stress concentrations caused by the sliding of a group of grains [58].

In the case of superplastic ceramics, cavity stringers were observed only in materials exhibiting large elongations (e.g., 700% in YTZP [30, 31]). In an extreme case, a 5 wt% SiO$_2$–3YTZP material that deformed to an exceptionally large strain (>1000% [32]) showed extensive cavity stringer formation along the tensile axis. The cavity stringers may result either from the extensive transverse interlinkage of cavities that mask the appearance of stringers, or from a genuine requirement of large strain deformation for the formation of cavity stringers [53]. In contrast, when the elongations to failure were less than 400%, there were no cavity stringers, and there was extensive cavity interlinkage perpendicular to the tensile axis. An example of such an observation is given in Figure 10.9 for a specimen pulled to an elongation to failure of ~150% at 1450 °C and a strain rate of 2.7×10^{-5} s^{-1}. The fracture tip is toward the left side of the micrograph, and the tensile axis is horizontal.

Quantitative data obtained from samples tested at 1550 °C and at various strain rates, plotted in the form of cavitation area fraction vs local true strain,

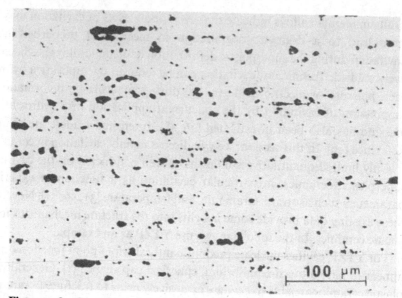

Figure 10.8   Cross-sectional micrograph from a YTZP specimen pulled to an elongation to failure of ~700% at 1550 °C and a strain rate of 8.3×10^{-5} s^{-1}.

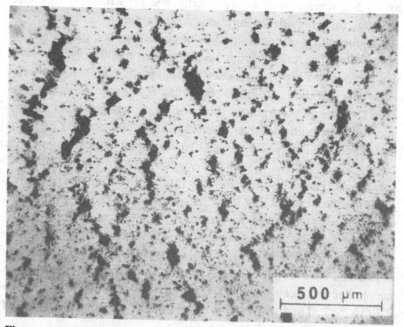

Figure 10.9   The absence of cavity stringers and cavity interlinkage perpendicular to the tensile axis in YTZP samples pulled to an elongation to failure of ~150% at 1450 °C and a strain rate of 2.7×10^{-5} s^{-1}.

are depicted in Figure 10.10. The experimental results demonstrate the striking capability of superplastic ceramics to tolerate cavitation, with levels of cavitation approaching about 30% at a strain rate of 8.3×10^{-5} s^{-1}, corresponding to an elongation to failure of ~700%. The shapes of the curves shown in Figure 10.10 tend to follow those for classical nucleation and growth phenomena, with an incubation period during which there is little cavitation, followed by significant cavity growth.

Figure 10.11 shows the variation of tensile elongation to failure with strain rate as well as the cavitation area fraction at the fracture tips of YTZP specimens tested at 1550 °C. It is clear that both the elongation to failure and cavitation area tend to follow a similar variation with strain rate; the highest level of cavitation was recorded on the specimen exhibiting the largest elongation to failure.

Concurrent cavitation can interrupt superplastic deformation to cause premature failure by the interlinkage of cavities in a direction transverse to the tensile axis. In actuality, it is not the total level of cavitation *per se* that is

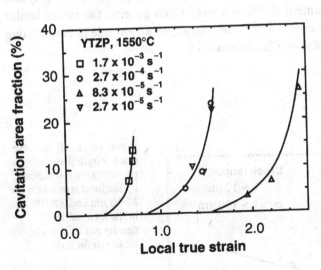

Figure 10.10
Quantitative data obtained from YTZP samples tested at 1550 °C and at various strain rates.

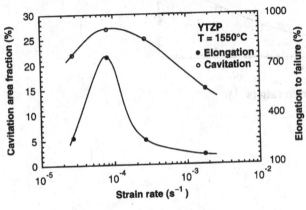

Figure 10.11   Variation of elongation to failure with strain rate, as well as the cavitation area fraction at the fracture tips of specimens tested at 1550 °C.

important, but rather the level of cavity interlinkage transverse to the tensile axis that must be considered. To illustrate this, Figure 10.12 shows the sizes, λ, and distribution of cracks in a direction perpendicular to the tensile axis. Shown in Figure 10.12 are the transverse crack length distributions per unit area normalized by the local true strain for $\lambda > 10 \ \mu m$ and $\lambda > 100 \ \mu m$, in the form of crack density per unit strain vs strain rate. Inspection of Figure 10.12 indicates that the density of large cracks ($\lambda > 100 \ \mu m$) attains a minimum value at a strain rate of $8.3 \times 10^{-5} \ s^{-1}$, corresponding to the maximum elongation to failure of ~700%. Not only are the densities of large cracks similar for specimens tested at strain rates of 2.7×10^{-5} and $2.7 \times 10^{-4} \ s^{-1}$, but also both the levels of cavitation and elongations to failure are similar for the two specimens (see Figure 10.10).

From the above observations, Schissler et al. [31] concluded that the tendency toward intergranular cavity interlinkage in a direction perpendicular to the tensile axis is the limiting factor for the tensile ductility of superplastic ceramics. This conclusion is consistent with the theory proposed by Kim et al. [12] and Chen and Xue [37]; namely, the flow stress, which governs the intergranular cavity formation and subsequent growth, plays a dominant role in determining the tensile elongation of superplastic ceramics.

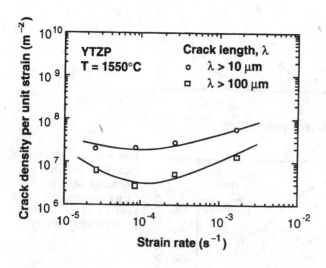

Figure 10.12 Transverse crack length distributions per unit area normalized by the local true strain for $\lambda > 10 \ \mu m$ and $\lambda > 100 \ \mu m$, in the form of crack density per unit strain versus strain rate.

10.3 Tensile ductility in superplastic intermetallic compounds

Although a number of intermetallic alloys, including Ti$_3$Al–TiAl [60–63], Ni$_3$Si [64–66], Ni$_3$Al [62, 67, 68], Ti$_3$Al [69–72], and Ni$_3$(Si,Ti) [73–75], have been demonstrated to be superplastic, only limited elongation data can be systematically evaluated for the functional dependence on grain size and flow stress. Among these materials, the tensile elongation behavior of boron-doped Ni$_3$Al [62, 67, 68] and of boron-doped and nondoped Ni$_3$(Si,Ti) [73–75] was analyzed. Figure 10.13 shows a plot of the true fracture strain as a function of the Zener–Hollomon parameter for these three intermetallic compounds. A linear correlation can be observed for the boron-doped Ni$_3$Al [67, 68] and for both the boron-doped and nondoped Ni$_3$(Si,Ti) [75]. The slopes of the curves in Figure 10.13 are particularly close to the predicted value (i.e., $q=7$) from the fracture mechanics model, which was initially developed for superplastic ceramics [12].

It has been shown that the superplastic elongation of the Ni$_3$(Si,Ti) decreases with increasing grain size (see Figure 7.6). An analysis of the grain-size dependence on the tensile elongation of Ni$_3$(Si,Ti) materials, tested at a constant strain rate, is presented in Figure 10.14 [73–75]. The resulting correlation is remarkably similar to that observed in superplastic ceramics. This limited analysis on the tensile elongation behavior of some intermetallic compounds indicates that failure mechanisms present in these materials are probably similar to those

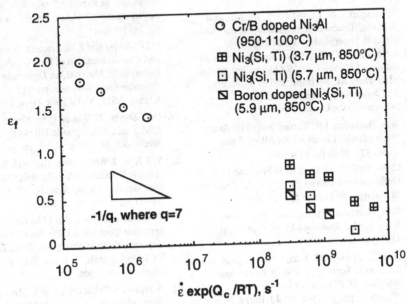

Figure 10.13 True fracture strain as a function of the Zener–Hollomon parameter for boron-doped Ni$_3$Al and for boron-doped and nondoped Ni$_3$(Si,Ti).

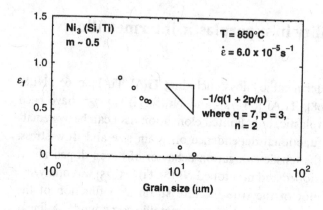

Figure 10.14 Grain-size dependence on the tensile elongation of $Ni_3(Si,Ti)$ tested at 850 °C and at a constant strain rate of $6 \times 10^{-5}\,s^{-1}$.

observed in superplastic ceramics. Additional research is certainly needed to provide insights on the fracture mechanisms of intermetallics. Up to the present time, cavitation in intermetallics during superplastic deformation has not yet been systematically studied.

References

1. E.W. Hart, 'Theory of the Tensile Test,' *Acta Metall.*, **15** (1967), pp. 351–355.

2. D.A. Woodford, 'Strain Rate Sensitivity as a Measure of Ductility,' *Trans. ASM*, **62** (1969), pp. 291–299.

3. Y. Takayama, N. Furushiro, T. Tozawa, and H. Kato, 'Effects of Specimen Shape on Flow Stress in Superplastic Materials,' in *International Conference on Superplasticity in Advanced Materials (ICSAM-91)*, pp. 69–74, ed. S. Hori, M. Tokizane, and N. Furushiro, The Japan Society for Research on Superplasticity, Osaka, Japan, 1991.

4. W.A. Backofen, I.R. Turner, and D.H. Avery, 'Superplasticity in an Al-Zn Alloy,' *Trans. ASM*, **57** (1964), pp. 980–990.

5. C. Rossard, 'Formation de la Striction Dans la Deformation a Chaud Par Traction,' *Revue de Metallurgie*, **63** (1966), pp. 225–235.

6. W.B. Morrison, 'Superplasticity of Low-Alloy Steels,' *Trans. ASM*, **61** (1968), pp. 423–434.

7. D.H. Avery and J.M. Stuart, 'The Role of Surfaces in Superplasticity,' in *Surface and Interfaces II, Physical and Mechanical Properties*, pp. 371–392, ed. J.J. Burke, N.L. Reed, and V. Weiss, Syracuse University Press, Syracuse, New York, 1968.

8. M.A. Burke and W.D. Nix, 'Plastic Instabilities in Tension Creep,' *Acta Metall.*, **23** (1975), pp. 793–798.

9. A.K. Ghosh and R.A. Ayres, 'On Reported Anomalies in Relating Strain-Rate Sensitivity (m) to Ductility,' *Metall. Trans.*, **7A** (1976), pp. 1589–1591.

10. O.D. Sherby and J. Wadsworth, 'Development and Characterization of Fine Grain Superplastic Material,' in *Deformation, Processing and Structure*, pp. 355–389, ed. G. Krauss, ASM, Metal Park, Ohio, 1984.

11. O.D. Sherby, B. Walser, C.M. Young, and E.M. Cady, 'Superplastic Ultra-High Carbon Steels,' *Scr. Metall.*, **9** (1975), pp. 569–574.

12. W.J. Kim, J. Wolfenstine, and O.D. Sherby, 'Tensile Ductility of Superplastic Ceramics and Metallic Alloys,' *Acta Metall. Mater.*, **39(2)** (1991), pp. 199–208.

13. F. Wakai and T. Nagano, 'The Role of Interface-Controlled Diffusion Creep on Superplasticity of Yttria-Stabilized Tetragonal ZrO_2 Polycrystals,' *J. Mater. Sci. Lett.*, **7** (1988), pp. 607–609.

14. F. Wakai, S. Sakaguchi, and Y. Matsuno, 'Superplasticity of Yttria-Stabilized Tetragonal ZrO_2 Polycrystals,' *Adv. Ceram. Mater.*, **1** (1986), pp. 259–263.

References

15. F. Wakai, Y. Kodama, S. Sakaguchi, N. Murayama, K. Izaki, and K. Niihara, 'A Superplastic Covalent Crystal Composite,' *Nature (London)*, **334(3)** (1990), pp. 421–423.

16. F. Wakai, Y. Kodama, S. Sakaguchi, and T. Nonami, 'Superplasticity of Hot Isostatically Pressed Hydroxyapatite,' *J. Am. Ceram. Soc*, **73(2)** (1990), pp. 257–260.

17. F. Wakai, Y. Kodama, N. Murayama, S. Sakaguch, T. Rouxel, S. Sato, and T. Nonami, 'Superplasticity of Functional Ceramics,' in *International Conference on Superplasticity in Advanced Materials (ICSAM–91)*, pp. 205–214, ed. S. Hori, M. Tokizane, and N. Furushiro, The Japan Society for Research on Superplasticity, Osaka, Japan, 1991.

18. F. Wakai, H. Okamura, N. Kimura, and P.G.E. Descamps, 'Superplasticity of Transition Metal Oxide-Doped Y-TZP at Low Temperatures,' in *Proc. 1st Japan Int'l SAMPE Symp.*, pp. 267–271, ed. N. Igata, K. Kimpara, T. Kishi, E. Nakata, A. Okura, and T. Uryu, Society for the Advancement of Materials and Process Engineering, 1989.

19. F. Wakai, 'Superplasticity of Non-Oxide Ceramics,' in *Superplasticity in Metals, Ceramics, and Intermetallics, MRS Proceeding No. 196R*, pp. 349–358, ed. M.J. Mayo, J. Wadsworth, and M. Kobayashi, Materials Research Society, Pittsburgh, PA, 1990.

20. F. Wakai, Y. Kodama, S. Sakaguchi, N. Murayama, H. Kato, and T. Nagano, 'Superplastic Deformation of Al_2O_3/ZrO_2 Duplex Composites,' in *MRS Intl. Meeting on Advanced Materials Vol 7 (IMAM–7, Superplasticity)*, pp. 259–266, ed. M. Doyama, S. Somiya, and R.P.H. Chang, Materials Research Soc., Pittsburgh, PA, 1989.

21. F. Wakai, Y. Kodama, and T. Nagano, 'Superplasticity of ZrO_2 Polycrystals,' *Japan J. Appl. Phy. Series 2, Lattice Defects in Ceramics*, (1989), p. 57.

22. T.G. Nieh, C.M. McNally, and J. Wadsworth, 'Superplastic Behavior of a Yttria-Stabilized Tetragonal Zirconia Polycrystal,' *Scr. Metall.*, **22** (1988), pp. 1297–1300.

23. T.G. Nieh, C.M. McNally, and J. Wadsworth, 'Superplasticity of Intermetallic Alloys and Ceramics,' *JOM*, **41(9)** (1989), pp. 31–35.

24. T.G. Nieh, C.M. McNally, and J. Wadsworth, 'Superplastic Behavior of a 20% Al_2O_3/YTZ Ceramic Composite,' *Scr. Metall.*, **23** (1989), pp. 457–460.

25. T.G. Nieh and J. Wadsworth, 'Superplastic Behavior of a Fine-Grained, Yttria-Stabilized, Tetragonal Zirconia Polycrystal (Y-TZP),' *Acta Metall. Mater.*, **38** (1990), pp. 1121–1133.

26. T.G. Nieh and J. Wadsworth, 'Superplasticity in Fine-Grained 20%Al_2O_3/YTZ Composite,' *Acta Metall. Mater.*, **39** (1991), pp. 3037–3045.

27. W.J. Kim, G. Frommeyer, O.A. Ruano, J.B. Wolfenstine, and O.D. Sherby, 'Superplastic Behavior of Iron Carbide,' *Scr. Metall.*, **23** (1989), pp. 1515–1520.

28. T. Hermanson, K.P.D. Lagerlof, and G.L. Dunlop, 'Superplastic Deformation of YTZP Zirconia,' in *Superplasticity and Superplastic Forming*, pp. 631–635, ed. C.H. Hamilton and N.E. Paton, TMS, 1988.

29. P. Gruffel, P. Carry, and A. Mocellin, 'Effects of Testing Conditions on Superplastic Creep of Alumina Doped with Ti and Y,' in *Science of Ceramics, Volume 14*, pp. 587–592, ed. D. Taylor, The Institute of Ceramics, Shelton, Stoke-on-Trent, UK, 1987.

30. Y. Ma and T.G. Langdon, 'An Investigation of the Mechanical Behavior of a Superplastic Yttria-Stabilized Zirconia,' in *Superplasticity in Metals, Ceramics, and Intermetallics, MRS Proceeding No. 196*, pp. 325–330, ed. M.J. Mayo, J. Wadsworth, and M. Kobayashi, Materials Research Society, Pittsburgh, PA, 1990.

31. D.J. Schissler, A.H. Chokshi, T.G. Nieh, and J. Wadsworth, 'Microstructural Aspects of Superplastic Tensile Deformation and Cavitation Failure in a Fine-Grained, Yttria Stabilized, Tetragonal Zirconia,' *Acta Metall. Mater.*, **39(12)** (1991), pp. 3227–3236.

32. Y. Yoshizaka and T. Sakuma, 'Improvement of Tensile Ductility in High-Purity Alumina due to Magnesia Addition,' *Acta Metall. Mater.*, **40** (1992), pp. 2943–2950.

33. K. Kajihara, Y. Yoshizawa, and T. Sakuma, 'Superplasticity in SiO_2-Containing Tetragonal Zirconia Polycrystal,' *Scr. Metall. Mater.*, **28** (1993), pp. 559–562.

34. X. Wu and I.-W. Chen, 'Exaggerated Texture and Grain Growth in a Superplastic SiAlON,' *J. Am. Ceram. Soc.*, **75(10)** (1992), pp. 2733–2741.

35. T. Rouxel, F. Wakai, and K. Izaki, 'Tensile Ductility of Superplastic Al_2O_3-Y_2O_3-Si_3N_4/SiC Composites,' *J. Am. Ceram. Soc.*, **75(9)** (1992), pp. 2363–2372.

36. J.-G. Wang and R. Raj, 'Mechanism of Superplastic Flow in a Fine-Grained Ceramic Containing Some Liquid Phase,' *J. Am. Ceram. Soc.*, **67**(6) (1984), pp. 399–409.

37. I.-W. Chen and L.A. Xue, 'Development of Superplastic Structural Ceramics,' *J. Am. Ceram. Soc.*, **73**(9) (1990), pp. 2585–2609.

38. W.J. Kim, J. Wolfenstine, O.S. Ruano, G. Frommeyer, and O.D. Sherby, 'Processing and Superplastic Properties of Fine-Grained Iron Carbide,' *Metall. Trans.*, **23A** (1992), pp. 527–535.

39. W.J. Kim, Ph.D. Thesis, Department of Materials Science and Engineering, Stanford University, (1993).

40. S.E. Dougherty, T.G. Nieh, J. Wadsworth, and Y. Akimune, 'Mechanical Properties of a 20%SiC Whisker-Reinforced Y-TZP at Elevated Temperatures,' *J. Mater. Res.*, **10**(1) (1995), pp. 113–118.

41. T. Kuroishi, K. Uno, and F. Wakai, 'Characterization of Superplastic ZrO_2-Toughened Al_2O_3 Prepared by Slip Casting,' in *MRS Intl. Meeting on Advanced Materials Vol 7 (IMAM-7, Superplasticity*, pp. 267–274, ed. M. Kobayashi and F. Wakai, Materials Research Soc., Pittsburgh, PA, 1989.

42. C.C. Bampton, A.K. Ghosh, and M.W. Mahoney, 'The Causes, Effects, and Control of Cavitation in Superplastic 7475 Aluminum Airframe Structures,' in *Superplasticity in Aerospace-Aluminum*, pp. 1–35, ed. R. Pearce and L. Kelly, Ashford Press, Curdridge, Southampton, Hampshire, 1985.

43. C.A. Henshall, J. Wadsworth, M.J. Reynolds, and A.J. Barnes, 'Design and Manufacture of a Superplastic-Formed Aluminum-Lithium Component,' *Materials and Design VIII*, (6) (1987), pp. 324–330.

44. H. Conrad, W.D. Cao, X.P. Lu, and A.F. Sprecher, 'Effect of Electric Field on Cavitation in Superplastic Aluminum Alloy 7475,' *Mater. Sci. Eng.*, **A138** (1991), pp. 247–258.

45. M. Li and S. Wu, 'Effect of External Electric Field on the Cavitation During the Superplastic Deformation of Duralumin LY12CZ,' *Scr. Metall. Mater.*, **31**(3) (1994), pp. 47–51.

46. C.C. Bampton, M.W. Mahoney, C.H. Hamilton, A.K. Ghosh, and R. Raj, 'Control of Superplastic Cavitation by Hydrostatic Pressure,' *Metall. Trans.*, **14A** (1983), pp. 1583–1591.

47. J. Pilling and N. Ridley, 'Cavitation in Superplastic Alloys and the Effect of Hydrostatic Pressure,' *Res. Mechanica*, **23** (1988), pp. 31–63.

48. K. Higashi and N. Ridley, 'Thermomechanical Processing and Superplasticity in a Commercial Copper-Base Alloy,' in *International Conference on Superplasticity in Advanced Materials (ICSAM-91)*, pp. 447–452, ed. S. Hori, M. Tokizane, and N. Furushiro, The Japan Society for Research on Superplasticity, Osaka, Japan, 1991.

49. A.H. Chokshi, 'Cavitation Behavior during Superplastic Deformation,' in *International Conference on Superplasticity in Advanced Materials (ICSAM-91)*, pp. 171–180, ed. S. Hori, M. Tokizane, and N. Furushiro, The Japan Society for Research on Superplasticity, Osaka, Japan, 1991.

50. A. Chokshi, D.J. Schissler, T.G. Nieh, and J. Wadsworth, 'A Comparative Study of Superplastic Deformation and Cavitation Failure in a Yttria Stabilized Zirconia and a Zirconia Alumina Composite,' in *Superplasticity in Metals, Ceramics, and Intermetallics, MRS Proceeding No. 196*, pp. 379–384, ed. M.J. Mayo, J. Wadsworth, and M. Kobayashi, Materials Research Society, Pittsburgh, PA, 1990.

51. A.H. Chokshi, T.G. Nieh, and J. Wadsworth, 'Role of Concurrent Cavitation in the Fracture of a Superplastic Zirconia-Alumina Composite,' *J. Am. Ceram. Soc.*, **74** (1991), pp. 869–873.

52. R.G. Fleck, C.J. Beevers, and D.M.R. Taplin, 'The Hot Fracture of an Industrial Copper-base Alloy,' *Metal Sci.*, **9** (1975), pp. 49–54.

53. A.H. Chokshi, 'An Experimental Study on the Alignment of Cavities in a Superplastic Commercial Copper Alloy,' *Metall. Trans.*, **18A** (1987), pp. 63–67.

54. H. Ishikawa, D.G. Bhat, F.A. Mohamed, and T.G. Langdon, 'Evidence of Cavitation in the Superplastic Zn–22%Al Eutectoid,' *Metall. Trans.*, **8A** (1977), pp. 523–525.

55. B.P. Kashyap and K. Tangri, 'Cavitation Behavior of an Al-Cu Eutectic Alloy During Superplastic Deformation,' *Metall. Trans.*, **20A** (1989), pp. 453–462.

56. K. Matsuki, Y. Ueno, M. Yamada, and Y. Murakami, 'Cavitation during Superplastic Deformation of an Al-Zn-Mg Alloy,' *J. JIM*, **11** (1977), pp. 1136–1144.

57. C.H. Caceres and D.S. Wilkinson, 'Large Strain Behavior of a Superplastic Copper Alloys-II. Cavitation and Fracture,' *Acta Metall.*, **32** (1984), pp. 423–434.

58. A.H. Chokshi and T.G. Langdon, 'The Influence of Rolling Direction on the Mechanical Behavior and Formation of Cavity Stringers in the Superplastic Zn–22%Al Alloy,' *Acta Metall. Mater.*, **37** (1989), pp. 715–724.

59. A.H. Chokshi and T.G. Langdon, 'The Nucleation and Growth of Cavities in a Superplastic Quasi-Single Phase Copper Alloy,' *Acta Metall. Mater.*, **38** (1990), pp. 867–877.

60. S.C. Cheng, J. Wolfenstine, and O.D. Sherby, 'Superplastic Behavior of Two-Phase Titanium Aluminides,' *Metall. Trans.*, **23A** (1992), pp. 1509–1513.

61. W.B. Lee, H.S. Yang, Y.-W. Kim, and A.K. Muhkerjee, 'Superplastic Behavior in a Two-Phase TiAl Alloy,' *Scr. Metall. Mater.*, **29** (1993), pp. 1403–1408.

62. H. Fukutomi, C. Harting, and H. Mecking, 'Microstructure in a TiAl Intermetallic Compound During High-Temperature Deformation,' *Z. Metallk.*, **81** (1991), pp. 272–280.

63. T. Maeda, M. Okada, and Y. Shida, 'Superplasticity in Ti-Rich TiAl,' in *International Conference on Superplasticity in Advanced Materials (ICSAM–91)*, pp. 311–316, ed. S. Hori, M. Tokizane, and N. Furushiro, The Japan Society for Research on Superplasticity, Osaka, Japan, 1991.

64. T.G. Nieh and W.C. Oliver, 'Superplasticity of a Nickel Silicide,' *Scr. Metall.*, **23** (1989), pp. 851–854.

65. S.L. Stoner and A.K. Muhkerjee. 'Superplasticity in Fine Grain Nickel Silicide,' in *International Conference on Superplasticity in Advanced Materials (ICSAM–91)*, pp. 323–328, ed. S. Hori, M. Tokizane, and N. Furushiro, The Japan Society for Research on Superplasticity, Osaka, Japan, 1991.

66. S.L. Stoner and A.K. Muhkerjee, 'Superplasticity in a Nickel Silicide Alloy-Microstructure and Mechanical Correlation,' *Mater. Sci. Eng.*, **A153** (1992), pp. 465–469.

67. M.S. Kim, S. Hanada, S. Wantanabe, and O. Izumi, 'Superplasticity in a Recrystallized Ni₃Al Polycrystal Doped with Boron,' *Mater. Trans. JIM*, **30**(1) (1989), pp. 77–85.

68. V.K. Sikka, C.T. Liu, and E.A. Loria, 'Processing and Properties of Powder Metallurgy Ni₃Al-Cr-Zr-B for Use in Oxidizing Environments,' in *Processing of Structural Metals by Rapid Solidification*, pp. 417–427, ed. F.H. Froes and S.J. Savage, Am. Soc. Metals, Metals Park, OH, 1987.

69. H.S. Yang, P. Jin, E. Dalder, and A.K. Muhkerjee, 'Superplasticity in a Ti₃Al-base Alloy Stabilized by Nb, V, and Mo,' *Scr. Metall. Mater.*, **25** (1991), pp. 1223–1228.

70. H.S. Yang, P. Jin, and A.K. Muhkerjee, 'Superplastic Properties of Ti₃Al,' *Mater. Sci. Eng.*, **A153** (1992), pp. 457–464.

71. A. Dutta and D. Banerjee, 'Superplastic Behavior in a Ti₃Al-Nb Alloy,' *Scr. Metall. Mater.*, **24** (1990), pp. 1319–1322.

72. A.K. Ghosh and C.-H. Cheng, 'Superplastic Deformation in Titanium Aluminides and Modeling of Transient Deformation,' in *International Conference on Superplasticity in Advanced Materials (ICSAM–91)*, pp. 299–310, ed. S. Hori, M. Tokizane, and N. Furushiro, The Japan Society for Research on Superplasticity, Osaka, Japan, 1991.

73. T. Takasugi, S. Rikukawa, and S. Hanada, 'Superplasticity in L1₂ Type Ni₃(Si,Ti) Intermetallics,' in *International Conference on Superplasticity in Advanced Materials (ICSAM–91)*, pp. 329–338, ed. S. Hori, M. Tokizane, and N. Furushiro, The Japan Society for Research on Superplasticity, Osaka, Japan, 1991.

74. T. Takasugi, S. Rikukawa, and S. Hanada, 'The Boron Effect on the Superplastic Deformation of Ni₃(Si,Ti) Alloys,' *Scr. Metall. Mater.*, **25** (1991), pp. 889–894.

75. T. Takasugi, S. Rikukawa, and S. Hanada, 'Superplastic Deformation in Ni₃(Si,Ti) Alloys,' *Acta Metall. Mater.*, **40** (1992), pp. 1895–1906.

Chapter 11

Internal-stress superplasticity (ISS)

In addition to the fine-structure superplasticity (FSS) described in the previous chapters, there is another type of superplasticity known as internal-stress super-plasticity (ISS) [1]. In these materials, in which internal stresses can be developed, considerable tensile plasticity can take place under the application of a low, externally applied stress. This is because internal-stress superplastic materials can have a strain-rate-sensitivity exponent as high as unity; i.e., they can exhibit ideal Newtonian viscous behavior. Such superplastic materials are believed to be deformed by a slip-creep mechanism.

There are many ways in which internal stresses can be generated. These include thermal cycling of composite materials, such as whisker- and particulate-reinforced composites, in which the constituents have different thermal expansion coefficients [2–9]; thermal cycling of polycrystalline pure metals or single-phase alloys that have anisotropic thermal expansion coefficients [10, 11]; and thermal cycling through a phase change [12–15]. In addition, pressure-induced phase changes have been cited as a possible source of superplastic flow in geological materials [16, 17]. For example, there is a phase transformation in the earth's upper mantle, because of pressure, from orthorhombic olivine to a spinel phase at a depth of about 400 km below the earth's surface [18]. And it is believed that internal-stress superplasticity, arising from the transformation stresses through pressure cycling (analogous to temperature cycling), leads to a mixed-phase region of low effective viscosity.

11.1 Whisker- and particle-reinforced composites

It has been shown that internal-stress superplasticity can be utilized to enhance the ductility of metal-matrix composites that are normally brittle [2]. For example, several whisker-reinforced metal-matrix composites (aluminum containing 20% SiC$_w$ alloy) were made ideally superplastic, i.e., Newtonian viscous in nature, during deformation under thermal cycling conditions [2, 7]. An example of the exceptional tensile ductility that can be achieved in this manner, in a whisker-reinforced 6061 aluminum alloy, is shown in Figure 11.1. Whereas the metal-matrix composite exhibits only 12% elongation under isothermal creep deformation at 450 °C, the same composite exhibits 1400% elongation if deformed under thermal cycling conditions (100 ↔ 450 °C at 100 s per cycle). Similar behavior was also demonstrated with a Zn–30 vol % Al$_2$O$_3$ particulate composite [3, 4]. Whereas this material exhibited essentially nil ductility when tested in tension at 300 °C, it exhibited Newtonian viscous behavior when deformed under thermal-cycling conditions, and elongations exceeding 150% were achieved. The concept of thermal-cycling superplasticity has also been used by Pickard and Derby in a SiC-reinforced, commercially pure aluminum composite [8]. In this case, the material was thermally cycled between approximately 130 to 450 °C, and it exhibited a tensile elongation of 150%, which is considerably higher than the 10% elongation under isothermal tensile deformation at 450 °C.

The basis of understanding the effect of internal stress on enhancing the ductility of metal-matrix composites has been proposed by Wu *et al.* [19]. During thermal cycling, internal stresses are developed at the interfaces between the metal matrix and the hard ceramic reinforcement. This is because the thermal expansion coefficient of the metal matrix is several times larger than that of the ceramic phase (e.g., coefficients of thermal expansion are 24×10^{-6} and

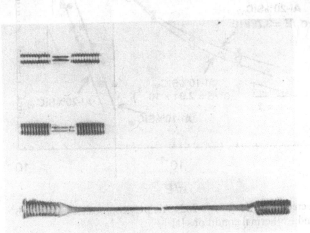

Figure 11.1 Exceptional tensile ductility obtained in a whisker-reinforced 6061 aluminum alloy by using a thermal-cycling technique. Undeformed sample (top), isothermally deformed sample (middle), and thermally cycled sample (bottom).

$4 \times 10^{-6}\ K^{-1}$ for Al and SiC, respectively). These internal stresses are relaxed by plastic deformation in the metal matrix to the value of the local interfacial yield stress of the material. It is this remaining local yield stress, which is defined as the internal stress, σ_i, that contributes to the low applied external stress and results in macroscopic deformation along the direction of the applied stress. A quantitative model is described in the next subsection. The creep behavior of a 2024 Al–SiC$_w$ composite under both thermal cycling and isothermal conditions is shown in Figure 11.2 [2]. The graph shows a plot of the diffusion-compensated creep rate as a function of the modulus-compensated stress. Three trends can be noted. First, the thermally cycled composites are much weaker than the iso-thermally tested composites at low applied stresses. Second, the thermally cycled composites have strain-rate-sensitivity exponents of unity at low stresses. Third, the thermally cycled samples and the isothermally tested samples yield data that converge at high stresses; this is expected since the internal stress generated by thermal cycling (a constant) will have a diminishing contribution to creep as the applied stress is increased.

It was also observed that the manner in which the silicon carbide whiskers rearranged during plastic deformation of the composite was dramatically differ-ent under thermal-cycling conditions compared to isothermal conditions [2, 6, 7]. For example, Wu *et al.* [2] have shown that regions devoid of whiskers in a 2024–20 vol % SiC$_w$ extruded composite become readily filled with whiskers

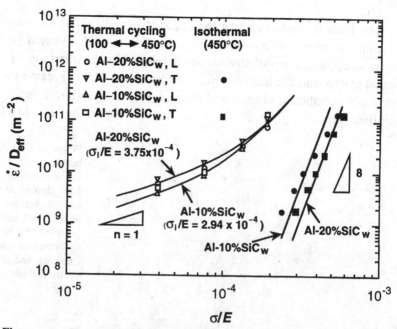

Figure 11.2   The creep behavior of two 2024 Al–SiC$_w$ composites under both thermal cycling and isothermal conditions [2].

when deformed in tension under thermal-cycling conditions. Another example showing the difference in the rearrangement of whiskers in the same extruded 2024–20 vol % SiC$_w$ composite was noted in tests conducted in compression under isothermal and thermal-cycling conditions [7]. This example is given in Figure 11.3. The sample that was deformed isothermally exhibited very limited reorientation of the whiskers toward the direction of compression flow. Furthermore, extensive surface cracks were observed in this sample. On the other hand, microstructural examination of the sample that was deformed under thermal-cycling conditions, to the same strain, showed that virtually none of the whiskers was oriented in the original longitudinal direction. Clearly, Newtonian viscous flow of the metal matrix accelerates the reorientation of the SiC whiskers. Also, no surface cracks were observed in the thermally cycled sample.

11.2 Anisotropic expanding polycrystalline materials

Other examples of internal-stress superplasticity are found in polycrystalline zinc and alpha uranium. In these two metals, the internal stress arises from the anisotropy of expansion coefficients present in these noncubic-structure materials. During thermal cycling, internal stresses are induced at the boundaries between adjacent grains.

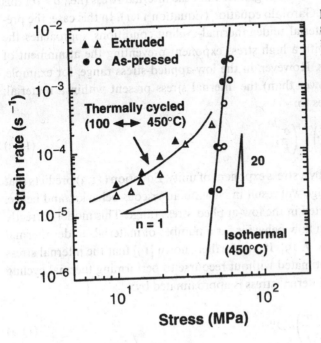

Figure 11.3 Example showing the difference in flow behavior in 2024–20 vol % SiC$_w$ composite, following either extrusion or pressing noted in tests conducted in compression under isothermal and thermal-cycling conditions.

A quantitative relation was developed by Wu *et al.* [19], to describe the thermal-cycling behavior of anisotropic polycrystalline materials and composites. The model begins with the Garofalo hyperbolic sine creep relation [20], which describes the steady-state creep rate of metals that are controlled by diffusion-assisted dislocation creep. This relation is given as:

$$\dot{\varepsilon} = \left(\frac{K}{\alpha^n}\right)\left(\frac{D_{eff}}{b^2}\right)\left(\frac{\sinh \alpha \sigma}{E}\right)^n \tag{11.1}$$

where $\dot{\varepsilon}$ is the creep rate, K and α are constants, D_{eff} is the effective diffusion coefficient, b is the Burgers vector, σ is the applied creep stress, E is Young's modulus, and n is the stress exponent. The value of n is typically in the range of 5 to 8. The theory of Wu *et al.* [19] uses Equation (11.1) on the assumption that the internal stresses generated during thermal cycling contribute to plastic flow in the following way. At any given time, half of the moving dislocations are aided by the presence of the internal stress, whereas the remaining half of the dislocations are opposed by the presence of the internal stress. The theory also assumes that these two groups of dislocations contribute to plastic flow independently of each other. The following equation results:

$$\dot{\varepsilon} = \frac{1}{2}\left(\frac{K}{\alpha^n}\right)\left(\frac{D_{eff}}{b^2}\right)\left\{\left[\sinh \alpha\left(\frac{\sigma+\sigma_i}{E}\right)\right]^n + \frac{|\sigma-\sigma_i|}{(\sigma-\sigma_i)}\left[\sinh \alpha\left|\frac{\sigma+\sigma_i}{E}\right|\right]^n\right\} \tag{11.2}$$

When the applied stress is much greater than the internal stress (i.e., $\sigma \gg \sigma_i$) this expression leads to the Garofalo equation (Equation 11.1). In this case, the prediction is that the material under thermal-cycling conditions approaches the isothermal behavior with a high stress exponent preventing the attainment of superplastic properties. However, in the low-applied-stress range, for example, on the order of (or lower than) the internal stress present within a material, Equation (11.2) reduces to:

$$\dot{\varepsilon} = Kn\left(\frac{D_{eff}}{b^2}\right)\left(\frac{\sigma_i}{E}\right)^{n-1}\left(\frac{\sigma}{E}\right) \tag{11.3}$$

which is characterized by a stress exponent of unity. Equation (11.3) predicts that deformation in this range will result in Newtonian viscous behavior, and hence, superplasticity is expected in the low-applied-stress range. This model correctly predicts the Newtonian flow behavior for a number of materials under thermal cycling conditions [3, 6, 7, 19]. It was further shown [19] that the internal stress term can, in fact, be estimated without recourse to performing thermal cycling tests. The value of the internal stress is approximated by:

$$\frac{\sigma_i}{E} = [2.2 \times 10^{-4}\left(\frac{b^2}{t}\right)KD]^{1/n} \tag{11.4}$$

where t is the time per cycle in seconds per cycle.

Equations (11.2) and (11.4) were used to predict the thermal-cycling behavior of polycrystalline Zn and α-U under different conditions of cycling and temperature ranges. The predictions are compared with experimental data in Figure 11.4. As can be seen, the thermal-cycling data are well predicted by Equation (11.2). At low values of the modulus-compensated stress, the model predicts, and the data demonstrate, the existence of ideal Newtonian viscous flow ($m=1$); under these experimental conditions, superplastic behavior is indeed observed [4, 10]. The thermal cycling data shown for the 2024–SiC$_w$ composite in Figure 11.2 were also predicted from Equation (11.2). It is evident in Figure 11.2 that the predicted curves fit quite well with the experimental data.

The dramatic increase in plastic strain during thermal cycling of a metal matrix composite has been successfully demonstrated to form MMC sheet. Shown in Figure 11.5. is a thin sheet of an MMC after bulge forming to a strain of 120% (with a normal tensile ductility of 7% at 350 °C) by cycling between 125 °C and 450 °C while applying a pressure of about 8 atmospheres [21]. This illustrates that thermal cycling is a promising method to make structural parts.

11.3 Materials undergoing polymorphic changes

Materials that undergo polymorphic changes with temperature can also exhibit Newtonian viscous behavior when tested under thermal-cycling conditions.

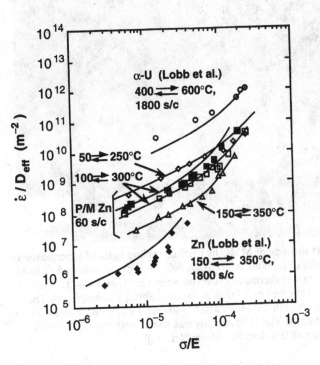

Figure 11.4 Comparison of the internal-stress creep model (Equations 11.2 and 11.4) with thermal cycling data for zinc and uranium (data from Refs. [10, 19]).

Among the early investigations on this subject on iron-based alloys were those by Sauveur [12], Wassermann [22], and more recently by de Jong and Rathenau [13], Clinard and Sherby [23], and Oelschlagel and Weiss [24]. The internal stress arising in this case is from the difference in molar volume between different

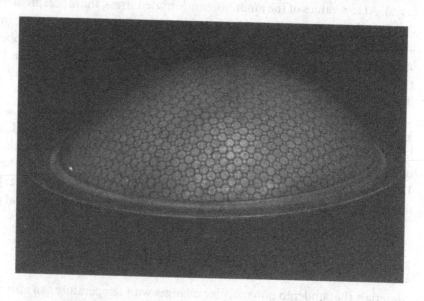

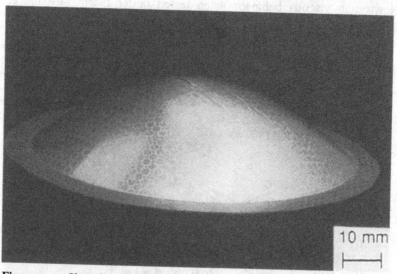

Figure 11.5   Sheet forming of MMC by thermal cycling induced superplasticity. Bulge tested samples of (a) 6061 Al–10 vol% SiC_p and (b) unreinforced 1100 Al alloy. The composite was thermally cycled between 125 °C and 450 °C, with a 6 minutes cycle, while stressed by a bulge pressure of about 8 atmospheres. The through-thickness strain in the center after 1140 cycles was about 120% with no sign of imminent failure. The 1100 Al alloy was tested isothermally at 350 °C and failed at a strain of less than 50% (from Ref. [21]).

phases during phase transformation. Such behavior is sometimes known as transformation plasticity. Application of ductility enhancement by cyclic phase transformation has been attempted with a number of polymorphic ceramics. Many of these ceramics are based on bismuth oxides. For example, it has been shown [15, 24–29] that high strain-rate-sensitivity exponents were obtained in many of the bismuth oxide-based materials under thermal-cycling conditions. But, it is pointed out that none of these ceramic systems has been convincingly shown to be superplastic because all tests were performed in compression. Since tensile tests were not performed, measurements of high values of the strain-rate-sensitivity exponent in compression are inconclusive evidence for superplasticity. As emphasized earlier, a high value of strain-rate-sensitivity exponent is a necessary but insufficient criterion for superplasticity. For example, we have pointed out in Chapter 6 that polycrystalline ceramics are susceptible to grain-boundary separation; therefore, it is possible that in the ceramic oxides studied for transformation plasticity, negligible ductility in tension may occur. This uncertainty should be clarified with appropriate additional experiments.

There exist other theoretical models [9] (known as the enhanced plasticity model or the repeatedly deforming model) to explain phase transformation superplasticity. In this type of model, the thermal stress combines with the applied load to exceed the material's yield stress, giving rise to a time-dependent strain which is biased according to the sense of the applied load. These models all have a form similar to that originally developed to describe the effect of the anisotropic expansion of polycrystalline uranium on creep strain [30]

$$\varepsilon_e = \frac{5}{6} \frac{\sigma}{\sigma_y} \Delta \alpha \Delta T \tag{11.5}$$

where $\Delta \alpha$ is the thermal expansion mismatch, σ_y is the uniaxial yield stress of the material, and ΔT is the thermal excursion.

For example, using the Levy-von Mises criterion and considering the biasing of internal strains by the externally applied stress σ, Greenwood and Johnson [31] derived a relationship for the strain $\Delta \varepsilon_t$ due to transformation plasticity:

$$\varepsilon_t = \frac{5}{3} \frac{\Delta V}{V} \frac{\sigma}{\sigma_y} \tag{11.6}$$

where $\Delta V/V$ is the fractional constrained volume change during the transformation and σ_y is the uniaxial yield stress of the weakest phase at the transformation temperature. Shown in Figure 11.6 are data obtained from thermal cycling of pure titanium. It is readily observed that the strain per cycle data are linearly dependent upon the applied stress, as predicted by Equation (11.6).

Interest in using phase transformation plasticity to manufacture structural parts is continuously high. Studies include that of Tozaki et al. on iron-based alloys [33], Furushiro et al. on commercially pure titanium [34], and Dunand and

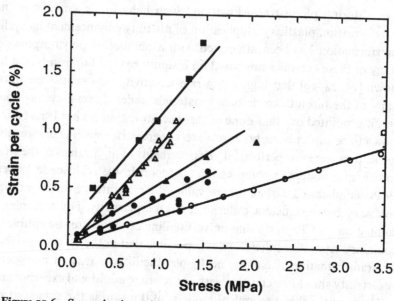

Figure 11.6   Superplastic strain per cycle upon various α–β–α cycling conditions of pure titanium as a function of applied stress (data from Ref. [32]).

Bedell on pure Ti and Ti-based composites [32]. It is particularly noted that Saotome and Iguchi [35, 36] have made *in situ* microstructural observations during transformation superplasticity in pure iron. They concluded that interphase boundary sliding was an important deformation mechanism. Such studies are worthy of pursuit, especially with respect to materials that are normally difficult to fabricate (such as polymorphic ceramics and highly ordered intermetallic compounds).

References

1. O.D. Sherby and J. Wadsworth, 'Internal Stress Superplasticity,' *Mater. Sci. Technol.*, **1** (1985), pp. 925.

2. M.Y. Wu, J. Wadsworth, and O.D. Sherby, 'Superplasticity in a Silicon Carbide Whisker Reinforced Aluminum Alloy,' *Scr. Metall.*, **18** (1984), pp. 773–776.

3. W.Y. Wu, Ph.D. Thesis, Dept. of Materials Science and Engineering, Stanford University, (1984).

4. M.Y. Wu, J. Wadsworth, and O.D. Sherby, 'Elimination of the Threshold Stress for Creep by Thermal Cycling in Oxide-Dispersion-Strengthened Materials,' *Scr. Metall.*, **21** (1987), pp. 1159–1164.

5. G. Gonzales-Doncel, R. McCann, A.P. Divecha, S.D. Karmarkar, S.H. Hong, and O.D. Sherby, 'Internal Stress Superplasticity in Metal Matrix Composite,' in *Proc. 19th SAMPE International Conf.*, pp. 619–629, SAMPE, Corvida, CA, 1987.

6. G. Gonzales-Doncel, S.D. Karmarkar, A.P. Divecha, and O.D. Sherby, 'Influence of Anisotropic Distribution of Whiskers on the Superplastic Behavior of Aluminum in a Back-Extruded 6061 Al–20% SiC$_w$ Composite,' *Comp. Sci. Technol.*, **35** (1989), pp. 105–120.

7. S.H. Hong, O.D. Sherby, A.P. Divecha, S.D. Karmarkar, and B.A. MacDonald, 'Internal

Stress Superplasticity in 2024 Al-SiC Whisker Reinforced Composites,' *J. Comp. Mater.*, **22** (1988), pp. 102–123.

8. S.M. Pickard and B. Derby, 'The Behavior of Metal Matrix Composites during Temperature Cycling,' *Mater. Sci. Eng.*, **A135** (1991), pp. 213–216.

9. B. Derby, 'The Mechanism of Internal Stress Superplasticity,' in *Superplasticity in Metals, Ceramics, and Intermetallics, MRS Proceeding No. 196*, pp. 115–120, ed. M.J. Mayo, J. Wadsworth, and M. Kobayashi, Materials Research Society, Pittsburgh, PA, 1990.

10. R.C. Lobb, E.C. Sykes, and R.H. Johnson, 'The Superplastic Behavior of Anisotropic Metals Thermally-Cycled under Stress,' *Met. Sci.*, **6** (1971), pp. 33–39.

11. M.Y. Wu, J. Wadsworth, and O.D. Sherby, 'Internal Stress Superplasticity in Anisotropic Polycrystalline Zinc and Uranium,' *Metall. Trans.*, **18A** (1987), pp. 451–462.

12. A. Sauveur, 'What is Steel? Another Answer.,' *Iron Age*, **113** (1924), pp. 581–583.

13. M. de Jong and G.W. Rathenau, 'Mechanical Properties of an Iron-Carbon Alloy During Allotropic Transformation in Pure Iron,' *Acta Metall.*, **9** (1961), pp. 714–720.

14. O.A. Ruano, J. Wadsworth, and O.D. Sherby, 'Enhanced Densification of White Cast Iron Powders by Cycling Phase Transformation under Stress,' *Metall. Trans.*, **13A** (1982), pp. 355–361.

15. L.A. Xue and R. Raj, 'Superplastic Deformation of Zinc Sulfide near Transformation Temperature (1020°C),' *J. Am. Ceram. Soc*, **72** (1989), pp. 1792–1796.

16. Y. Maehara and T.G. Langdon, 'Review: Superplasticity in Ceramics,' *J. Mater. Sci.*, **25** (1990), pp. 2275–2286.

17. T.G. Nieh, J. Wadsworth, and F. Wakai, 'Recent Advances in Superplastic Ceramics and Ceramic Composites,' *Inter. Mater. Rev.*, **36**(4) (1991), pp. 146–161.

18. C.M. Sung and R.G. Burns, 'Kinetics of High Pressure Phase Transformation: Implications to the Evolution of the Olivine-Spinel Transformation in the downgoing Lithosphere and its Consequences on the Dynamics of the Mantle,' *Tectonophy.*, **31** (1976), pp. 1–31.

19. M.Y. Wu, J. Wadsworth, and O.D. Sherby, 'Internal Stress Superplasticity in Anisotropic Polycrystalline Zinc and Uranium,' *Metall. Trans.*, **18A** (1987), pp. 451–462.

20. F. Garofalo, 'An Empirical Relation Defining the Stress Dependence of Minimum Creep Rate in Metals,' *Trans. TMS-AIME*, **227** (1963), pp. 351–356.

21. Y.C. Chen, G.S. Daehn, and R.H. Wagoner, 'The Potential for Forming Metal Matrix Composite Components via Thermal Cycling,' *Scr. Metall. Mater.*, **24** (1990), pp. 2157–2162.

22. G. Wasserman, 'Untersuchungen an Eisen-Nickel-Legierung uber die Verformbarkeit wahrend der γ–α Unwandlung,' *Archiv Fur der Eisenhutt.*, **7** (1937), pp. 321–325.

23. F. Clinard and O.D. Sherby, 'Strength of Iron During Allotropic Transformation,' *Acta Metall.*, **12** (1964), pp. 911–919.

24. D. Oelschlagel and V. Weiss, 'Superplasticity of Steels During the Ferrite-Austenite Transformation,' *Trans. Am. Soc. Metals*, **59** (1966), pp. 143–154.

25. J.R. Smythe, R.C. Bradt, and J.H. Hoke, 'Transformation Superplasticity in the Bi_2O_3-Sm_2O_3 Eutectoid System,' *J. Amer. Ceram. Soc.*, **58** (1975), pp. 381–382.

26. J.R. Smythe, R.C. Bradt, and J.H. Hoke, 'Isothermal Deformation in Bi_2O_3-Sm_2O_3,' *J. Mater. Sci.*, **12** (1977), pp. 1495–1502.

27. L.A. Winger, R.C. Bradt, and J.H. Hoke, 'Transformational Superplasticity of Bi_2WO_6 and Bi_2MoO_6,' *J. Amer. Ceram. Soc.*, **63**(5–6) (1980), pp. 291–294.

28. C.A. Johnson, R.C. Bradt, and J.H. Hoke, 'Transformational Plasticity in Bi_2O_3 Eutectoid System,' *J. Amer. Ceram. Soc.*, **58**(1–2) (1975), pp. 37–40.

29. C.A. Johnson, J.R. Smythe, R.C. Bradt, and J.H. Hoke, 'Transformational Superplasticity in Pure Bi_2O_3 and the Bi_2O_3-Sm_2O_3 Eutectoid System,' in *Deformation of Ceramic Materials*, ed. R.C. Bradt and R.E. Treisler, Plenum Press, New York, 1975, pp. 443–453.

30. A.C. Roberts and A.H. Cottrell, 'Creep of α-Uranium during Irradiation with Neutrons,' *Phil. Mag.*, **1** (1956), pp. 711–717.

31. G.W. Greenwood and R.H. Johnson, 'The Deformation of Metals under Small Stresses during Phase Transformation,' *Proc. R. Soc. Lond. A*, **283** (1965), pp. 403–422.

32. D.C. Dunand and C.M. Bedell, 'Transformation-Mismatch Superplasticity in Reinforced and Unreinforced Titanium,' *Acta Mater.*, **44** (1995), pp. 1063–1076.

33. H. Tozaki, Y. Uesugi, T. Okada, and I. Tamura, 'Analysis of Phase Transformation Superplasticity by Using Continuum Mechanics,' *J. JIM*, **50** (1986), pp. 56–62.

34. N. Furushiro, H. Kuramoto, Y. Takayama,
    and S. Hori, 'Fundamental Characteristics of
    the Transformation Superplasticity in a
    Commercially-Pure Titanium,' *Trans. ISIJ,* **27**
    (1987), pp. 725–729.

35. Y. Saotome and N. Iguchi, 'In-Situ
    Microstructural Observations and Micro-Grid
    Analyses of Transformation Superplasticity in
    Pure Iron,' *Trans. ISIJ,* **27** (1987), pp.
    696–704.

36. Y. Saotome and N. Iguchi. 'Microstructrual
    Changes and Straining Behavior in
    Transformation Superplasticity in Iron and
    Steels,' in *Superplasticity in Metals, Ceramics,
    and Intermetallics, MRS Proceeding No. 196,*
    pp. 105–114, ed. M.J. Mayo, J. Wadsworth,
    and M. Kobayashi, Materials Research
    Society, Pittsburgh, PA, 1990.

Chapter 12

Other possible superplasticity mechanisms

In addition to the descriptions of superplastic behavior given in the previous chapters, there are other examples of superplastic or potential superplastic mechanisms. These mechanisms, including deformation processes controlled by lattice dislocations (with a strain-rate-sensitivity exponent, $m=0.33$) and by Newtonian viscous flow ($m=1$), will be discussed. In addition, the phenomenon of apparent large tensile elongations observed at ultrahigh strain rates ($>10^3$ s^{-1}), will also be described, although the exact deformation mechanism is not yet clear.

12.1 Class I superplasticity in coarse-grained materials

Class I solid solutions are a group of alloys in which the glide segment of the glide/climb dislocation creep process is rate controlling because solute atoms impede dislocation motion [1–3]. The plastic flow behavior of this group of alloys can be described by [4]:

$$\dot{\varepsilon}=\frac{0.33kTD_s\sigma^3}{G^4b^5e^2c},$$
(12.1)

where k is Boltzman's constant, T is the absolute temperature, D_s is the diffusion coefficient of the solute, σ is the applied stress, G is the shear modulus, b is the length of Burgers vector, c is the concentration of the solute, and e is the atomic size difference between the solute and solvent atoms. This group of alloys is of interest in the coarse-grained condition for superplastic studies because, as a result of the glide-controlled creep mechanism, they have an intrinsically high-strain-rate sensitivity of about $m=0.33$ (over certain temperature and strain-rate ranges) and, therefore, are expected to exhibit high elongations [5, 6]. The intrinsic nature of the high-strain-rate sensitivity is important because it suggests that complex thermomechanical processing, such as that needed for developing fine-grained superplastic alloys, is unnecessary in Class I solid solutions.

Because the strain-rate sensitivity in Class I solid solutions is not as high as that found in classical superplastic materials (i.e., $m=0.33$ vs $m=0.5$ to 1.0), the elongations to failure are relatively modest, i.e., about 200 to 400%. It is probable that some early reports of superplasticity in coarse-grained alloys (e.g., Al–Mg [3] and W–Re [7]), are in fact the result of Class I solid-solution behavior.

Table 12.1 shows that coarse-grained Class I alloys based on W, Nb, and Al exhibit large tensile elongations at elevated temperatures ($T/T_m>0.4$). It is particularly pointed out that Class I solid-solution behavior is commonly observed in refractory metals. This is because refractory metals usually form complete solid solutions (or exhibit extended mutual solid solubility) with other refractory metals [8]. Alloys are listed in Table 12.1 that exhibit a strain-rate sensitivity of $m\approx0.33$ and that are also believed to fulfill the criteria for Class I solid solutions. These criteria include atomic size mismatch* and modulus [1, 2] as well as chemical diffusion considerations [9].

Other materials showing $m=0.33$ include: In–Pb, Pb–In, Au–Ni, NaCl–KCl, Fe–Al–C, In–Hg, Pb–Sn, Ni–Sn, Mg–Al, Al–Cu–Mg, In–Sn, Pb–Cd, β-brass, Ti–Al–V, In–Cd, Cu–Sn, Ni–Au, Al–Cu, and Al–Mg–0.1Zr [1, 2, 20, 21].

Figures 12.1 and 12.2 show examples of creep behavior exhibiting a stress exponent of $n=3$ ($m=0.33$) in Nb-based alloys, including Nb–10Hf–1Ti (C103), Nb–5.8Hf–5.6W (C129), and Nb–19.2Hf–5.6W (WC3009), and the concurrent, uniform, large tensile elongation for the Nb–10Hf–1Ti alloy [10]. For the case of C103, other typical features of Class I solid-solution behavior have also been found. These include the following: creep curves of the C103 alloy do not exhibit primary creep, and a large elongation is observed in relatively coarse-grained (75–100 μm) samples at intermediate temperatures and strain rates (thus removing the likelihood of significant grain-boundary sliding as the mechanism, as is the case in classical superplasticity). In addition, the measurement of activation energy led to the conclusion that Hf atoms, rather than Ti atoms, were responsible for the solute effects on dislocation motion that lead to control by dislocation glide [10]. Similar behavior was also observed in other Nb-based alloys. A detailed review of Class I behavior in Nb-based alloys is given in Reference [8].

Creep deformation behavior exhibiting a value of $n=3$ ($m=0.33$) can be

---

*The atomic size factor is defined as: $|\Omega|=|\Omega_B-\Omega_A|$, where Ω_B is the effective atomic volume of a solute B in a matrix A, and Ω_A is the effective atomic volume of solution A.

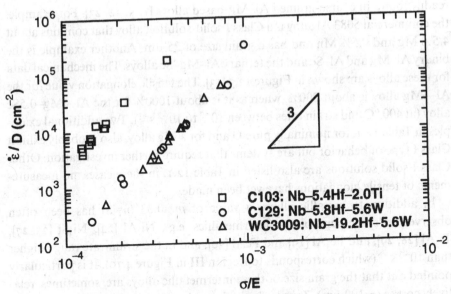

Figure 12.1 Diffusion-compensated strain rate vs modulus-compensated stress for C103, C129, and WC3009 alloys from several studies exhibiting $n=3$ ($m=0.33$) (data from Ref. [10]).

Table 12.1 Alloys exhibiting Class I solid-solution behavior

Alloys/reference	m	d (μm)	Maximum elongation (%)
W–33 at. % Re [7]	0.2–0.3	50–400	260
Nb–10Hf–1Ti [10]	0.33	75	>125
Nb–5V–1.25Zr [11]	0.26	ASTM #8	170
Nb–29Ta–8W–0.65Zr–0.32C [12]	0.28	NA[a]	>80
Al–5456 (Al–5Mg-based) [3]	0.33	20	153
Al–4Mg–0.5Sc [13]	0.3	0.5[b]	>1020
Al–2 to 4% Ge [14]	0.3–0.6	100–200	200–260
Al–2.2Li–0.5Sc [15]	0.32	3	500
Al–Ca–Zn [16]	0.25	2	NA[a]
Mg–5.5 Zn–0.5Zr [17]	0.33	NA[a]	~100
β–U (Ag, Al, Zn impurities) [18]	0.33	100	680
Fe$_3$Al-Ti [19]	0.33	100	333

[a]NA=Not available.

[b]Subgrain size.

readily found in coarse-grained Al–Mg-based alloys [1, 5, 22, 23]. For example, the commercial 5083 Al alloy is a Class I solid-solution alloy that contains about 4.5% Mg and 0.5% Mn and has a grain size of 25 μm. Another example is the binary Al–Mg and Al–Sc and the ternary Al–Mg–Sc alloys. The mechanical data for these alloys are shown in Figure 12.3 [13]. The tensile elongation value for the Al–4Mg alloy is about 200%, whereas it is about 1000% for the Al–4Mg–0.5Sc alloy (at 400 °C and strain rates between 10^{-3} to 10^{-2} s^{-1}). Two additional examples in Table 12.1, for nominally pure U and for a Mg alloy, also probably exhibit Class I type of behavior but are systems that require further investigation. Other Class I solid solutions are also listed in Table 12.1; in these cases, no measurements of tensile elongations have yet been made.

In addition to metallic alloys, a value of $m \approx 0.33$ ($n \approx 3$) has been often observed in some superplastic intermetallics, e.g., Ni_3Al [24], Ni_3Si [25–27], Ti_3Al [28, 29], and Fe_3Al [19] and FeAl [30], and in the strain-rate range higher than 10^{-3} s^{-1} (which corresponds to Region III in Figure 4.10). It is particularly pointed out that the grain size of these intermetallic alloys are, sometimes, relatively coarse ($\sim$100 μm). Tensile elongation values, depending on the materials,

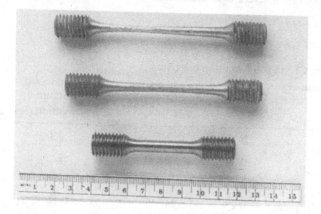

Figure 12.2 Large, uniform, tensile plastic strains of C103 (over 100%) (from Ref. [10]).

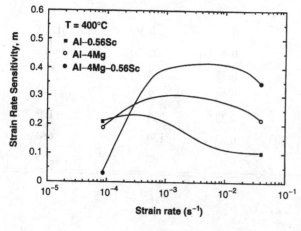

Figure 12.3 Strain-rate sensitivity as a function of strain rate for some of the Al–(Mg)–0.5Sc alloys at 400 °C (from Ref. [13]).

are typically over 150%. Yang *et al.* [31] have argued that, when disorder is introduced into an ordered solid by the glide of a dislocation, the steady-state velocity is limited by the rate at which chemical diffusion can reinstate order behind the glide dislocation. In such cases, the deformation is controlled by a viscous glide process.

There is yet another possible explanation. Because intermetallics are often nonstoichiometric, the solute drag process may be a result of the nonstoichiometry of these alloys. In addition, intermetallic alloys usually contain certain amounts of interstitial impurities, e.g., oxygen, nitrogen, and carbon. The presence of these impurities may also contribute to the drag process [32]. Yaney and Nix [33] have suggested that the addition of solute atoms is not the only way to cause dislocations to move in a viscous manner. Lattice friction effects may also reduce glide mobility, particularly in covalently bonded solids (such as Ge and Si) and ordered intermetallic compounds. In this latter group, strong repulsive forces exist between atoms of like character. Thus, the deformation behavior of ordered intermetallics, in which lattice friction effects limit glide mobility, is expected to be quite similar to that of Class I solid solutions.

Although extended ductility is achievable with Class I solid-solution alloys, it is to be pointed out that such materials generally do not have high strength at low temperatures. These materials are not heat treatable. Moreover, a superplastically formed part usually is in its softest state. Therefore, these materials are used mostly as secondary structural components.

12.2 Viscous creep mechanisms for superplasticity

At least three creep mechanisms yield a stress exponent, n, of unity and therefore behave in a Newtonian viscous manner (i.e., $m=1$). These mechanisms are the two diffusional-based mechanisms of Nabarro–Herring (N-H) and Coble creep, and the slip-creep mechanism of Harper–Dorn (H-D) creep. It should be noted that Coble creep has often been considered and incorporated into models of superplastic flow [34–36]. Furthermore, because some studies [37–39] suggest that H-D creep usually predominates over N-H creep, the potential for H-D creep to yield superplastic behavior is emphasized here.

H-D creep is a slip-creep mechanism that has been the subject of great interest in recent years [37–41]. The mechanism yields a stress exponent of unity and is usually observed at high temperatures and modest strain rates. A deformation mechanism map is shown in Figure 12.4 in which grain size compensated by the Burgers vector is plotted on the abscissa and modulus-compensated stress is plotted on the ordinate at a fixed homologous temperature. It is noted that H-D creep dominates at relatively coarse grain sizes and low stresses.

Because H-D creep has, to the present time, only been observed at low stresses, the associated low creep rates have not permitted measurements of strain to failure. The fact that ideal Newtonian viscous behavior is found in materials deforming by the H-D creep mechanism leads to the intriguing possibility of extensive plastic flow in these pure metals and alloys. This could be of technological importance if H-D creep can be developed at commercially interesting strain rates, i.e., $\dot{\varepsilon} > 10^{-3}$ s^{-1}. The likelihood of such a development depends upon an understanding of the physical basis of H-D creep. It is currently believed that the origin of H-D creep is the same as that obtained in ISS materials (described in Chapter 11). The internal stress in the case of H-D creep arises from the presence of random, stationary dislocations existing within subgrains. The equation for H-D creep is similar to Equation (11.3). The key physical parameters controlling the strain rate at which H-D creep can be expected to dominate include high dislocation density, low stacking fault energy, and high self-diffusivity.

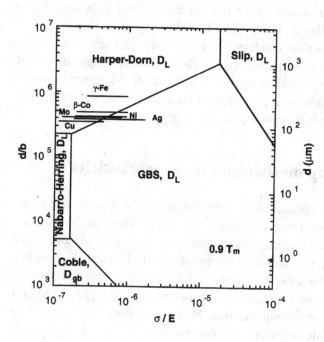

Figure 12.4 A deformation mechanism map showing grain size compensated by the Burgers vector vs modulus-compensated stress at a fixed homologous temperature.

12.3 Ultrahigh-strain-rate superplasticity

The final example of a viscous-like creep mechanism is that found at extremely high-strain-rates, i.e., up to approximately 10^3 to 10^5 s^{-1}. Such high strain rates are in excess of those used in high rate-forming operations such as explosive forming and dynamic compaction. The evidence for possible superplastic behavior at strain rates of 10^4 to 10^5 s^{-1} has been reported [42, 43]. In this work, flash X-rays were taken of Al and Cu shape-charged liners undergoing large apparent tensile elongations. In these experiments, a shape-charged liner (i.e., a hollow cone) undergoes a shape change to a long rod, in the solid state, at extremely high strain rates of up to 10^5 s^{-1}; this is demonstrated in Figure 12.5. The X-ray data support the contention that deformation is occurring in the solid state, i.e., well below the melting point of the metals.

Notably, there are a number of studies of deformation at high strain rates in

Figure 12.5 Flash radiograph of an aluminum jet (from Ref. [42, 43]).

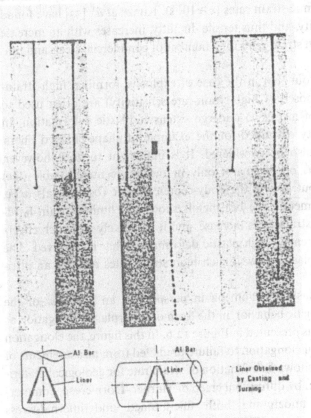

which high-strain-rate sensitivities have been reported, for example, in LiF crystals [44] and in composite materials [45]. Raw data on metals from the work by Follansbee [46] suggest that apparently high-strain-rate sensitivities are found at ultrahigh strain rates. The interpretation of such data, however, has been discussed by Follansbee, who has questioned whether such high-strain-rate sensitivities are real or are instead the result of testing techniques as well as structural changes with strain rate. On the other hand, it has been argued that adiabatic heating during high-strain-rate deformation may cause dynamic recrystallization and, thus, the subsequent observation of extensive ductility [47–50]. For example, it has been demonstrated that deformation at an extremely high strain rate can result in the formation of nanocrystalline grains [51, 52]. In addition, temperature rise resulting from adiabatic heating during dynamic deformation can, sometimes, be several hundred degrees (e.g., over 500 °C in Cu [49]). As a result of the temperature rise, and the fact that the material has a nanocrystalline structure, grain boundary sliding can readily take place in an extremely fast rate (recall that: $\dot{\varepsilon} \propto d^{-3}$, where $p=2-3$). This leads to the observation of extensive tensile ductility at dynamic strain rates ($\dot{\varepsilon} \geq 10^3$ s). Kunze et al. [53] have found that strain-rate-sensitivity, and thus tensile ductility, increases with an increase in strain rate at very high strain rates for a number of commercial steels and titanium alloys.

It is worth pointing out that in the case of explosive forming, high-strain-hardening rates (as opposed to high-strain-rate sensitivity) are often used to explain the capability of metals to undergo extensive plastic deformation. In the case of the plasticity described for the example of shape-charged liners, this viewpoint should also be considered. It is important to note, however, that the phenomenon is observed not only in the face-centered cubic (fcc) metals of Al and Cu, but also in the body-centered cubic (bcc) metals α-Fe, Ta, and W. Unlike fcc metals, the bcc metals show very limited strain hardening characteristics as strain rates increase, and it is unlikely that such effects can account for the extremely high plastic deformation that is observed. The area of potential superplastic flow at such high strain rates remains an intriguing one.

To place the superplastic phenomena in perspective, an overview of the various types of superplastic behavior, in the form of superplastic elongation as a function of strain rate, is presented in Figure 12.6. In this figure, the elongation to failure, or the potential elongation to failure predicted from measurements of strain-rate sensitivity, is shown as a function of strain rate for geological materials, materials that deform by diffusional creep or Harper–Dorn creep, commercial aluminum alloys undergoing both fine-grained and internal-stress superplasticity, the superplastic YTZP ceramic, and mechanically alloyed aluminum and nickel alloys. In addition, the predicted range of superplastic behavior in nanophase (grain size approximately 10 μm) materials and the possible

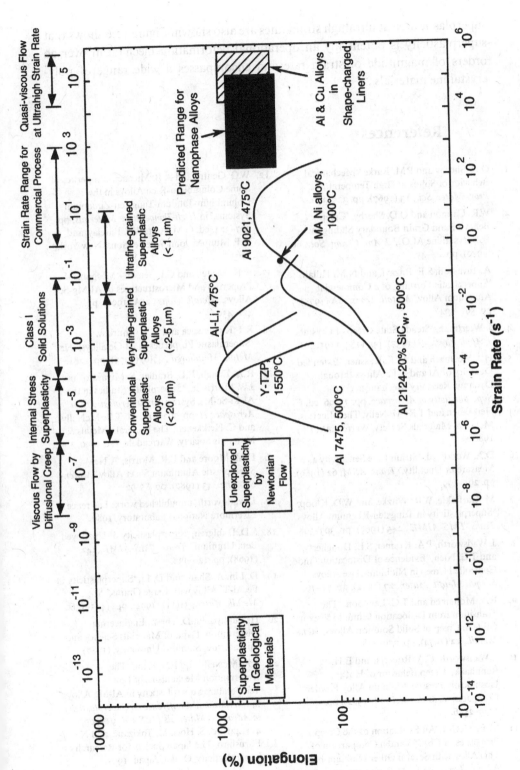

Figure 12.6 Overview of the various types of superplastic behavior in the form of superplastic elongation as a function of strain rate.

superplastic area at ultrahigh strain rates are also shown. Figure 12.6 shows that superplasticity is potentially an operational deformation mechanism over 20 orders of magnitude of strain rate and encompasses a wide range of polycrystalline materials.

References

1. O.D. Sherby and P.M. Burke, 'Mechanical Behavior of Solids at High Temperatures,' *Prog. Mater. Sci.*, **13** (1967), pp. 325–390.

2. W.R. Cannon and O.D. Sherby, 'Creep Behavior and Grain Boundary Sliding in Polycrystalline Al$_2$O$_3$,' *J. Am. Ceram. Soc.*, **60** (1970), pp. 44–47.

3. A. Tavassoli, S.E. Razavi, and N.M. Fallah, 'Superplastic Forming of a Commercial Aluminum Alloy,' *Metall. Trans.*, **6A** (1975), pp. 591–594.

4. J. Weertman, 'Steady State Creep of Crystals,' *J. Appl. Phys.*, **28** (1957), pp. 1185–1191.

5. H.J. McQueen and M.E. Kassner, 'Extended Ductility of Al and α-Fe Alloys through Dynamic Recovery Mechanisms,' in *Superplasticity in Aerospace*, pp. 77–96, ed. C. Heikkenen and T.R. McNelly, The Minerals, Metals & Materials Society, Warrendale, PA, 1988.

6. D.A. Woodford, 'Strain Rate Sensitivity as a Measure of Ductility,' *Trans. ASM*, **62** (1969), pp. 291–299.

7. M. Garfinkle, W.R. Witzke, and W.D. Klopp, 'Superplasticity in Tungsten-Rhenium Alloys,' *Trans. TMS-AIME.*, **245** (1969), pp. 303–308.

8. J. Wadsworth, P.A. Kramer, S.E. Dougherty, and T.G. Nieh, 'Evidence of Dislocation Glide Controlled Creep in Niobium-Base Alloys,' *Scripta Metall. Mater.*, **27** (1992), pp. 71–76.

9. F.A. Mohamed and T.G. Langdon, 'The Transition from Dislocation Climb to Viscous Glide in Creep of Solid Solution Alloys,' *Acta Metall.*, **22** (1974), pp. 779–788.

10. J. Wadsworth, C.A. Roberts, and E.H. Rennhack, 'Creep Behavior of Hot Isostatically Pressed Niobium Alloy Powder Compacts,' *J. Mater. Sci.*, **17** (1982), pp. 2539–2546.

11. M.E. McCoy, 'An Evaluation of the Creep Properties of Cb-753 and a Comparison of This Alloy with Several Other Niobium-base Alloys,' *Trans. ASM*, **59** (1966), pp. 277–287.

12. W.O. Gentry and A.B. Michael, 'Properties of Some Columbium-Rich Alloys in the Columbium-Tatalum-Tungsten-Zirconium Systems,' in *High Temperature Materials*, pp. 307–324, ed. G.M. Ault, W.F. Barclay, and H.P. Munper, John Wiley & Sons, New York, 1963.

13. R.R. Sawtell and C.L. Jensen, 'Mechanical Properties and Microstructures of Al-Mg-Sc Alloys,' *Metall. Trans.*, **21A** (1990), pp. 421–430.

14. R.I. Kuznetsova and N.N. Zhurkov, 'Superplastic Properties in Al-Ge alloys,' *Fiz. Metal., Metalloved.*, **47**(6) (1979), pp. 1281.

15. R.A. Emigh, E.L. Bradley, S. Miyasato, and J.W. Morris Jr., 'Superplastic Studies in the Al-Li-Sc-Mg System,' in *Superplasticity in Aerospace II*, pp. 303–315, ed. T.R. McNelly and C. Heikkenen, The Minerals, Metals & Materials Society, Warrendale, PA, 1990.

16. D.M. Moore and L.R. Morris, 'A New Superplastic Aluminum Sheet Alloy,' *Mater. Sci. Eng.*, **43** (1980), pp. 85–92.

17. J. Wadsworth, unpublished work, Lawrence Livermore National Laboratory, 1989.

18. S.D. Dahlgren, 'Superplasticity of Unalloyed Beta Uranium,' *Trans. TMS-AIME*, **242** (1968), pp. 126–132.

19. D. Lin, A. Shan, and D. Li, 'Superplasticity in Fe$_3$Al-Ti Alloy with Large Grains,' *Scr. Metall. Mater.*, **31**(11) (1994), pp. 1455–1460.

20. H. Fukuyo, Ph.D. Thesis, Engineering Dissertation, Dept. of Materials Science and Engineering, Stanford University, (1987).

21. T.R. McNelly and P.N. Kalu, 'The Deformation Mechanisms of Low Temperature Superplasticity in Al-Mg Alloys,' in *International Conference on Superplasticity in Advanced Materials (ICSAM-91)*, pp. 413–421, ed. S. Hori, M. Tokizane, and N. Furushiro, The Japan Society for Research on Superplasticity, Osaka, Japan, 1991.

22. H.J. McQueen and M.E. Kassner, 'Behavior

of Al-Mg Alloys in Hot Working and Superplasticity,' in *Superplasticity in Aerospace II*, pp. 77–96, ed. T.R. McNelly and C. Heikkenen, The Minerals, Metals & Materials Society, Warrendale, PA, 1990.

23. E. Taleff, G.A. Henshall, D.R. Lesuer, and T.G. Nieh, 'Warm Formability of Aluminum-Magnesium Alloys,' in *Aluminum Alloys Vol. I – Their Physical and Mechanical Properties (ICAA4)*, pp. 338–345, ed. T.H. Sanders Jr. and E.A. Starke Jr., The Georgia Institute of Technology Press, 1994.

24. J. Mukhopadhyay, G.C. Kaschner, and A.K. Muhkerjee, 'Superplasticity and Cavitation in Boron Doped Ni_3Al,' in *Superplasticity in Aerospace II*, pp. 33–46, ed. T.R. McNelly and C. Heikkenen, The Minerals, Metals & Materials Society, Warrendale, PA, 1990.

25. T.G. Nieh and W.C. Oliver, 'Superplasticity of a Nickel Silicide,' *Scr. Metall.*, **23** (1989), pp. 851–854.

26. S.L. Stoner and A.K. Muhkerjee, 'Superplasticity in Fine Grain Nickel Silicide,' in *International Conference on Superplasticity in Advanced Materials (ICSAM-91)*, pp. 323–328, ed. S. Hori, M. Tokizane, and N. Furushiro, The Japan Society for Research on Superplasticity, Osaka, Japan, 1991.

27. T. Takasugi, S. Rikukawa, and S. Hanada, 'Superplastic Deformation in $Ni_3(Si,Ti)$ Alloys,' *Acta Metall. Mater.*, **40** (1992), pp. 1895–1906.

28. H.S. Yang, P. Jin, E. Dalder, and A.K. Muhkerjee, 'Superplasticity in a Ti_3Al-base Alloy Stabilized by Nb, V, and Mo,' *Scr. Metall. Mater.*, **25** (1991), pp. 1223–1228.

29. A. Dutta and D. Banerjee, 'Superplastic Behavior in a Ti_3Al-Nb Alloy,' *Scr. Metall. Mater.*, **24** (1990), pp. 1319–1322.

30. D. Li, A. Shan, Y. Liu, and D. Lin, 'Study of Superplastic Deformation in an FeAl Based Alloy with Large Grains,' *Scr. Metall. Mater.*, **33** (1995), pp. 681–685.

31. H.S. Yang, W.B. Lee, and A.K. Muhkerjee, 'Superplastic Characteristics of Two Titanium Aluminides: γ-TiAl and α_2-Ti_3Al,' in *Structural Intermetallics*, pp. 69–76, ed. R. Darolia, J.J. Lewandowski, C.T. Liu, P.L. Martin, D.B. Miracle, and M.V. Nathal, The Minerals, Metals & Materials Science, Warrendale, PA, 1993.

32. T.G. Nieh, T.C. Chou, J. Wadsworth, D. Owen, and A.H. Chokshi, 'Creep of a Niobium Beryllide, Nb_2Be_{17},' *J. Mater. Res.*, **8**(4) (1993), pp. 757–763.

33. D.L. Yaney and W.D. Nix, 'Mechanisms of Elevated-Temperature Deformation in the B2 Aluminides NiAl and CoAl,' *J. Mater. Sci.*, **23** (1988), pp. 3088–3098.

34. *Superplastic Forming of Structural Alloys*, edited by N.E. Paton and C.H. Hamilton, TMS-AIME, Warrendale, PA, 1982.

35. J.K. Gregory, J.C. Gibeling, and W.D. Nix, 'High Temperature Deformation of Ultra-Fine-Grained Oxide Dispersion Strengthened Alloys,' *Metall. Trans.*, **16A** (1985), pp. 777–787.

36. J. Karch, R. Birringer, and H. Gleiter, 'Ceramics Ductile at Low Temperature,' *Nature*, **330** (1987), pp. 556–558.

37. M.Y. Wu and O.D. Sherby, 'Unification of Harper-Dorn and Power Law Creep through Consideration of Internal Stress,' *Acta Metall.*, **32**(9) (1984), pp. 1561–1572.

38. O.A. Ruano, J. Wadsworth, and O.D. Sherby, 'Harper-Dorn Creep in Pure Metals,' *Acta Metall.*, **36**(4) (1988), pp. 1117–1128.

39. O.A. Ruano, J. Wadsworth, and O.D. Sherby, 'Harper-Dorn and Power Law Creep in Fe–3wt%Si,' *Scr. Metall.*, **22** (1988), pp. 1907–1910.

40. J.N. Wang, 'On the Transition from Power Law Creep to Harper-Dorn Creep,' *Scr. Metall. Mater.*, **29** (1993), pp. 733–736.

41. J.N. Wang, 'Comments on the Internal Stress Model of Harper-Dorn Creep,' *Scr. Metall. Mater.*, **29** (1993), pp. 1267–1270.

42. F. Jamet, 'Investigation of Shaped Charge Jets Using Flash X-Ray Diffraction,' in *8th International Symposium on Ballistics*, pp. VI-v6, ed. W.G. Reinecke, AVCO System Division, Wilmington, MA, 1984.

43. F. Jamet, *La Diffraction Instantanee*, Report CO 227/84, Institut St. Louis, France, 1984.

44. J.J. Gilman, 'The Strength of Ceramic Crystals,' in *The Physics and Chemistry of Ceramics*, ed. C. Klingsberg, Grodon and Breach, New York, 1963, pp. 240–265.

45. J. Lankford, 'Temperature, Strain Rate and Fibre Orientation Effects in the Compressive Fracture of SiC Fibre-Reinforced Glass-Matrix Composites,' *Composites*, **18**(2) (1987), pp. 145–154.

46. P.S. Follansbee, ed. *High-Strain-Rate Deformation of FCC Metals and Alloys*, pp.

451–479, ed. P.S. Follansbee, Marcel Dekker, Inc., New York, 1986.

47. M. Hatherly and A.S. Malin, 'Shear Bands in Deformed Metals,' *Scr. Metall.*, **18** (1984), pp. 449–454.

48. A.H. Chokshi and M.A. Meyers, 'The Perspects for Superplasticity at High Strain Rates: Preliminary Considerations and an Example,' *Scr. Metall. Mater.*, **24** (1990), pp. 605–610.

49. K.S. Vecchio, U. Andrade, M.A. Meyers, and L.W. Meyer, 'Microstructural Evolution in High Strain, High Strain-Rate Deformation,' in *Shock Compression of Condensed Matter 1991*, pp. 527–530, ed. S.C. Schmidt, R.D. Dick, J.W. Forbes, and D.G. Tasker, Elsevier Science Publishers B.V., 1992.

50. S.A. Atroshenko, Y.I. Meshcheryakov, E.V. Nesterova, and V.V. Rybin, 'On Dynamic Recrystallization at the Localized Sgear Bands Induced by Impact Loading,' *Phy. Met. Metallogr.*, **75**(4) (1993), pp. 427–436.

51. H. Chen, Y. He, G.J. Shiflet, and S.J. Poon, 'Plastic Deformation-Induced Nanocrystalline Aluminum in Al-Based Amorphous Alloys, Mater. Res. Soc. Symp. Proc. Vol. 321,' in *Crystallization and Related Phenomena in Amorphous Materials*, pp. 251–256, ed. C.V. Thompson, Mater. Res. Soc., Pittsburgh, PA, 1994.

52. S.G. Pskhie, S.Y. Korostelev, and V.I. Vorobyov, 'Formation of Nanocrystalline Structure by Shock Wave Progagation Through Amorphous Materials,' in *Shock Compression of Condensed Matter 1991*, pp. 157–159, ed. S.C. Schmidt, R.D. Dick, J.W. Forbes, and D.G. Tasker, Elsevier Science Publishers B.V., 1992.

53. *Metallurgical Applications of Shock-Wave and High-Strain-Rate Phenomena*, pp. 481, ed. H.D. Kunze, L.E. Murr, L.P. Staudhammer, and M.A. Meyers, Marcel Dekker, Inc., New York, 1986.

Chapter 13

Enhanced powder consolidation through superplastic flow

A practical method for superplastic forming of bulk material is through the use of powder metallurgy methods. The approach here is to achieve net-shaped products, with high density, by compaction of both metal and ceramic powders, using fine-structure or internal-stress superplasticity (FSS and ISS) methods. Studies have been performed by Ruano et al. [1] on the use of ISS in enhancing the densification of white cast iron powders. Caligiuri [2] and Isonishi and Tokizane [3], on the other hand, used FSS to enhance the densification of ultra-high carbon steel (UHCS) powders. In addition, Allen [4] used FSS to consolidate Ni-based IN–100 powders (Gatorizing™* process). For ceramics, Kellett et al. [5] used FSS to extrude fine zirconia powders and Wakai et al. [6] to perform the bulge forming of YTZP pipes directly from powders. Panda et al. [7] Akmoulin et al. [8], Uchic et al. [9], and Kwon et al. [10] have sinter forged nanometer zirconia, titania, and alumina powders.

13.1   ISS compaction of white cast iron powders

The advent of new technologies centered on fine powders often requires development of methods of enhancing densification wherein the fine structures present in such powders are retained. Low temperatures must be used to achieve this goal, but this usually requires the application of high pressures if a high density is to be achieved. High pressures are often a limiting factor in the manufacture

* Trademark of United Technologies Corporation.

of powder products. One method of enhancing densification of powders is by accelerating plastic flow through the generation of internal stresses during warm pressing. As previously described, one technique for generating internal stress is through the use of multiple, solid-state phase transformations.

Ruano *et al.* [1] showed that the densification of rapidly solidified white cast iron powders can be enhanced by multiple phase transformations. The basis of this result is that, during phase transformation, volume changes occur that generate internal stresses. These internal stresses assist plastic flow and result in enhanced pressure-sintering kinetics. An example of such a result is shown in Figure 13.1. The densification kinetics of white cast irons are shown as a function of applied stress under both isothermal and thermal-cycling conditions. The results demonstrate that thermal cycling, under stress, is an important factor in enhancing densification. For example, under a very low, externally applied stress of 6.9 MPa, a density of 90% is found for one cycle but increases to 95% after 10 cycles. By contrast, densities of much less than 80% are found for the iso-thermally warm-pressed samples at both 650 and 775 °C. Transformation cycling is also seen to enhance the densification of white cast iron powders at high applied stresses. For example, at 20 MPa, the density is about 95% for one cycle but increases to 99% after 10 cycles. Without transformation cycling, a density of only 90.5% is found after warm pressing for 0.5 hour at 775 °C and less than 80% is found after warm pressing for 0.5 hour at 650 °C. The results shown in Figure 13.1 indicate that high densification can be achieved in a short time by using transformation cycling under small applied stresses. Utilization of thermal cycling to achieve high densification of powders under hot isostatic pressing conditions, in principle, appears worthy of exploration.

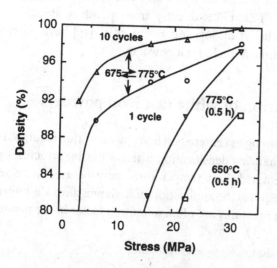

Figure 13.1 Influence of multiple phase transformations (internal-stress superplasticity) on the densification of white cast iron powders as a function of stress. High densities are achieved at low applied stresses under thermal-cycling conditions.

13.2   FSS compaction of ultrahigh carbon steel powders

Studies on the pressure-sintering kinetics of metal powders and, specifically, the kinetics of pressure sintering of ultrahigh carbon steel (1.6%C) powders containing either coarse-grained (nonsuperplastic) structure or fine-grained (superplastic) structure have been carried out [2, 11]. For comparison, the pressure sintering kinetics of iron powders were also investigated. To interpret the densification mechanisms of powders under an applied stress, Sherby *et al.* [11] investigated the creep behavior of the ultrahigh carbon steels containing coarse and fine microstructures. The creep rate of the fine-grained steel at a given low stress was found to be higher than the creep rate of the coarse-grained steel. The densification of coarse- and fine-grained UHCS powders was found to parallel their creep behavior. The fine-grained powders densified more readily than the coarse-grained powders. From these results, it was concluded that a superplastic microstructure enhances densification and permits the pressing of powders into high-density compacts at intermediate temperatures and low pressures. From the densification and creep studies, Caligiuri developed an equation showing that the intermediate-stage densification rate, $\dot{\rho}$, can be related to the steady-state creep rate, $\dot{\varepsilon}_{ss}$, at a given temperature by:

$$\dot{\rho} = A \left(\frac{1-\rho}{\rho} \right)^n \dot{\varepsilon}_{ss} \qquad\qquad (13.1)$$

where A is a constant (of value 57 for UHCS), ρ is the density, and n is the stress exponent. Thus, for a fixed value of relative density, the densification rate at a given intermediate temperature and low pressure is directly proportional to the steady-state creep rate of the material and to the stress exponent of the material. Through Equation (13.1), the densification behavior of a material can be predicted by simply knowing its creep properties at the temperature and pressure of interest.

Since superplastic materials generally creep faster than nonsuperplastic materials, at intermediate temperatures and low stresses, Equation (13.1) predicts that superplastic materials will densify faster than nonsuperplastic materials during pressing. Furthermore, since $(1-\rho)/\rho$ is usually less than unity, the relationship between densification rate and creep rate predicts that, through the term $[(1-\rho)/\rho]^n$, superplastic materials (low stress exponent) will densify faster than nonsuperplastic materials (high stress exponent) at the same creep rate. This prediction has been verified experimentally. For example, creep experiments were performed to determine the stress, at 650 °C, at which a superplastic 1.6%C steel has the same steady-state creep rate as a nonsuperplastic commercially pure iron. (This stress was determined to be 43.1 MPa at 650 °C.) Powders of these materials were then compacted at 43.1 MPa. The results of these experiments are shown in Figure 13.2. It is evident that the 1.6%C superplastic powders densify more rapidly than the iron powders. Tirosh and Miller [12] have developed a mechanics-plasticity model that describes quite well the results of Caligiuri.

13.3   FSS consolidation of Ni-based superalloy powders

The Pratt and Whitney Company has developed the well-known Gatorizing process for forging jet engine parts. This technique incorporates the advantages offered by superplasticity to forge high-temperature nickel-base superalloys that are normally difficult to forge [13].

Gatorizing utilizes the low flow stresses of the alloys at high temperatures to produce forgings with extremely close tolerances. This allows a decrease in the initial billet weight and subsequent lowering of machining costs. Forging pressures are less than those required for conventional forgings and allow the use of lighter forging equipment. Surface finishes on many forgings are sufficient to eliminate final machining, and uniform mechanical properties are obtained in the finished forgings. Alloys which can be forged using the Gatorizing technique include IN–100, INCO–901, A286, Astroloy, B–1900, and Waspalloy [4, 13, 14].

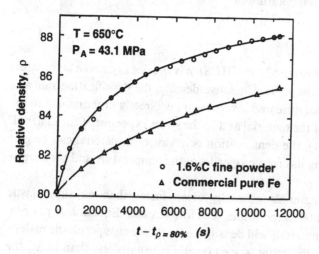

Figure 13.2 Relative density–time curves for fine-grained 1.6% UHCS powders and commercially pure iron powders compacted at 650 °C. A constant applied pressure of 43.1 MPa was used, at which stress the creep rates of the two dissimilar materials are identical.

An example of the complexity of forgings which can be produced is shown in Figure 13.3.

The precision forging of titanium alloys under superplastic conditions has also been explored [15]. In this case, Inconel dies made by precision forging was used. The nickel-base alloy dies retain sufficient strength for prolonged times at elevated temperatures to permit isothermal forging of titanium alloys in the superplastic temperature ranges. Thus, slower deformation rates as compared to normal forging practice are allowed and press forces are reduced, permitting complex die cavities to be filled without the customary preforging operations. In a manner similar to that in the consolidation of Fe-based powders, proper control of the powder properties and, thus, material flow during thermomechanical processing are critical to the success of the technique.

13.4 FSS extrusion and sinter forging of ultrafine ceramic powders

The recent advances in ceramic powder processing technology have greatly improved the quality of ceramic powders; high-purity ceramics of sub-

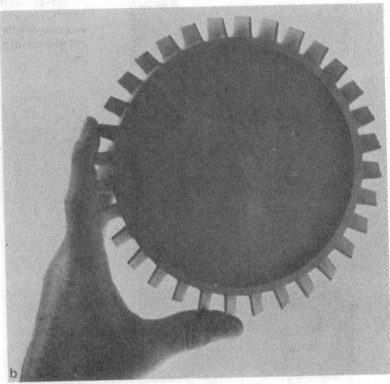

Figure 13.3 Turbine disk forged from IN–100.

micrometer particle size and more consistent microstructures can be obtained routinely. This, combined with the rapidly gained fundamental knowledge from superplastic ceramics, has promoted the concept of hot-forming bulk ceramics directly from powders. Kellett et al. [5, 16], for example, have demonstrated that submicrometer-sized (~0.23 μm) 3YTZP powders can be extruded superplastically through graphite dies, with an 8 to 1 reduction in area ratio (or a true strain of –2.2), at 1500 °C in vacuum to near full density. This result is illustrated in Figure 13.4. The extrusion behavior of the powders was reasonably described using the uniaxial constitutive creep equation of 3YTZP and a slab analysis [5].

Another useful powder processing technique that has been demonstrated is sinter forging nanocrystalline powders at low temperatures. Sinter forging is a net-shape forming technique where powder preforms are compressed in such a way that the preform consolidates to full density and obtains the desired shape in one step. Panda et al. [7] and Yamana et al. [17] have applied the technique to form bulk ZrO_2, stabilized by various amount of Y_2O_3, from fine powders (~25 nm) to near full density (96.5–99.5%) at 1400 °C at a strain rate faster than 10^{-3} s^{-1}. It was found that the densification rate (or pore closure) was related to

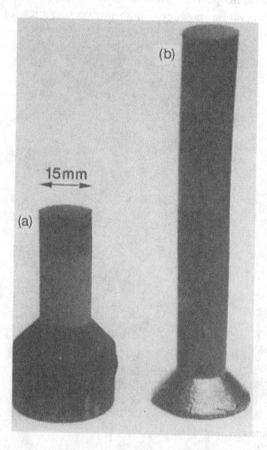

Figure 13.4 Fine YTZP powders extruded at 1500 °C with a reduction in area ratio of 8. With (a) and without (b) graphite liner.

the applied strain rather than to the applied hydrostatic pressure. In other words, the powder compact requires a critical amount of compressive strain to consolidate to full density, irrespective of the strain rate or the stress at which that strain is applied [7].

In addition to zirconia, Uchic et al. [9] have sinter forged nanocrystalline TiO_2 powder at temperatures below $0.5\ T_m$, where T_m is the absolute melting point of TiO_2. In contrast to fine zirconia powder, however, the densification rate of nanocrystalline TiO_2 was found to be related to applied hydrostatic stress, which can be described by a diffusional model proposed by Venkatachari and Raj [18]. Apparently, there is still no general view on the densification mechanism of nanocrystalline powders.

References

1. O.A. Ruano, J. Wadsworth, and O.D. Sherby, 'Enhanced Densification of White Cast Iron Powders by Cycling Phase Transformation under Stress,' *Metall. Trans.*, **13A** (1982), pp. 355–361.

2. R.D. Caligiuri, Ph.D. Thesis, Dept. of Materials Science and Engineering, Stanford University, (1977).

3. K. Isonishi and M. Tokizane, 'Hot Pressing of a Gas Atomized Ultrahigh Carbon Steel Powder and Superplasticity,' *J. JIM*, **49** (1985), pp. 149–158.

4. M.M. Allen, 'Iso-Forging of Powder Metallurgy Superalloys for Advanced Turbine Engine Applications'. pp. 5-1-5-15, Agard Conference Proc. No. 200, 1976.

5. B.J. Kellett, C. Carry, and A. Mocellin, 'High-Temperature Extrusion Behavior of a Superplastic Zirconia-Based Ceramic,' *J. Amer. Ceram. Soc*, **74**(7) (1990), pp. 1922–1927.

6. F. Wakai, S. Sakaguchi, K. Kanayama, H. Kato, and H. Onishi, 'Hot Work of Yttria-Stabilized Tetragonal ZrO_2 Polycrystals,' in *Ceramic Materials and Components for Engines, Proc. 2nd International Symposium*, pp. 315–322, ed. W. Bunk and H. Hausner, Verlag Deutsche Keramische Gesellschaft, 1986.

7. P.C. Panda, J. Wang, and R. Raj, 'Sinter-Forging Characteristics of Fine-Grained Zirconia,' *J. Amer. Ceram. Soc.*, **71**(12) (1988), pp. C-507-C-509.

8. I.A. Akmoulin, M. Djahazi, N.D. Buravova, and J.J. Jonas, 'Superplastic Forging Procedures for Manufacture of Ceramic Yttria Stabilized Tetragonal Zirconia Products,' *Mater. Sci. Technol.*, **9** (1993), pp. 26–33.

9. M. Uchic, H.J. Hofler, W.J. Flick, R. Tao, P. Kurath, and R.S. Averback, 'Sinter-Forging of Nanophase TiO_2,' *Scr. Metall. Mater.*, **26** (1992), pp. 791–796.

10. O.-H. Kwon, C.S. Nordahl, and G.L. Messing, 'Submicrometer Transparent Alumina by Sinter Forging Seeded γ-Al_2O_3 Powders,' *J. Am. Ceram. Soc.*, **78**(2) (1995), pp. 491–494.

11. O.D. Sherby, R.D. Caligiuri, E.S. Kayali, and R.A. White, 'Fundamentals of Superplasticity and its Applications,' in *Advanced Metal Processing*, pp. 131–171, ed. J.J. Burke, R. Mehrabian, and V. Weiss, Plenum Press, 1981.

12. J. Tirosh and A.K. Miller, 'Damage Evolution and Rupture in Creeping of Porous Materials,' *Inter. J. Solids and Structure*, **24**(6) (1988), pp. 567–580.

13. J.B. Moore, J. Tequesta, and R.L. Athey, *U.S. Patent 3,519,503*, 1970.

14. R.L. Athey and J.B. Moore, 'Development of IN–100 Powder Metallurgy Discs for Advanced Jet Engine Applications,' in *Powder Metallurgy for High-Performance Applications*, pp. 281–293, ed. J.J. Burke and V. Weiss, Syracuse University Press, 1972.

15. T.J. Headley, D. Kalish, and E.E. Underwood, 'The Current Status of Applied Superplasticity,' in *Ultrafine Grain Metals*, pp. 325–353, ed. J.J. Burke and V. Weiss, Syracuse University Press, 1970.

16. B.J. Kellett, C. Carry, and A. Mocellin,
    'Extrusion of Tet-ZrO$_2$ at Elevated
    Temperatures,' in *Superplasticity and
    Superplastic Forming*, pp. 625–630, ed. C.H.
    Hamilton and N.E. Paton, The Minerals,
    Metals, and Materials Society, Warrendale,
    PA, 1988.

17. Y. Yamana, Y. Y., S. Nakamura, K. Kitagawa,
    T. Yoshimura, T. Mano, and Y. Shintani,
    'Concurrent Sintering and Forging of YTZP,'
    J. Ceram. Soc. Japan, **97** (1989), pp. 758–762.

18. K.R. Venkatachari and R. Raj, 'Superplastic
    Flow in Fine-Grained Alumina,' *J. Am.
    Ceram. Soc.*, **69(2)** (1986), pp. 135–138.

Chapter 14

Superplastic forming and diffusion bonding

Concurrent superplastic forming and diffusion bonding (SPF/DB) has been recognized as a viable manufacturing technology that can result in both cost and weight savings compared with conventional manufacturing methods [1–4]. An important property of fine-grained superplastic materials is that they can often be readily bonded in the solid state, either to themselves or to other nonsuperplastic materials. This type of bonding is possible because of the presence of many grain boundaries in fine-grained materials that act as short-circuit paths for diffusion [5]. Grain-boundary migration is also faster in fine-grained than in coarse-grained materials [6]. Also, the bonding pressure is lower in fine-grained materials than in coarse-grained materials; the creep rate of the former material is faster than the latter. The same materials, if coarse-grained, may not readily bond under identical conditions of pressure, temperature, and time.

At superplastic temperatures (typically $>0.75T_m$, where T_m is the absolute melting point of the material), diffusion processes, and particularly grain-boundary diffusion, usually are rapid, permitting extensive boundary migration and void shrinkage occurring at interfaces. Based upon available data, as shown in Figure 14.1, Hwang and Balluffi [7] suggested that grain-boundary diffusivity, D_b, can be expressed by:

$$D_b = 1.0 \times 10^{-4} \exp(-9.35T_m/T) \text{ m}^2\text{s}^{-1} \tag{14.1}$$

for $1.0 < T_m/T < 2.4$ (or $0.42 < T/T_m < 1.0$), where T_m is the melting point of the material. At $T = 0.75T_m$, the value of D_b is estimated to be about 4×10^{-10} m^{2}s^{-1}; i.e., atoms diffuse about 10 μm s^{-1}. Up to the present, many studies have been devoted to the mechanistic description of diffusion bonding, especially on the kinetics of void shrinkage at bond interfaces [8–14]. However, the basic under-standing of surface condition requirements (e.g., chemical composition, oxide state, solute distribution, and roughness) and their effects on the quality of diffu-sion bonds are still limited. As a result, large variations in bond properties are often encountered; this hinders the general acceptance of diffusion bonding as a viable manufacturing technique. In the following, some examples of SPF/DB of both metals and ceramics are discussed.

14.1   Metals

14.1.1   Titanium

Superplastic titanium sheets have been used to make intricate parts by the SPF/DB method. Weisert and coworkers [3, 4] pioneered processing by SPF/DB

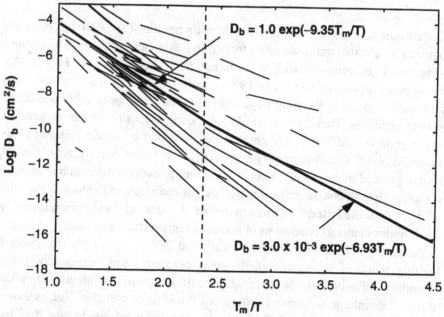

Figure 14.1   Arrhenius plot of grain-boundary diffusivities in a number of poly-crystalline metals (data from Ref. [7]).

in many applications. An example of the procedure used in processing components is shown schematically in Figure 14.2. The greatest advantage of processing components in this way is considered to be the contribution from diffusion bonding. The solid-state bond created between two superplastic titanium sheets is so well formed that it is usually impossible to detect the original interface, even at high magnification. There are a number of papers describing SPF/DB of Ti alloys, and in particular, Ti–6Al–4V [3, 15–19].

One of the reasons that a good solid-state bond can be readily established between two titanium sheets is attributed to the fact that Ti has extensive solid solubility (~34 at. % in α-Ti) for oxygen at high temperatures (greater than 900 °C) where superplasticity takes place. At diffusion-bonding temperatures, the Ti surface oxide is dissolved, resulting in extensive interdiffusion across the bond interface, and giving rise to a good metallurgical bond. However, this favorable situation does not normally exist in other metal alloys.

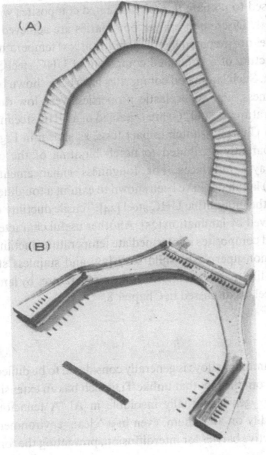

(A)

(B)

Figure 14.2 Nacelle center-beam frame of alloy Ti–6Al–4V, demonstrating the unusual forming capabilities of a superplastic material. Eight such frames form the structure separating the engines on the B–1 aircraft. Formerly (B), each frame was made by forming eight separate pieces of the alloy that then had to be joined with 96 fasteners. The superplastic forming route, A (carried out at 920 °C), which also uses diffusion bonding, results in considerable weight and cost savings.

14.1.2 Iron and steels

Diffusion bonding of Fe- or Ni-based alloys is relatively easy, mainly because iron and nickel oxides are thermodynamically less stable than most of the other metal oxides. For example, diffusion bonding of a microduplex ($\alpha-\gamma$) stainless steel has been successfully performed by uniaxial [20] and isostatic [21] hot-pressing techniques. Bonding parameters, particularly temperature and pressure, used in these studies are within the range for superplasticity in the alloy ($T\sim1000$ °C and $P\sim20$ MPa). A vacuum environment, however, was used during bonding. The lap shear strength of the bond is similar to that of the parent metal subjected to the same thermal treatment. Although not performed, it is envisioned that concurrent SPF/DB of the stainless steel can be made readily. It is interesting to note that a SPF/DB technique was used to fabricate hollow spheres from two dissimilar superplastic alloys Ti6Al4V and SUS304, in which SUS304 is nonsuperplastic [22].

Another example of solid-state diffusion bonding is fine-grained ultrahigh carbon (UHC) steels. In this case, the ease of solid-state bonding of fine-grained UHC steels has been used to prepare ferrous laminated composites with sharp interfaces between layers. Discrete interlayer boundaries are achieved because the laminated composites are prepared by roll bonding at low temperatures (e.g., 650 °C). The microstructure of a laminated composite of UHC steel/Fe–3Si is presented in Figure 14.3. Such laminated composites have been shown to exhibit unusual impact, toughness, and superplastic properties. Very low ductile-to-brittle transition temperatures (-140 °C) are obtained in a UHC steel/mild steel laminated composite in Charpy V-notch impact tests, as shown in Figure 14.4 [23]. This result is apparently attributed to notch blunting of the crack by delamination at the layer interfaces. For toughness enhancement, UHC steel/brass (70Cu–30Zn) laminates have been shown to exhibit a toughness value over three times that of the monolithic UHC steel [24]. Tensile ductility of UHC steels can also be improved by lamination [25]. Another useful characteristic of the UHC steel laminated composites is intermediate temperature ductility. Thus, it is possible to make nonsuperplastic mild steel [26] and stainless steel [27] behave in a superplastic-like manner at intermediate temperatures by lamination to superplastic UHC steel, as discussed in Chapter 8.

14.1.3 Aluminum

Diffusion bonding of aluminum alloys is generally considered to be difficult. The difficulty mainly arises from the fact that unlike Ti, which has an extensive solid solubility for oxygen, oxygen is virtually insoluble in Al. A tenacious, thin surface oxide forms readily on aluminum, even in a 'clean' environment. This thin oxide acts as an effective barrier for interdiffusion, preventing the formation of a good metallurgical bond between two samples. Despite this difficulty, many attempts have been made to join advanced superplastic Al alloys, particularly Al–Li alloys [28–41] and Al 7475 [30, 33, 37, 42, 43].

Two techniques are generally used for diffusion bonding of Al–Li alloys. The first technique is to sputter clean sample surfaces in an ultrahigh vacuum prior to diffusion bonding *in situ* [28]. This technique can produce strong bonds but is technologically impractical. The second technique is to combine mechanical and

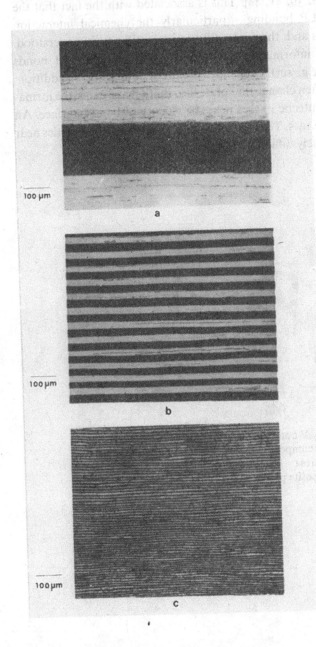

Figure 14.3
Microstructure of a laminated composite of UHC steel/Fe–3Si containing (a) 25, (b) 250, and (c) 2500 individual layers.

chemical methods to clean sample surfaces, followed by plating a thin layer ($\sim 10 \ \mu$m) of transient liquid phase (TLP) forming elements, and then performing diffusion bonding in an ambient atmosphere. The TLP elements must also be highly soluble in Al to minimize the formation of brittle intermetallics. Elements such as Zn and Cu satisfy these requirements. Presently, TLP bonding techniques have been the preferred methods for bonding Al structures. Experimental results indicate that diffusion bonds produced by these techniques are often inconsistent [38, 39, 41, 44]. This is associated with the fact that the exact mechanism for TLP bonding – particularly the chemical interaction between various elements and, thus, interdiffusion – is still poorly understood. Thus, the most critical information to produce good metallurgical bonds between two Al samples (e.g., surface chemistry) cannot be assessed. In addition, the interlayer materials often change the solute distribution and cause the formation of grain-boundary eutectic phases near the vicinity of bond interface. An example is given in Figure 14.5. The control of solute redistribution kinetics near a bond interface is extremely difficult.

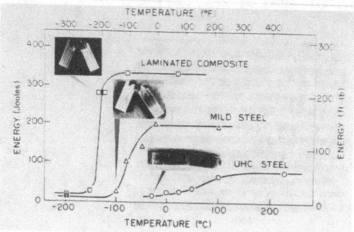

Figure 14.4 Charpy V-notch impact properties of a laminated composite vs those of its component metals. Very low ductile-to-brittle transition temperatures (-140 °C) are obtained in a UHC steel/mild steel laminated composite produced by a diffusion-bonding technique.

Several studies have been conducted on diffusion bonding of superplastic 7475 Al [30, 33, 37, 42, 43]. In contrast to Al–Li alloys, no interlayer (or coating) was used for the diffusion bonding of 7475 Al. Good bonds can be obtained from samples requiring only mechanical polishing and chemical cleaning prior to bonding. The success is apparently related to the particular type of oxides formed on the surface of 7475 Al. Although many studies have been conducted on the diffusion bonding and the subsequent bond strength of 7475 Al, unfortunately, no detailed analysis has yet been performed to identify the oxide states and their roles during diffusion bonding [33, 42, 43]. As a result of this insufficiency, the bond shear strength of 7475 Al was noted to suffer large variations. Therefore, although concurrent SPF/DB is now a well-established manufacturing route for the production of components from Ti sheet alloy, it has not yet been consistently demonstrated for Al alloys. This is mainly because of an insufficient understanding of the surface conditions that are required to produce successful diffusion bonding. Nonetheless, in view of the potential application of Al structures, this is certainly an high-payoff area of research.

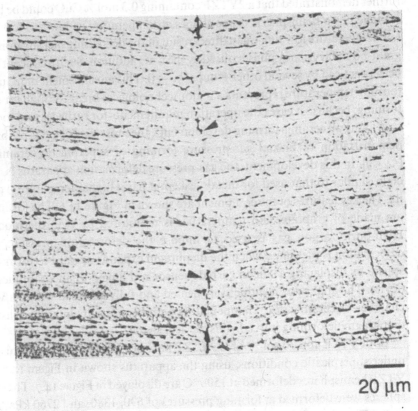

Figure 14.5 Grain-boundary eutectic formation in 8090 Al samples diffusion bonded using a copper interlayer.

14.2   Ceramics

14.2.1   Superplastic forming

In the case of ceramic bulk forming, Kellett and Lange [45] have hot forged dense 20 vol % ZrO_2/Al_2O_3 to 60% reduction in thickness at 1500 °C. Akmoulin *et al.* [46] performed upset forging of porous YTZP up to 90% strain at 1300 °C using graphite tools. The densified YTZP was further machined into surgical blades. Wu and Chen [47] have conducted hot extrusion (extrusion ratios from 4 to 9) of two dense ceramics, i.e., 10 wt% $Al_2O_3/2YTZP$ and $Bi_2Sr_2CaCu_2O_{8+x}$ super-conductors. For Al_2O_3, Carry and Mocellin [48] demonstrated bar bending and inverse extrusion of fine-grained Al_2O_3 at 1500 and 1600 °C.

In the case of ceramic sheet forming, many demonstration parts have been formed superplastically from fine-grained 3YTZP. For example, Wakai [49] successfully formed a YTZP sheet sample at 1450 °C in air using SiC tools. The superplastically formed parts have a smooth surface finish with the excellent dimensional accuracy required for aerospace applications. Wu and Chen [50] further demonstrated that a 2YTZP containing 0.3 mol % CuO could be biaxially stretched into hemispheric dome shapes at temperature as low as 1150 °C. The low-temperature forming was a direct result of the decreasing melting point of the grain-boundary glassy phase caused by the CuO addition. Such a concept has also been practiced by several other groups [51–53]. This advancement of reduction in forming temperature represents a major technological breakthrough for ceramic forming, because it eases the operation requirements for furnaces and tools.

Among all ceramic forming developments, the most significant is probably the demonstration of biaxial gas-pressure forming of superplastic ceramic sheet [54–59]. In this development, the gas-pressure deformation apparatus is, in principle, similar to that used for metals, except it has a higher temperature capability. The key features of the equipment are shown in Figure 14.6. The forming dies are made of graphite, and temperature is monitored by twin thermocouples. High-purity argon gas is used to impose the deformation pressure. *In situ* deformation is measured when the diaphragm expands upward to form a hemisphere by displacing a silicon carbide sensor rod linked to an strain measuring device. The apparatus is inductively heated and fully enclosed within a vacuum chamber. A ceramic disc, clamped at its periphery, results in an unconstrained diaphragm deforming into a hemispherical shape.

Both YTZP and 20 vol % $Al_2O_3/YTZP$ have been deformed into hemispheres, under superplastic conditions, using the apparatus shown in Figure 14.6. Three YTZP hemispheres deformed at 1500 °C are displayed in Figure 14.7. These hemispheres were deformed at forming pressures of 690, 1380, and 2760 kPa, respectively. The superplastic forming behavior for YTZP sheet is summarized in the deformation–time plots of Figure 14.8. The data indicate that there are three distinct regions of behavior as deformation progresses for each of the experimental conditions examined. Initially, the height of the deforming dome increases quite rapidly. This stage is followed by a period of apparent steady-state deformation. Finally, as the height approaches that of a hemisphere, the deformation rate increases again. This three-stage behavior appears to be similar to the creep curve of metal alloys deformed under a constant value of uniaxial stress, but the interpretation is quite different. This is because, although the applied forming pressure remains constant throughout the test, the resultant stress acting in the deforming shell varies continuously during the course of deformation. The relationship between applied forming pressure and resultant shell stress can be explained through a consideration of the pressure–curvature relationship for a spherical shell:

$$\sigma = \frac{P\rho}{2t}, \tag{14.2}$$

where σ is the principal tangential stress acting in a shell, having a wall of thickness t and radius ρ, and P is the applied gas pressure. For the experiments of this

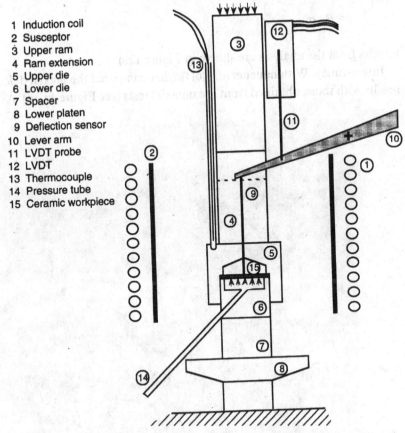

1 Induction coil
2 Susceptor
3 Upper ram
4 Ram extension
5 Upper die
6 Lower die
7 Spacer
8 Lower platen
9 Deflection sensor
10 Lever arm
11 LVDT probe
12 LVDT
13 Thermocouple
14 Pressure tube
15 Ceramic workpiece

Figure 14.6 Gas-pressure forming apparatus for ceramics. Upper die (5) and lower die (6). Argon gas is admitted through an integral niobium pressure tube (14); the ceramic disc (15) deforms upward, causing the silicon carbide deflection sensor (9) to move. Temperature is monitored by twin thermocouples (13).

study, P remains constant, ρ and t (and, therefore, σ) vary during the course of the test. Qualitatively, Equation (14.2) predicts a high flow stress σ at the beginning of a test (when ρ is very high) and at the end of a test (as t decreases). The result of this variation in flow stress is a high forming rate at both low and high dome heights. Because t and ρ are interdependent, the second stage of deformation occurs at a more or less constant rate, as a decreasing radius of curvature is balanced by a decreasing shell thickness. Therefore, the three-stage behavior shown in Figure 14.8 not only results from the creep of the material, but also from the nature of constant-pressure biaxial forming.

From the data in Figure 14.8, and assuming that (1) the volume of the thin-walled deforming shell remains constant and (2) the thickness of the shell decreases uniformly during deformation, Wittenauer et al. [56] performed additional analyses and were able to show that the average strain rate $\dot{\varepsilon}$ of the deforming YTZP sheet is proportional to the nominal flow stress raised to the nth power, i.e.,

$$\dot{\varepsilon} = B \cdot \sigma^n. \tag{14.3}$$

Results from the analysis are shown in Figure 14.9.

Interestingly, Wittenauer et al. [56] further compared the biaxial deformation results with those obtained from the uniaxial tests (see Figure 14.10). As shown,

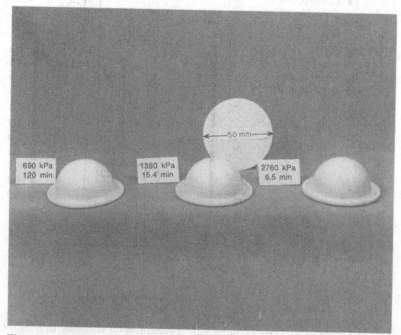

Figure 14.7   Hemispheres made from superplastic YTZP by gas pressure at 1500 °C.

although the strain rate from biaxial forming tests is generally faster than that from uniaxial tests, both data are in reasonable agreement. The generally higher strain rate for the biaxial forming tests was attributed mainly to the fact that the thickness of the sheet was assumed to decrease uniformly during deformation. In fact, the actual thickness during these experiments was less than the calculated thickness, because the stress for the forming experiments of the present study are actually higher than those predicted by Equation (14.2).

Using the biaxial gas-pressure forming technology, it is possible to make intricate, net-shaped parts from superplastic ceramic sheet. Figure 4.11 shows examples of a cone-on-cylinder geometry, a hat section, and a hemisphere. The articles shown were all formed at 1550 °C, using a forming pressure of 690 kPa, and requiring forming times of 20 to 90 min. From Figure 14.11, it is clear that the manufacturing of net-shaped ceramic articles is now technologically feasible. A multitude of shape geometries, as determined by the shape of the die, are possible with this process.

In addition to superplastic YTZP, several other ceramics, such as superplastic 20 vol %Al_2O_3/YTZP [60], Fe/Fe_3C [57], silicon [58], and even a YTZP/C–103 alloy hybrid structure [55] have been gas-pressure formed using the apparatus shown in Figure 14.6. The advantage of gas-pressure forming over biaxial punch forming is the avoidance of possible chemical and mechanical (friction) interactions between sample and die during high-temperature forming operations.

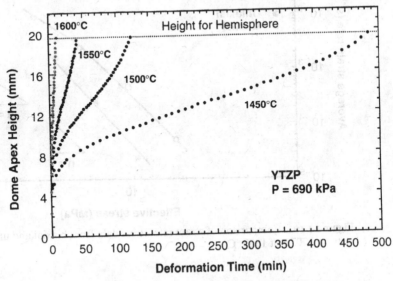

Figure 14.8 Deformation–time data for superplastic YTZP gas-pressure formed at 690 kPa.

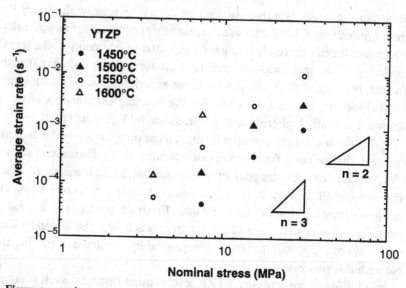

Figure 14.9   Average strain rate vs nominal stress for YTZP. The data suggest that the stress exponent at low strain rates has a value of about 3.

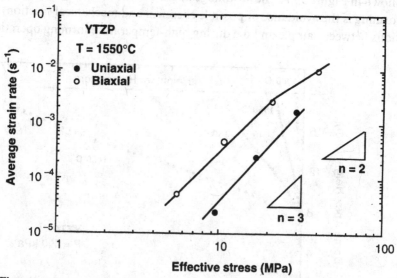

Figure 14.10   Comparison of strain rate–stress data from biaxial and uniaxial tests on YTZP at 1550 °C.

14.2.2 Diffusion bonding

Despite the fact that concurrent SPF/DB has been successfully applied for metals, these two processes have not yet been incorporated into a single operation for ceramics. Even in the area of diffusion bonding alone, only limited studies have been carried out with superplastic ceramics by Wakai and colleagues [61–65] and Ye and Dominguez-Rodriguez [66]. These superlastic ceramics include 3YTZP, Al_2O_3/YTZP composites with various amounts of Al_2O_3, and fine-grained mullite. For example, for Al_2O_3/YTZP composites containing various amounts of Al_2O_3, successful bonding can be produced at superplastic temperatures, i.e., 1450–1500 °C. The bond strength, measured by bend tests, was found to depend on the bonding temperature and the compositions of the bond materials (similar and dissimilar). Generally, the larger the difference in thermal expansion between the two bonding materials, the more difficult it is for bonding and the weaker the bond strength. A maximum bond strength of over 1300 MPa was recorded in a 20%Al_2O_3/YTZP–20%Al_2O_3/YTZP bond produced at 1470 °C; this compares favorably with a strength of 2200 MPa for the base material [62].

Another interesting study on the concurrent SPF/DB of a YTZP ceramic disc and the Nb-based refractory alloy C–103 (nominal composition Nb–10Hf–1Ti) to manufacture a net-shaped, metal–ceramic hybrid structure has been carried

Figure 14.11 Intricately shaped, net shaped parts from superplastic ceramic sheet by gas pressure forming.

out [55, 58]. Alloy C–103 is a single-phase Class I solid-solution alloy (see Section 12.1), which is commonly used in spacecraft propulsion systems [67]. The net-shaped YTZP/C–103 laminates produced offer possible applications such as fluidics and integral cooling passages in high thermal flux structures. The above studies are all exploratory in nature. Considering the relatively high cost of diffusion bonding processes, the materialization of these technologies will probably be only for high pay-off applications, such as aerospace structures.

References

1. C.H. Hamilton, 'Superplastic Forming and Diffusion Bonding of Titanium Alloys,' in *Titanium Science and Technology*, pp. 621–647, ed. R.T. Jaffee and H.M. Burke, Plenum Press, New York, 1973.

2. J.R. Williamson, 'Aerospace Applications of SPF and SPF/DB,' in *Superplastic Forming of Structural Alloys*, ed. N.E. Paton and C.H. Hamilton, TMS-AIME, Warrendale, Pennsylvania, 1982, pp. 291–306.

3. E.D. Weisert and G.W. Stacher, 'Concurrent Superplastic Forming/Diffusion Bonding of Titanium,' in *Superplastic Forming of Structural Alloys*, pp. 273–289, ed. N.E. Paton and C.H. Hamilton, TMS-AIME, Warrendale, PA, 1982.

4. E.D. Weisert, 'Hollow Titanium Turbofan Blades,' in *Superplasticity in Aerospace*, pp. 315–330, ed. C. Heikkenen and T.R. McNelly, TMS-AIME, Warrendale, PA, 1988.

5. O.D. Sherby, J. Wadsworth, R.D. Caligiuri, L.E. Eiselstein, B.C. Snyder, and R.T. Whalen, 'Superplastic Bonding of Ferrous Laminates,' *Scr. Metall.*, **13** (1979), pp. 941–946.

6. W. Hu, D. Ponge, and G. Gottstein, 'Origin of Grain Boundary Motion during Diffusion Bounding by Hot Pressing,' *Mater. Sci. Eng.*, **A190** (1995), pp. 223–229.

7. J.C.M. Hwang and R.W. Balluffi, 'On Possible Temperature Dependence of the Activation Energy for Grain Boundary Diffusion in Metals,' *Scr. Metall.*, **12** (1978), pp. 709–714.

8. B. Derby and E.R. Wallach, 'Theoretical Model for Diffusion Bonding,' *Met. Sci.*, **16** (1982), pp. 49–56.

9. B. Derby and E.R. Wallach, 'Diffusion Bonding: Development of Theoretical Model,' *Met. Sci.*, **18** (1984), pp. 427–413.

10. G. Garmong, N.E. Paton, and A.S. Argon, 'Attainment of Full Interfacial Contact during Diffusion Bonding,' *Metall. Trans.*, **6A** (1975), pp. 1269–1279.

11. W.H. King and W.A. Owczarski, 'Diffusion Welding of Commercial Pure Titanium,' *Weld. Res. Supp.*, (1967), pp. 289s–298s.

12. Y. Takahashi and K. Inoue, 'Recent Void Shrinkage Models and Their Applicability to Diffusion Bonding,' *Mater. Sci. Tech.*, **8** (1992), pp. 953–964.

13. Z.X. Guo and N. Ridley, 'Modelling of Diffusion Bonding of Metals,' *Mater. Sci. Technol.*, **3** (1987), pp. 945–953.

14. R. Raj and M.F. Ashby, 'Intergranular Fracture at Elevated Temperature,' *Acta Metall.*, **23** (1975), pp. 653–666.

15. J. Pilling, D.W. Livesey, J.B. Hawkyard, and N. Ridley, 'Solid State Bonding in Superplastic Ti–6Al–4V,' *Metal Sci.*, **18** (1987), pp. 117–122.

16. P.-J. Winkler, 'Diffusion Bonding and Superplastic Forming, Two Complementary Manufacturing Techniques,' in *Superplasticity and Superplastic Forming*, pp. 491–506, ed. C.H. Hamilton and N.E. Paton, The Minerals, Metals & Materials Soc., 1988.

17. W. Ji, Z. Wang, and F. Song, 'Development of SPF and SPF/DB Techniques for Ti and Al Alloys and Their Applications,' in *Superplasticity in Metals, Ceramics, and Intermetallics, MRS Proceeding No. 196*, pp. 173–180, ed. M.J. Mayo, J. Wadsworth, and M. Kobayashi, Materials Research Society, Pittsburgh, PA, 1990.

18. H.E. Friedrich and P.-J. Winkler, 'Fundamental Questions Concerning the Applications of Superplastic Forming and Superplastic Forming/Diffusion Bonding,' in

International Conference on Superplasticity in Advanced Materials (ICSAM-91), pp. 601–610, ed. S. Hori, M. Tokizane, and N. Furushiro, The Japan Society for Research on Superplasticity, Osaka, Japan, 1991.

19. Z. Li and H. Guo, 'Superplastic Forming/Diffusion Bonding Combined Technology of Curved Panel of Titanium Alloy,' in International Conference on Superplasticity in Advanced Materials (ICSAM-91), pp. 705–708, ed. S. Hori, M. Tokizane, and N. Furushiro, The Japan Society for Research on Superplasticity, Osaka, Japan, 1991.

20. Y. Komizo and Y. Maehara, 'Diffusion Bonding of a Superplastic Microduplex Stainless Steel,' Trans. Jpn. Wel. Soc., 19 (1988), pp. 83–91.

21. N. Ridley, M.T. Salehi, and J. Pilling, 'Isostatic Diffusion Bonding of Microduplex Stainless Steel,' Mater. Sci. Tech., 8 (1992), pp. 791–795.

22. T.H. Chuang, S.Y. Chang, J.H. Cheng, H.P. Kao, and S.E. Hsu, 'Diffusion Bonding/Suerplastic Forming of Ti6Al4V/SUS304/Ti6Al4V,' in International Conference on Superplasticity in Advanced Materials (ICSAM-91), pp. 661–668, ed. S. Hori, M. Tokizane, and N. Furushiro, The Japan Society for Research on Superplasticity, Osaka, Japan, 1991.

23. D.W. Kum, T. Oyama, J. Wadsworth, and O.D. Sherby, 'The Impact Properties of Laminated Composites Containing Ultrahigh Carbon (UHC) Steels,' J. Mech. Phys. Solids, 31 (1983), pp. 173–186.

24. C.K. Syn, D.R. Lesuer, K.L. Cadwell, O.D. Sherby, and K.R. Brown, 'Microstructures and Deformation Properties from Room Temperature to 1400°C of an Al_2O_3-Ni_3Al Composite,' in Developments in Ceramic and Metal-Matrix Composites, pp. 85–96, ed. K. Upadhya, The Minerals, Metals & Materials Society, Warrendale, PA, 1992.

25. C.K. Syn, D.R. Lesuer, J. Wolfenstine, and O.D. Sherby, 'Layer Thickness Effect on Ductile Tensile Fracture of Ultrahigh Carbon Steel-Brass Laminates,' Metall. Trans., 24A (1993), pp. 1647–1653.

26. B.C. Snyder, J. Wadsworth, and O.D. Sherby, 'Superplastic Behavior in Ferrous Laminated Composites,' Acta Metall, 32 (1984), pp. 919–923.

27. G. Daehn, D.W. Kum, and O.D. Sherby,

'Superplasticity of a Stainless Steel Clad Ultrahigh Carbon Steel,' Metall. Trans., 17A (1986), pp. 2295–2298.

28. D.V. Dunford and P.G. Partridge, 'Shear and Peel Strengths of Diffusion Bonded Al-Alloys,' in Superplasticity in Aerospace-Aluminum, pp. 257–284, ed. R. Pearce and L. Kelly, Ashford Press, Curdridge, Southampton, Hampshire, 1985.

29. J. Harvey, P.G. Partridge, and C.L. Snooke, 'Diffusion Bonding and Testing of Al-Alloy Lap Shear Test Pieces,' J. Mater. Sci., 20 (1985), pp. 1009–1014.

30. T.G. Nieh, LMSC Independent Research and Development Report, LMSC-F070359, pp.7–301–7–321, Lockheed Missiles and Space Co., 1986.

31. D.V. Dunford and P.G. Partridge, 'The Peel Strength of Diffusion Bonded Joints between Clad Al-Alloy Sheets,' J. Mater. Sci., 22 (1987), pp. 1790–1798.

32. P.G. Partridge and D.V. Dunford, 'On the Testing of Diffusion-boned Overlap Joints between Clad Al-Zn-Mg Alloy (7010) Sheet,' J. Mater. Sci., 22 (1987), pp. 1597–1608.

33. J. Pilling and N. Ridley, 'Solid State Bonding of Superplastic AA7475,' Mater. Sci. Technol., 3 (1987), pp. 353–359.

34. D.V. Dunford and P.G. Partridge, 'Strength and Fracture Behavior of Diffusion-bonded Joints in Al-Li (8090) Alloy,' J. Mater. Sci., 25 (1990), pp. 4957–4964.

35. Y. Huang, J. Cui, and L. Ma, 'Diffusion Bonding of Superplastic 7075 Aluminum Alloy,' in Superplasticity in Metals, Ceramics, and Intermetallics, MRS Proceeding No.196, pp. 137–142, ed. M.J. Mayo, J. Wadsworth, and M. Kobayashi, Materials Research Society, Pittsburgh, PA, 1990.

36. D.S. McDarmaid, P.G. Partridge, and A. Wisbey, 'A Review of the Mechanical Prperties after Superplastic Forming and Diffusion Bonding,' in International Conference on Superplasticity in Advanced Materials (ICSAM-91), pp. 621–632, ed. S. Hori, M. Tokizane, and N. Furushiro, The Japan Society for Research on Superplasticity, Osaka, Japan, 1991.

37. M. Hirohashi, Y. Park, and H. Asanuma, 'Superplastic Forming and Diffusion Bonding of Aluminum Alloy 7475 Sheet in Air,' in International Conference on Superplasticity in Advanced Materials (ICSAM-91), pp. 649–654, ed. S. Hori, M. Tokizane, and N.

Furushiro, The Japan Society for Research on Superplasticity, Osaka, Japan, 1991.

38. D.V. Dunford, C.J. Gilmore, and P.G. Partridge, 'Transient Liquid Phase Diffusion Bonding of 8090 Al-Li Alloy,' in *Aluminum-Lithium V*, pp. 1057–1062, ed. P.J. Winkler, DMS, MBB, 1992.

39. N. Ridley and D.W. Livesey, 'Diffusion Bonding of Superplastic Al-Li (8090) Using a Zinc Interlayer,' in *Aluminum-Lithium V*, pp. 1063–1068, ed. P.J. Winkler, DMS, MBB, 1992.

40. P.J. Winkler, T. Heinrich, R. Keyte, G.J. Mahon, and R.A. Rick, 'Bonding and Superplastic Forming of Al-Li Alloy AA8090 for Commercial Applications,' in *Aluminum-Lithium V*, pp. 1069–1074, ed. P.J. Winkler, DMS, MBB, 1992.

41. A. Sunwoo, 'Diffusion Bonding of Aluminum Alloy, 8090,' *Scr. Metall. Mater.*, **31**(3) (1994), pp. 407–412.

42. J. Kennedy, 'Diffusion Bonding and Superplastic Forming of 7475 Aluminum Alloy,' in *Superplasticity and Superplastic Forming*, pp. 523–527, ed. C.H. Hamilton and N.E. Paton, The Minerals, Metals & Materials Science Soc., 1988.

43. A. Sunwoo and R. Lum, 'Superplastic Deformation Enhanced Diffusion Bonding of Aluminum Alloy 7475,' *Scr. Metall. Mater.*, **33** (1995), pp. 639–644.

44. T.G. Nieh, unpublished work, Lawrence Livermore National Laboratory, 1993.

45. B.J. Kellett and F.F. Lange, 'Hot Forging Characteristics of Transformation-Toughened Al_2O_3/ZrO_2 Composites,' *J. Mater. Res.*, **3**(3) (1988), pp. 545–551.

46. I.A. Akmoulin, M. Djahazi, N.D. Buravova, and J.J. Jonas, 'Superplastic Forging Procedures for Manufacture of Ceramic Yttria Stabilized Tetragonal Zirconia Products,' *Mater. Sci. Technol.*, **9** (1993), pp. 26–33.

47. X. Wu and I.-W. Chen, 'Hot Extrusion of Ceramics,' *J. Am. Ceram. Soc.*, **75**(7) (1992), pp. 1846–1853.

48. C. Carry and A. Mocellin, 'Example of Superplastic Forming Fine-Grained Al_2O_3 and ZrO_2 Ceramics,' in *High Tech Ceramics*, pp. 1043–1052, ed. P. Vincenzini, Elsevier Science Publishers, Amsterdam, 1987.

49. F. Wakai, 'A Review of Superplasticity in ZrO_2-Toughened Ceramics,' *Brit. Ceram. Trans. J.*, **88** (1989), pp. 205–208.

50. X. Wu and I.W. Chen, 'Superplastic Bulging

of Fine-Grained Zirconia,' *J. Am. Ceram. Soc.*, **73**(3) (1990), pp. 746–749.

51. N. Kimura, H. Okamura, and J. Morishita, 'Preparation of Low-Y_2O_3 TZP by Low-Temperature Sintering,' in *Advances in Ceramics, Vol. 24A*, pp. 183–191, ed. S. Somiya, N. Yamamoto, and H. Yanagida, American Ceramic Society, Westerville, Ohio, 1988.

52. Y. Yoshizaka and T. Sakuma, 'Effect of Grain-Boundary Glassy Phase on Superplastic Deformation in ZrO_2–2.5mol% Y_2O_3,' in *Proc. 1st Japan Int'l SAMPE Symp.*, pp. 272–277, ed. N. Igata, K. Kimpara, T. Kishi, E. Nakata, A. Okura, and T. Uryu, Society for the Advancement of Materials and Process Engineering, 1989.

53. F. Wakai, H. Okamura, N. Kimura, and P.G.E. Descamps, 'Superplasticity of Transition Metal Oxide-Doped Y-TZP at Low Temperatures,' in *Proc. 1st Japan Int'l SAMPE Symp.*, pp. 267–271, ed. N. Igata, K. Kimpara, T. Kishi, E. Nakata, A. Okura, and T. Uryu, Society for the Advancement of Materials and Process Engineering, 1989.

54. J.P. Wittenauer, T.G. Nieh, and J. Wadsworth, 'A First Report on Superplastic Gas-Pressure Forming of Ceramic Sheet,' *Scripta Metall. Maters.*, **26** (1992), pp. 551–556.

55. J.P. Wittenauer, T.G. Nieh, and J. Wadsworth. 'Superplastic Forming of a Zirconia Ceramic Sheet,' in *Processing and Fabrication of Advanced Materials*, pp. 541–557, ed. J.J. Morris, TMS, Warrendale, PA, 1993.

56. J.P. Wittenauer, T.G. Nieh, and J. Wadsworth, 'Superplastic Gas-Pressure Deformation of Y-TZP Sheet,' *J. Am. Ceram. Soc.*, **76** (1993), pp. 1665–1672.

57. O.D. Sherby, unpublished work, Stanford University, 1993.

58. T.G. Nieh and J. Wadsworth, 'Gas-Pressure Forming of Ceramic Sheet,' in *Transaction of the Materials Research Society of Japan Vol 16B – Composites, Grain Boundaries and Nanophase Materials*, pp. 1061–1064, ed. M. Sakai, M. Kobayashi, T. Suga, R. Watanabe, Y. Ishida, and K. Niihara, Elsevier Science, Netherland, 1994.

59. K. Higashi, N. Ikegami, and S. Tanimura, 'The Simulation and the Bulge Forming of Superplastic YTZP,' in *Transaction of the Materials Research Society of Japan Vol 16B – Composites, Grain Boundaries and Nanophase Materials*, pp. 1065–1068, ed. M. Sakai, M. Kobayashi, T. Suga, R. Watanabe, Y.

Ishida, and K. Niihara, Elsevier Science, Netherland, 1994.

60. T.G. Nieh and J. Wadsworth, 'Biaxial Gas-Pressure Forming of a Superplastic Al$_2$O$_3$/YTZP,' *J. Mater. Eng. Performance*, **3(4)** (1994), pp. 496–500.

61. T. Nagano, H. Kato, and F. Wakai, 'Solid State Bonding of Superplastic Ceramics,' in *MRS Intl. Meeting on Advanced Materials Vol 7 (IMAM–7) – Superplasticity*, pp. 285–292, ed. M. Kabayashi and F. Wakai, Materials Research Society, Pittsburgh, PA, 1989.

62. T. Nagano, H. Kato, and F. Wakai, 'Diffusion Bonding of Zirconia/Alumina Composites,' *J. Amer. Ceram. Soc.*, **73(11)** (1990), pp. 3476–3480.

63. T. Nagano, H. Kato, and F. Wakai, 'Diffusion Bonding of Ceramics: Mullite, ZrO$_2$-Toughened Mullite,' *J. Mater. Sci.*, **26** (1990), pp. 4985–4990.

64. F. Wakai, S. Sakaguchi, K. Kanayama, H. Kato, and H. Onishi, 'Hot Work of Yttria-Stabilized Tetragonal ZrO$_2$ Polycrystals,' in *Ceramic Materials and Components for Engines, Proc. 2nd International Symposium*, pp. 315–322, ed. W. Bunk and H. Hausner, Verlag Deutsche Keramische Gesellschaft, 1986.

65. T. Nagano and F. Wakai, 'Superplastic Diffusion Bonding,' *in Transaction of the Materials Research Society of Japan Vol 16B – Composites, Grain Boundaries and Nanophase Materials*, pp. 1069–1074, edited by M. Sakai, M. Kobayashi, T. Suga, R. Watanabe, Y. Ishida, and K. Niihara, Elsevier Science, Netherland, 1994.

66. J. Ye and A. Dominguez-Rodriguez, 'Joining of Y-TZP Parts,' *Scr. Metall. Mater.*, **33** (1995), pp. 441–445.

67. J. Wadsworth, T.G. Nieh, and J.J. Stephens, 'Recent Advances in Aerospace Refractory Metal Alloys,' *Inter. Mater. Rev.*, **33(3)** (1988), pp. 131–151.

Chapter 15

Commercial examples of superplastic products

15.1 Titanium alloys

Superplastic forming of titanium alloys has been widely used by the aerospace industry. A well-known example of superplastic forming carried out at Rockwell International using a Ti–6Al–4V alloy was shown in Figure 14.2 [1]. The component shown was a nacelle center-beam frame. (A number of such parts formed a proposed structure in the B–1 aircraft.) In this example, a single superplastic forming and diffusion bonding operation was designed to replace a production route which had involved the forming of eight separate pieces of the same alloy, which then had to be joined together with 96 fasteners. Estimated cost savings of 55% and weight savings of 33% were estimated using this fabrication route compared with the conventional production technique. Superplastically formed Ti–6Al–4V was also used for service doors and panels for Airbus aircraft [2], missile fins [3], turbofan blades [4], and turbine disks [5].

For non-aerospace applications, the most significant commercial product is probably golf club heads, shown in Figure 15.1. A titanium golf club head offers a light weight, large volume, and a wide sweet spot area. It is now produced in both Japan [6] and China [7]. In Japan, the head is made of SP–700 Ti alloy which has the nominal composition of Ti–4.5Al–3V–2Fe–2Mo. The alloy, marketed by Nippon Kokan (NKK), features a low SPF temperature (as low as 775 °C) for the maximum formability, and can be age-hardened [8].

Another interesting product is cookware. A certain type of titanium alloys specifically alloyed with aluminum shows very high electric resistivity, low heat capacity, and generates heat effectively under a high frequency magnetic field, for example 22 kHz. These features make the titanium cookware very attractive even though the titanium alloy itself is not a good heat conductor. In this case, superplastic forming has been chosen as an effective forming method. To further reduce the production cost, a specially designed driving sheet for the forming operation (Figure 15.2) has been adapted to save up to 20% of the titanium alloy sheet. Figure 15.3 shows an example of superplastically formed Ti–6Al–4V skillet.

15.2 Nickel alloys

A similar and equally impressive demonstration of a superplastic forming operation (Gatorizing™), developed by Pratt and Whitney [9–11], is shown in Figure 15.4. Here, a superplastic, nickel-based superalloy (IN–100), prepared by powder-metallurgy processing, is formed into the desired shape using a two-step forming operation. In Figure 15.4, the billet is made by hot isostatic pressing of powders made fine-structured by atomization. First, the billet is warm-pressed in a die to the configuration of a disk (center of Figure 15.4). Second, this intermediate part is superplastically formed into the complex disk with turbine blades (right side of Figure 15.4). The fine structure of the IN–100 superalloy is an advantage in the body of the disk during operation but a disadvantage in the blades, where a high temperature is encountered in service. For this reason, the part of the disk containing the blades is selectively heat-treated after forming to create a coarse grain size in the blades. Such a treatment considerably improves the creep-resistance characteristics of the blades, but leaves a tough, fatigue-resistant, fine-grained material in the body of the disk.

Intense efforts have also been devoted to developing superplasticity in Inconel 718, one of the most widely used superalloys [12]. The new alloy, 718SPF, is now commercially available. Shown in Figure 15.5 are examples of aircraft gas turbine components (exhaust mixer nozzle and noise suppressor) made from 718SPF alloy by Murdock, Inc., USA.

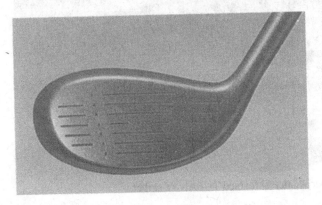

Figure 15.1 Golf club head made from SP–700 Ti alloy by Yamaha.

Figure 15.2   Specially designed driving sheet for the forming of titanium cookwares.

Figure 15.3   Superplastically formed Ti–6Al–4V skillet.

15.3 Iron alloys

Several potential applications of the superplastic forming of ultrahigh carbon (UHC) steels have been suggested by Lesuer *et al.* [13]. Four components made from UHC steels are shown in Figure 15.6. All were formed in single operations. The back-extruded tube and trunion (Figures 15.6(a) and 15.6(b)) were shaped at conventional forming rates (~10 s^{-1}) and illustrate the high formability of fine-grained UHC steels at high strain rates. The temperatures of forming were 750 °C for the tube and 700 °C for the trunion. The height reduction for the trunion was approximately 4:1. The UHC steels used for both components contained 1.25 wt% carbon and 1.6 wt% aluminum. The guided missile aft-closure (Figure 15.6(c)) was forged at a temperature (815 °C) and strain rate (10^{-3} s^{-1}) at which the UHC steels (1.6% carbon) exhibit superplasticity. The material for this component was obtained from fine-structure powders prepared via liquid atomization. The dome on this component has a uniform wall thickness, illustrating uniform flow during forming. The subscale compressor hub (Figure 15.6(d)) was forged at near superplastic conditions (750 °C; strain rate of 10^{-3} s^{-1}) from a 40-mm-high, 76-mm-diam cylinder. The component was formed to demonstrate the die-filling capability of UHC steels by manufactur-

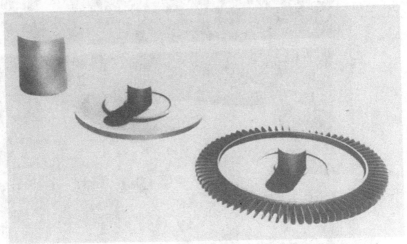

Figure 15.4 Net-shaped forming of an ultrafine-grain, nickel-based alloy by superplastic forming in two stages. (Left) Original powder-metallurgy IN–100 billet. (Center) Powder-metallurgy billet pressed into disk shape. (Right) Disk-shape billet superplastically pressed into disk and turbine blades. (Courtesy of J. Moore and R. Athey, Pratt and Whitney, Florida.)

ing a complex part to net shape. Figure 15.7 illustrates some bulk formed components made in Russia which were probably also formed under high strain-rate-sensitivity value (*m*) conditions [14]. The sample on the left is a relatively coarse-grained Fe-Al alloy formed isothermally at 800 °C into an remarkably intricate electrical connector component (about 2 min to form the part).

In passenger airplanes, lavatories must be designed so that many functions work effectively in their small spaces. A sink deck is one of the parts installed in

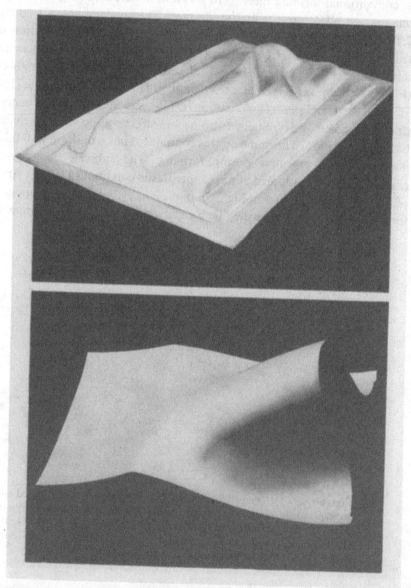

Figure 15.5   Aircraft gas turbine components, exhaust mixer nozzle (top) and noise suppressor (bottom), made from superplastic Inconel 718 (from Ref. [12]).

the lavatory, and sometimes a complicated three-dimensional shape is required on the sink deck. Shown in Figure 15.8 is a sink deck with a dust box lid installed in the Boeing 737. The sink deck was formed by SPF using a duplex stainless steel, SuperDux64 (nominal composition in weight %: Fe–25Cr–6.5Ni–3.2 Mo–0.14Ni). The typical dimension of a sink deck is 1.1 m in width, 0.35 m in depth, and 0.270 m in height.

15.4 Aluminum alloys

Superplastic forming technique has been succesfully, if not the most successfully, applied on commercial aluminum alloys. In fact, aluminum alloys have been used

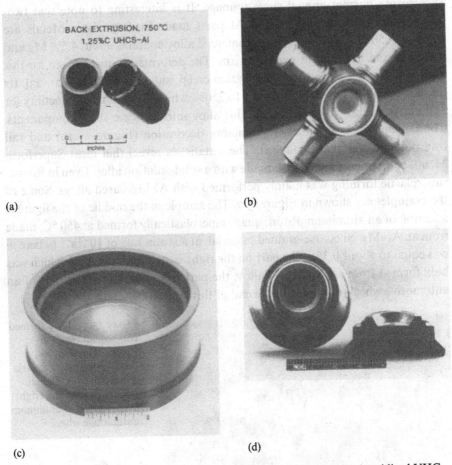

(a)

(b)

(c)

(d)

Figure 15.6 Components that have been formed from fine-grained spheroidized UHC steels: (a) a back extrusion made at the Raychem Corporation (courtesy of Dieter Stockel); (b) a trunion forged at the General Motors Technology Center (courtesy of W. Mueller); (c) a forged guided missile aft-closure; and (d) a subscale compressor hub. (Courtesy of Bryant Walker and Roy Athey from Pratt and Whitney, Florida.)

commercially for superplastic forming operations since 1973. The first super-plastically formed aluminum component was probably the prototype amenity panel made from SUPRAL 100 (Al–6%Cu–0.4%Zr) for the Concord supersonic aircraft. Many thousands of components have been formed annually by Superform Metals in England and Superform USA [15–17]. Structural components include aircraft ejector seats (SUPRAL 100), aircraft fuselage fairings (SUPRAL 150), electronic enclosures (2004 Al), and under-carriage door panels (8090 Al). Rail transport is now the largest user of superplastic aluminum in Europe.

Superform Metals has been gas-pressure-forming complex sheet shapes mainly from four different composition aluminum alloys (2004, 7475, 8090 and 5083), as shown in Figure 15.9. Two of these alloys (2004 and 5083) are used much more extensively than the others. Their combined usage exceeds 95% of Superform's current annual sheet tonnage. It is interesting to note that two-thirds of all gas-pressure formed Al parts made by Superform Metals are made with 5083 Al [17–20]. This commercial alloy contains about 4.5% Mg and 0.5% Mn and has a grain size of 25 μm. The deformation mechanism for this alloy is solute-drag-controlled dislocation creep, and m is about 0.3 [21–23]. Its tensile ductility is typically about 200 to 250%, which gives sufficient ductility for Superform Metals to manufacture this alloy into intricate sheet components. Components such as panels for building decoration (Figure 15.10) and rail transport are routinely produced. These statistics reveal that most Superform Metals components are, in fact, made with a solid-solution alloy. Even in Russia, superplastic forming was mainly performed with Al-Mg based alloys. Some of the examples are shown in Figure 15.7. The sample in the middle of the figure is a section of an aluminum piston, quasi-superplastically formed at 450 °C, made from an Al–Mg–Si coarse-grained material, at a strain rate of $10^{-2}\,s^{-1}$ (where m was equal to about 0.35). The part on the right is an aluminum alloy which was bulk formed from an Al–Mg–Si alloy; the part was made for use as part of an automotive wheel (discussed in the end of this section).

Figure 15.7 Bulk-formed component made in Russia. The samples include (left) steel electric connector, (middle) aluminum piston, and (right) cross section of aluminum automobile wheel.

As pointed out in Chapter 12, although extended ductility is achievable with Class I solid-solution alloys, such materials generally do not have high strength at low temperatures. Therefore, these materials are used mostly as secondary structural components. In primary structural components, high to ultrahigh strengths are desired at low temperature. In this case, ultrafine-grained materials with ultrafine second phases are natural choices. These high-strength aluminum alloys include 2004 Al [15–17], 7475 Al [24–26], and Al–Li alloys 8090 [26, 27], which are fine-grained and exhibit m values in the order of 0.5. Figure 15.11 shows a prototype parachute holder for an advanced fighter produced by superplastic forming a specially processed Al–Li alloy 2090 (Al–2.6%Cu–2.4%Li–0.18%Zr) by Superform USA. Figure 15.12 shows the oil inlet production parts formed superplastically from SUPRAL 100 for the U.S. Navy aircraft P3 Orion. The complexity of the parts is apparent.

Five different sheet forming methods are in regular use at Superform. They are as follows.

1. Simple female forming in which the heated superplastic sheet is clamped around its edge and stretched into the heated cavity tool using gas pressure.

2. Drape forming where the heated and clamped superplastic sheet is stretched into a cavity containing one or more male form blocks.

3. Male forming in which gas pressure and tool movement are combined enabling deeper more uniform thickness parts to be made.

4. Back pressure forming (BPF) utilizes gas pressure on both sides (front and back) of the deforming superplastic sheet which produces a

Figure 15.8 Sink deck made from SuperDux64 stainless steel for passenger aircraft.

Al alloys:

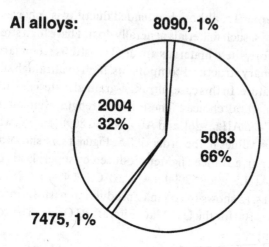

8090, 1%

2004
32%

5083
66%

7475, 1%

Figure 15.9 Two-thirds of all gas-pressure-formed parts made by Superform Metals are made with 5083 Al (data in 1993).

Figure 15.10 Panels made from 5083 Al for building decoration.

Figure 15.11 A complex shape produced by superplastic forming a specially processed Al–Li alloy 2090.

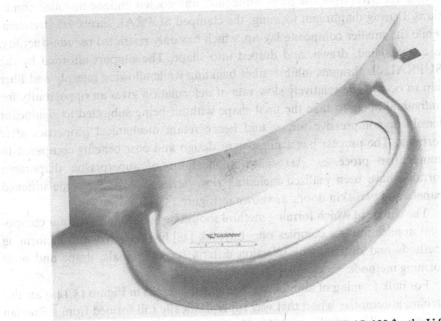

Figure 15.12 Oil inlets formed superplastically from SUPRAL 100 for the U.S. Navy aircraft P3 Orion.

hydrostatic confining pressure capable of suppressing cavitation. Gas control creates a positive pressure differential enabling forming to be achieved. BPF is usually applied to female and drape forming methods when structural applications demand cavitation levels be contained below 0.5% volume fraction.

5. Diaphragm forming uses one or two 'slave' sheets of superplastic alloy to urge a smaller unclamped blank to be augmentedly drawn and draped into tool contact. The process can shape nonsuperplastic metals having limited formability at room temperature and in certain cases superplastic alloys are shaped this way when limited thickness variation is demanded. This process has also been successfully applied to shaping advanced thermoplastic graphite composites, such as polyethylether ketone (or PEEK) [15]. In this technology, SUPRAL 100 alloy is used in a sacrificial forming mode at the lower end of its superplastic temperature range (300–400 °C). These low temperatures are coincident with the melt and fabrication temperatures for PEEK and other polymers.

In the forming operation, reinforcement fibers and plies are placed with the polymeric matrix in position between two cold SUPRAL aluminum sheets (diaphragms). The interlaminar space is evacuated before the PEEK matrix melts, and the 'hot' sandwich is then shaped by one of the forming techniques (i.e., simple female forming, female drape forming or male forming). The form is then removed from the press while 'hot' and cooled under controlled conditions. During diaphragm forming, the clamped SUPRAL sheets are stretched, while the smaller composite lay-up, which has only restricted pseudo-ductility, is consolidated, drawn, and draped into shape. The support afforded by the SUPRAL diaphragms inhibits fiber buckling while allowing interply and fiber slip to occur. The relatively slow rate of deformation gives an opportunity for individual fibers to take the final shape without being subjected to significant tensile or compressive forces, and hence retain mechanical properties after forming. The process has a number of design and cost benefits compared to competitive processes. Aerospace applications of superplastic diaphragm forming have been realized including ribs, fairings, channels, blade-stiffened panels, and two-skin doors, as shown in Figure 15.13.

The choice of which forming method should be used for a particular component application is a complex one. Laycock [28] has reviewed various forming methods and outlined the selection criteria for the female, drape and male forming methods.

For bulk forming of aluminum, the right-hand part in Figure 15.14 is an aluminum automotive wheel that was superplastically roll-formed from a Russian aluminum alloy (similar to 6010 Al). The roll-forming involves the following operations.

1. A round disc was sliced from an extruded aluminum alloy cylindrical bar with the required starting blank diameter. The large extrudate has previously been processed to an ultrafine grain size.

2. The aluminum blank was isothermally forged in a conventional hydraulic press (~1000 ton) to the final wheel disc shape with a simple rim preform. This operation takes 2–4 min at 400–500 °C.

3. The final rim shape was isothermally roll-formed on a lathe-type machine. This operation takes 3–5 min at 400–500 °C.

The formed wheel was subsequently heat-treated and surface-finished for appearance and corrosion resistance. The above sequence is illustrated in Figure 15.14. The roll-forming technology is expected to have a broad range of commercial applications. In addition to automotive wheels, it can be used to manufacture other axisymmetric products such as large spherical accumulators for hydraulic earth-moving equipment, large hollow shafts, large axle hubs, and large brake drums and rotors. An additional benefit of the superplastic forming process derives from the basic nature of the process, which is incremental, localized forming. This allows scale-up of the workpiece to any size that physically fits within the traversing limits of the forming rolls. Forming loads are not increased by part dimension increases – only the overall forming time. Therefore, scale up to very large axisymmetric components is feasible.

Figure 15.13 Various diaphragm formed parts, including thermoplastic matrix composites and metals for aerospace applications.

Готовое изделие   Заготовка под   Заготовка
                        раскатку   под штамповку

Figure 15.14 Superplastic roll formed aluminum automotive wheels made from a fine-grained Russian aluminum alloy (similar to 6010 Al).

References

1. C.H. Hamilton and G.W. Stacher, 'Superplastic Forming of Ti–6Al–4V Beam Frames,' *Metal Prog.*, **109**(3) (1976), pp. 34–37.

2. P. -J. Winkler, 'Superplasticity in Use: A Critical Review of its Status, Trends and Limits,' in *Superplasticity in Metals, Ceramics, and Intermetallics, MRS Proceeding No. 196R*, pp. 123–136, ed. M.J. Mayo, J. Wadsworth, and M. Kobayashi, Materials Research Society, Pittsburgh, PA, 1990.

3. Z. Li and H. Guo, 'Superplastic Forming/Diffusion Bonding Combined Technology of Curved Panel of Titanium Alloy,' in *International Conference on Superplasticity in Advanced Materials (ICSAM-91)*, pp. 705–708, ed. S. Hori, M. Tokizane, and N. Furushiro, The Japan Society for Research on Superplasticity, Osaka, Japan, 1991.

4. E.D. Weisert, 'Hollow Titanium Turbofan Blades,' in *Superplasticity in Aerospace*, pp. 315–330, ed. C. Heikkenen and T.R. McNelly, TMS-AIME, Warrendale, PA, 1988.

5. J. Hai, J. Dai, and S. Chen, 'Development of Superplastic Forming Technology in China,' in *Superplasticity and Superplastic Forming*, pp. 571–582, ed. C.H. Hamilton and N.E. Paton, The Minerals, Metals & Materials Science Soc., 1988.

6. K. Osada and H. Yoshida, 'Recent Applications of Superplastic Materials in Japan,' in *Superplasticity in Advanced Materials – ICSAM-94*, pp. 715–724, ed. T.G. Langdon, Trans Tech Publications Ltd, Switzerland, 1994.

7. Y. Wang, personal communication, Central Iron and Steel Research Institute, Beijing, China, 1995.

8. M. Ishikawa, O. Kuboyama, M. Niikura, and C. Ouchi, 'Microstructure and Mechanical Properties Relatioship of β-rich α-β Titanium Alloy; SP-700,' in *Titanium '92, Science and Technology, Vol. 1*, pp. 141–148, ed. F.H. Froes and I.L. Caplin, TMS, Warrendale, PA, 1992.

9. J.B. Moore, J. Tequesta, and R.L. Athey, *U.S. Patent 3,519,503*, 1970.

10. R.L. Athey and J.B. Moore, 'Development of IN-100 Powder Metallurgy Discs for Advanced Jet Engine Applications,' in *Powder Metallurgy for High-Performance Applications*, pp. 281–293, ed. J.J. Burke and V. Weiss, Syracuse University Press, 1972.

11. M.M. Allen, 'Iso-Forging of Powder Metallurgy Superalloys for Advanced Turbine Engine Applications'. pp. 5–1–5–15, Agard Conference Proc. No. 200, 1976.

12. G.D. Smith and H.L. Flower, 'Superplastic Forming of Alloy 718,' *Adv. Mater. Proc.*, **145**(4) (1994), pp. 32–34.

13. D.R. Lesuer, C.K. Syn, A. Goldberg, J. Wadsworth, and O.D. Sherby, 'The Case for Ultrahigh Carbon Steels as Structural Materials,' *JOM*, **45(8)** (1993), pp. 40–46.

14. O.D. Sherby, T.G. Nieh, and J. Wadsworth, 'Overview on Superplasticity Research on Small-Grained Materials,' in *Materials Science Forum Vols. 170–172*, pp. 13–22, ed. T.G. Longdon, Trans Tech Publications, Switzerland, 1994.

15. A.J. Barnes, 'Commercial Superplastic Aluminum Alloys – Opportunities and Challenges,' in *Superplasticity in Aerospace-Aluminum*, pp. 424–447, ed. R. Pearce and L. Kelly, Ashford Press, Curdridge, Southampton, Hampshire, 1985.

16. A.J. Barnes, 'Advances in Superplastic Aluminum Forming,' in *Superplasticity in Aerospace*, pp. 301–313, ed. C. Heikkenen and T.R. McNelly, TMS-AIME, Warrendale, PA, 1988.

17. R. Grimes, M.J. Stowell, and B.M. Watts, 'The Forming Behavior of Commercially Available Superplastic Aluminum Alloys,' in *Superplasticity in Aerospace*, pp. 97–113, ed. C. Heikkenen and T.R. McNelly, TMS-AIME, Warrendale, PA, 1988.

18. M. Kobayashi, M. Miyagawa, N. Furushiro, K. Matsuki, and Y. Nakazawa, 'State of Fundamental and Industrial Research in Superplasticity in Japan,' in *Superplasticity*, pp. 12.1–12.15, ed. B. Baudelet and M. Suery, Centre National de la Recherche Scientifique, Paris, Grenoble, France, 1985.

19. B.J. Dunwoody, R.J. Stracey, and A.J. Barnes, 'Mechanical Properties of 5083 SPF After Superplastic Deformation,' in *Superplasticity in Metals, Ceramics, and Intermetallics, MRS Proceeding No.196*, pp. 161–166, ed. M.J. Mayo, J. Wadsworth, and M. Kobayashi, Materials Research Society, Pittsburgh, PA, 1990.

20. P. Fernandez, 'A New Superplastic Aluminum Alloy, FORMALL^R 570,' in *International Conference on Superplasticity in Advanced Materials (ICSAM–91)*, pp. 675–680, ed. S. Hori, M. Tokizane, and N. Furushiro, The Japan Society for Research on Superplasticity, Osaka, Japan, 1991.

21. H. Imamura and N. Ridley, 'Superplastic and Recrystallization Behavior of a Commercial Al-Mg Alloy 5083,' in *International Conference on Superplasticity in Advanced Materials (ICSAM–91)*, pp. 453–458, ed. S. Hori, M. Tokizane, and N. Furushiro, The Japan Society for Research on Superplasticity, Osaka, Japan, 1991.

22. J.S. Vetrano, C.A. Lavender, C.H. Hamilton, M.T. Smith, and S.M. Bruemmer, 'Superplastic Behavior in a Commercial 5083 Aluminum Alloy,' *Scr. Metall. Mater.*, **30** (1994), pp. 565–570.

23. R. Verma, A.K. Ghosh, S. Kim, and C. Kim, 'Grain Refinement and Superplasticity in 5083 Al,' *Mater. Sci. Eng.*, **A191** (1995), pp. 143–150.

24. J.A. Wert, N.E. Paton, C.H. Hamilton, and M.W. Mahoney, 'Grain Refinement in 7075 Aluminum by Thermomechanical Processing,' *Metall. Trans.*, **12A** (1981), pp. 1267–1276.

25. B. Plege and C. Schleinzer, 'Development of a SPF-Aluminum Aircraft Structure,' in *International Conference on Superplasticity in Advanced Materials (ICSAM–91)*, pp. 801–806, ed. S. Hori, M. Tokizane, and N. Furushiro, The Japan Society for Research on Superplasticity, Osaka, Japan, 1991.

26. T. Tsuzuku, A. Takahashi, and A. Sakamoto, 'Applications of Superplastic Forming for Aerospace Components,' in *International Conference on Superplasticity in Advanced Materials (ICSAM–91)*, pp. 611–620, ed. S. Hori, M. Tokizane, and N. Furushiro, The Japan Society for Research on Superplasticity, Osaka, Japan, 1991.

27. H.E. Friedrich and P.-J. Winkler, 'Fundamental Questions Concerning the Applications of Superplastic Forming and Superplastic Forming/Diffusion Bonding,' in *International Conference on Superplasticity in Advanced Materials (ICSAM–91)*, pp. 601–610, ed. S. Hori, M. Tokizane, and N. Furushiro, The Japan Society for Research on Superplasticity, Osaka, Japan, 1991.

28. D.B. Laycock, 'Superplastic Forming of Sheet Metal,' in *Superplastic Forming of Structural Alloys*, ed. N.E. Paton and C.H. Hamilton, The Metallurgical Society of AIME, Warrendale, PA, 1982, pp. 257–272.

Index